TONY DAWSON obtained his degree at the University of Southampton and has since taught in various Surrey schools and colleges. He was Head of the Mathematics Department at Woking Grammer School for Boys and is now the Deputy Principal of Woking Sixth Form College.

ROD PARSONS graduated from Bristol University and completed a post-graduate certificate in Education at Southampton University. Since then he has taught in schools and sixth form colleges in Surrey. He has taught both modern and traditional syllabuses and has written for S.M.P. and lectured to teachers at S.M.P. conferences. He was Head of Mathematics Department at Woking Sixth Form College and is now Vice Principal at Esher Sixth Form College.

key facts

GCE A-Level Passbooks

APPLIED MATHEMATICS, E. M. Peet, B.A.

BIOLOGY, H. Rapson, B.Sc.

CHEMISTRY, J. E. Chandler, B.Sc.
and R. J. Wilkinson, Ph.D.

ECONOMICS, R. Maile, B.A.

GEOGRAPHY, R. Bryant, B.A.,
R. Knowles, M.A. and J. Wareing, B.A., M.Sc.

PHYSICS, J. R. Garrood, M.A., Ph.D.

PURE MATHEMATICS, R. A. Parsons, B.Sc.
and A. G. Dawson, B.Sc.

GCE A-Level Passbook

Pure and Applied Mathematics

R. A. Parsons, B.Sc.
and
A. G. Dawson, B.Sc.

key facts

Published by
Charles Letts and Co Ltd
London, Edinburgh and New York

First published 1980 by Intercontinental Book Productions

Published 1982 by Charles Letts & Co Ltd
Diary House, Borough Road, London SE1 1DW

1st edition, 4th impression

Made and printed by Charles Letts (Scotland) Ltd
ISBN 0 85097 389 9

Contents

Introduction

This book is a notebook for students of Pure and Applied Mathematics at advanced level. The Authors have intended to cover most of the topics that are studied at advanced level both for modern and traditional syllabuses. The differences between modern and traditional syllabuses are now decreasing and the topics in this book include most of those that an advanced level student would encounter, no matter which syllabus he or she is following.

At advanced level a student usually covers a little of each topic at a time, and as the course progresses the topics become interdependent and more related to each other. For ease of reference and clarity this book is written as twenty different chapters, although there are cross-references between the chapters.

Each topic is explained thoroughly with formulae derived as though the student were doing the work himself. Frequent examples are included to enable the student to assimilate a process thoroughly and to understand its applications.

Various approaches to learning and teaching each topic have been given so that students are not handicapped by seeing the work presented in a completely different way. Indeed a new method of presentation can often lead to greater understanding.

The style of examinations for the different Examining Boards can vary and students are best advised to study past examination papers which are obtainable from the Examination Boards. Some examinations at advanced level include multi-choice papers, with papers of short or long questions. Other papers have a compulsory section A of short questions with a choice of longer questions in section B.

Key facts and definitions are emphasized in bold print and these appear in the index at the back of the book. There is a short summary at the end of each chapter, but the student is advised by the authors to consult the full text whenever possible.

Chapter 1
Algebra

If $A = 2x + 1$ and $B = x + 3$, then

$A + B = 3x + 4,$

$A - B = (2x + 1) - (x + 3) = x - 2,$

$A \times B = (2x + 1)(x + 3) = 2x^2 + 7x + 3,$

$$A \div B = \frac{2x+1}{x+3} = \frac{2x+6-5}{x+3} = \frac{2x+6}{x+3} - \frac{5}{x+3} = 2 - \frac{5}{x+3}.$$

A is an example of a linear function since the graph of $y = 2x + 1$ or $f(x) = 2x + 1$ is a straight line. A is also a **polynomial** of degree 1, while $A \times B = 2x^2 + 7x + 3$ is a polynomial of degree 2 since the highest power of x occurring in the expression is x^2. It may be helpful when drawing the graph of $y = 2x^2 + 7x + 3$ to know its **factors** are $2x + 1$ and $x + 3$.

Example 1 Draw the graph of $g(x) = x^2 - 3x + 2$ $= (x - 1)(x - 2)$.

Table of values

x	-1	0	1	2	3	4
g(x)	6	2	0	0	2	6

$g(3) = 2$

2 is the value of

$g(x)$ when $x = 3$

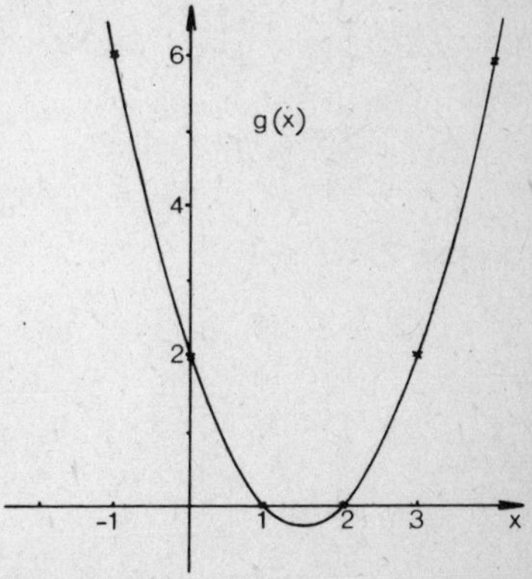

Figure 1

$x - 1$ and $x - 2$ are the factors of $g(x) = x^2 - 3x + 2$. $x = 1$ and $x = 2$ are **zero values** (or **zeros**) of $g(x)$ since $g(1) = 0$ and $g(2) = 0$. $\{g(1)$ is the value of $g(x)$ when $x = 1\}$.

$x = 1$ and $x = 2$ are the **solutions** (or **roots**) of the equation $g(x) = x^2 - 3x + 2 = 0$.

Polynomials are added and subtracted by considering their **corresponding terms** (powers of x).

If $g(x) = x^2 - 3x + 2$ and $h(x) = 2x + 1$,

$$\begin{aligned} g(x) + h(x) &= x^2 - x + 3; \\ g(x) - h(x) &= x^2 - 3x + 2 - (2x + 1) \\ &= x^2 - 5x + 1. \end{aligned}$$

Polynomials are multiplied by multiplying out the brackets.

$$\begin{aligned} g(x) \times h(x) &= (x^2 - 3x + 2)(2x + 1) \\ &= x^2(2x + 1) - 3x(2x + 1) + 2(2x + 1) \\ &= 2x^3 + x^2 - 6x^2 - 3x + 4x + 2 \\ &= 2x^3 - 5x^2 + x + 2. \end{aligned}$$

Long multiplication process

$$\begin{array}{rrrr} & & x^2 - 3x & + 2 \\ \times & & 2x & + 1 \\ \hline 2x^3 & - 6x^2 & + 4x & \\ & + x^2 & - 3x & + 2 \\ \hline 2x^3 & - 5x^2 & + x & + 2 \end{array}$$

or

coefficients

$$\begin{array}{rrrr} & 1 & -3 & 2 \\ \times & & 2 & 1 \\ \hline 2 & -6 & 4 & 0 \\ & 1 & -3 & 2 \\ \hline 2 & -5 & 1 & 2 \end{array}$$

Example 2 Multiply $x^3 + 2x^2 + 3x + 4$ by $x^2 - x + 2$.

$$\begin{array}{rrrrrr} & & 1 & 2 & 3 & 4 \\ & & \times & 1 & -1 & 2 \\ \hline 1 & 2 & 3 & 4 & 0 & 0 \\ & -1 & -2 & -3 & -4 & 0 \\ & & 2 & 4 & 6 & 8 \\ \hline 1 & 1 & 3 & 5 & 2 & 8 \end{array}$$

$$\begin{aligned} &(x^3 + 2x^2 + 3x + 4)(x^2 - x + 2) \\ &= x^5 + x^4 + 3x^3 + 5x^2 + 2x + 8. \end{aligned}$$

Dividing polynomials

This is more involved. It is best achieved by setting the polynomials out as a division process.

$$\begin{array}{r} 21 \text{ R. } 1 \\ 12\overline{)253} \\ \underline{24} \\ 13 \\ \underline{12} \\ 1 \end{array} \qquad \text{divisor}\overline{)\ \ \text{dividend}\ \ }^{\text{quotient + remainder}} \qquad \begin{array}{r} 2x + 1 \text{ R. } 1 \\ x + 2\overline{)2x^2 + 5x + 3} \\ \underline{2x^2 + 4x } \\ 0 \quad + \quad x + 3 \\ \underline{x + 2} \\ \text{R.} \quad 1 \end{array}$$

This can be summarized.

$$253 = 12 \times 21 \text{ R. } 1 \qquad 2x^2 + 5x + 3 = (x + 2)(2x + 1) + 1.$$

$$\frac{253}{12} = 21 + \frac{1}{12} \qquad \frac{2x^2 + 5x + 3}{x + 2} = 2x + 1 + \frac{1}{x + 2}.$$

Another method is to divide by inspection.

Example 3 Divide $2x^3 - 5x^2 + x + 2$ by $x + 2$.
The quotient is quadratic so

$2x^3 - 5x^2 + x + 2 = (x + 2)(ax^2 + bx + c) + \text{R}.$

$a = 2$ to give $2x^3$, so $2 \times ax^2$ gives $4x^2$.

To give a total of $-5x^2$, b must now be -9. $b = -9$ gives $-18x$, so $c = 19$ to give $19x$ and $+x$ in total. $2c = 38$ so the remainder is -36.

$$2x^3 - 5x^2 + x + 2 = (x + 2)(2x^2 - 9x + 19) - 36.$$

The method of inspection is more useful when factorizing a polynomial.

Example 4 Divide $2x^3 - 5x^2 + x + 2$ by $x - 1$.

$$\begin{array}{r} 2x^2 - 3x - 2 \\ x - 1\overline{)2x^3 - 5x^2 + x + 2} \\ \underline{2x^3 - 2x^2 } \\ -3x^2 + x \\ \underline{-3x^2 + 3x } \\ -2x + 2 \\ \underline{-2x + 2} \\ 0 \end{array}$$

There is no remainder which shows that $x - 1$ is a factor of $2x^3 - 5x^2 + x + 2$.

There is a simple way to find the remainder when a polynomial is divided by a linear factor. It is known as the **remainder theorem**.

Remainder theorem

If $f(x)$ is divided by $x - a$ then the remainder is $f(a)$.

Example 5 From example 3 above, if $f(x) = 2x^3 - 5x^2 + x + 2$,

$$f(-2) = 2(-2)^3 - 5(-2)^2 + (-2) + 2 = -36.$$

In example, 4, if $g(x) = 2x^3 - 5x^2 + x + 2$,

$$g(1) = 2 - 5 + 1 + 2 = 0,$$

showing a remainder of 0.

Proof of remainder theorem

$f(x)$ when divided by $x - a$ gives a quotient of $Q(x)$ and a remainder of R, which means

$$f(x) = (x - a)Q(x) + R.$$

$$x = a \text{ gives } \quad f(a) = 0 \times Q(a) + R \quad \text{so} \quad R = f(a).$$

This is most useful when factorizing a polynomial, for if $F(a) = 0$ then $x - a$ is a factor of $F(x)$. This is known as the factor theorem.

Example 6 Factorize $f(x) = 2x^3 - 5x^2 + x + 2$.

Possible factors are $x + 1$, $x - 1$, $x + 2$, $x - 2$, $2x + 1$, $2x - 1$, by examining the first and last terms.

$f(1) = 2 - 5 + 1 + 2 = 0$, $\quad f(-1) = -2 - 5 - 1 + 2 = -6$,
$f(2) = 16 - 20 + 2 + 2 = 0$, $\quad f(-2) = -16 - 20 - 2 + 2 = -36$,
$f(-\frac{1}{2}) = -\frac{1}{4} - 1\frac{1}{4} - \frac{1}{2} + 2 = 0 \Rightarrow 2x + 1$ is a factor of $f(x)$.

There are 3 linear factors $x - 1$, $x - 2$, $2x + 1$, so

$$2x^3 - 5x^2 + x + 2 = (x - 1)(x - 2)(2x + 1)$$

and we need not try $(2x - 1)$.

If one linear factor, say $x - 1$ can be found it is better to divide this into $f(x)$ to get

$$2x^3 - 5x^2 + x + 2 = (x - 1)(2x^2 - 3x - 2),$$

either by the division process or by inspection.

Then $\qquad 2x^2 - 3x - 2 = (x - 2)(2x + 1).$

It is easy to see whether a quadratic expression will factorize by using the formula for solving quadratic equations.

NB The factors are unique:

$$\begin{aligned} 2x^3 - 5x + x + 2 &= (x-2)(2x^2 - x + 1) \\ &= (x-2)(x-1)(2x+1). \end{aligned}$$

Dividing a cubic polynomial by a quadratic will, in general, leave a remainder of a linear polynomial, unless it divides exactly.

Example 7 Divide $2x^3 - 5x^2 + x + 2$ by $x^2 + x + 1$.

$$\begin{array}{r l} & 2x - 7 \\ x^2 + x + 1 \big) & 2x^3 - 5x^2 + x + 2 \\ & 2x^3 + 2x^2 + 2x \\ \hline & -7x^2 - x + 2 \\ & -7x^2 - 7x - 7 \\ \hline & +6x + 9 \end{array}$$

$$\begin{aligned} &2x^3 - 5x^2 + x + 2 \\ &= (x^2 + x + 1)(2x - 7) \\ &\quad + 6x + 9. \end{aligned}$$

Special factors

$\mathbf{x^2 - a^2 = (x - a)(x + a)}$ $\quad f(x) = x^2 - a^2. \quad f(\pm a) = 0.$
$x - a$ and $x + a$ are both factors.

$\mathbf{x^2 + 2ax + a^2 = (x + a)^2}$ and $\mathbf{x^2 - 2ax + a^2 = (x - a)^2}$.

$$\mathbf{x^3 - a^3 = (x - a)(x^2 + ax + a^2)}$$

and
$$\mathbf{x^3 + a^3 = (x + a)(x^2 - ax + a^2)}.$$

$$\mathbf{x^4 - a^4 = (x^2 - a^2)(x^2 + a^2) = (x - a)(x + a)(x^2 + a^2)}.$$

Roots of equations

Example 8 The quadratic equation $x^2 - 3x + 2 = 0$ has two solutions (or roots) $x = 1$ or $x = 2$.

$$x^2 - 3x + 2 = 0.$$

So $\quad (x-1)(x-2) = 0$ and $x - 1 = 0$ or $x - 2 = 0$.

Thus $\quad x = 1$ or $x = 2$.

Sum of roots $= 3$; $\quad$ product of roots $= 2$.

If $x^2 + px + q = 0$ factorizes to $(x - \alpha)(x - \beta) = 0$ then the roots are α and β, and $(x - \alpha)(x - \beta) = x^2 - (\alpha + \beta)x + \alpha\beta$.

Sum of roots $= \alpha + \beta = -p$; $\quad$ product of roots $= \alpha\beta = q$.

In general, if $ax^2 + bx + c = 0$ has roots α and β, then

$$\alpha + \beta = -\frac{b}{a} \quad \text{and} \quad \alpha\beta = +\frac{c}{a}.$$

Consider the cubic equation:

$$ax^3 + bx^2 + cx + d = 0. \qquad (a \neq 0) \qquad (1)$$

If the roots are α, β, γ then $(x - \alpha)(x - \beta)(x - \gamma) = 0$,

$$(x - \alpha)\{x^2 - (\beta + \gamma)x + \beta\gamma\} = 0,$$

$$x^3 - (\alpha + \beta + \gamma)x^2 + (\alpha\beta + \beta\gamma + \alpha\gamma)x - \alpha\beta\gamma = 0.$$

Comparing with (1), after dividing through by a:

$$\alpha + \beta + \gamma = -\frac{b}{a}, \qquad \alpha\beta + \beta\gamma + \alpha\gamma = +\frac{c}{a}, \qquad \alpha\beta\gamma = -\frac{d}{a}.$$

Example 9 $2x^3 - 5x^2 + x + 2 = 0, \quad (x - 1)(x - 2)(2x + 1) = 0.$

Roots are $\alpha = 1, \quad \beta = 2, \quad \gamma = -\frac{1}{2}$.

Sum of roots, $\alpha + \beta + \gamma = 2\frac{1}{2} = -\frac{b}{a}$;

product of roots, $\alpha\beta\gamma = -1 = -\frac{d}{a}$.

Sum of roots in pairs, $\alpha\beta + \beta\gamma + \alpha\gamma = 2 - 1 - \frac{1}{2} = \frac{1}{2} = +\frac{c}{a}$.

Partial fractions

$$\frac{1}{x+1} + \frac{1}{x+2} = \frac{(x+2) + (x+1)}{(x+1)(x+2)} = \frac{2x+3}{(x+1)(x+2)}.$$

$$\frac{1}{x-1} - \frac{1}{x+1} = \frac{(x+1) - (x-1)}{(x+1)(x-1)} = \frac{2}{x^2-1}.$$

It is useful when studying graphs and integration to perform the reverse operation, that is to express $\dfrac{5}{(x+2)(x-3)}$ in simpler fractions.

$$\frac{5}{(x+2)(x-3)} \equiv \frac{a}{x+2} + \frac{b}{x-3} \quad \text{for all values of } x.$$

Multiplying through by $(x + 2)(x - 3)$ gives

$$5 = a(x - 3) + b(x + 2).$$

When $x = 3$, $5 = 5b$ and $b = 1$; when $x = -2$, $5 = -5a$ and $a = -1$.

Thus $$\frac{5}{(x+2)(x-3)} = \frac{-1}{x+2} + \frac{1}{x-3},$$

and the expression is said to be in partial fractions.

Example 10 Express $\dfrac{11x+12}{(2x+3)(x+2)(x-3)}$ in partial fractions.

$$\frac{11x+12}{(2x+3)(x+2)(x-3)} \equiv \frac{a}{2x+3} + \frac{b}{x+2} + \frac{c}{x-3}.$$

Multiplying through by $(2x+3)(x+2)(x-3)$ gives

$$11x + 12 \equiv a(x+2)(x-3) + b(2x+3)(x-3) + c(2x+3)(x+2)$$

$$x = 3 \Rightarrow 45 = 45c \Rightarrow c = 1; \qquad x = -2 \Rightarrow -10 = 5b \Rightarrow b = -2;$$

$$x = -1\tfrac{1}{2} \Rightarrow a = 2.$$

So $$\frac{11x+12}{(2x+3)(x+2)(x-3)} = \frac{2}{2x+3} - \frac{2}{x+2} + \frac{1}{x-3}.$$

The previous working shows that

$$a = \frac{11x+12}{(x+2)(x-3)} \quad \text{when} \quad x = -1\tfrac{1}{2},$$

$$b = \frac{11x+12}{(2x+3)(x-3)} \quad \text{when} \quad x = -2;$$

and $$c = \frac{11x+12}{(2x+3)(x+2)} \quad \text{when} \quad x = 3.$$

This indicates a simple process for finding a, b and c. To find a, the numerator of $2x + 3$ in the partial fractions form of $F(x) = \dfrac{11x+12}{(2x+3)(x+2)(x-3)}$, work out $F(-1\tfrac{1}{2})$ ignoring the factor $(2x + 3)$.

To find b, work out $F(-2)$ ignoring the factor $(x + 2)$.
To find c, work out $F(3)$ ignoring the factor $(x - 3)$.

$(x + 2)$ in the denominator can be covered up and $F(-2)$ evaluated without it. This covering-up rule gives a quick easy way for finding linear partial fractions. Care must be taken when the denominator contains repeated factors like $(x + 3)^2$, or quadratic factors like $x^2 + 1$ which have no linear factors.

If the degree of the numerator is greater than or equal to the degree of the denominator, it is necessary to divide the denominator into the numerator first. If the denominator contains factors like $x^2 + 1$ which do not factorize, a general linear expression, $ax + b$, must be sought. The following example illustrates these points.

Example 11 $\dfrac{x^3 + x^2 + 5x + 7}{(x + 1)(x^2 + 1)} = 1 + \dfrac{4x + 6}{(x + 1)(x^2 + 1)}$ by division.

$\dfrac{4x + 6}{(x + 1)(x^2 + 1)} = \dfrac{a}{x + 1} + \dfrac{bx + c}{x^2 + 1}$; $a = 1$ by covering up.

$4x + 6 \equiv x^2 + 1 + (x + 1)(bx + c);$
$b = -1$ and $c = 5$ (by comparing coefficients).

$$\frac{x^3 + x^2 + 5x + 7}{(x + 1)(x^2 + 1)} = 1 + \frac{1}{x + 1} + \frac{-x + 5}{x^2 + 1}.$$

Repeated factors

Example 12 $\dfrac{x + 1}{(x + 2)^2} = \dfrac{x + 2 - 1}{(x + 2)^2} = \dfrac{x + 2}{(x + 2)^2} - \dfrac{1}{(x + 2)^2}$

$$= \frac{1}{x + 2} - \frac{1}{(x + 2)^2}.$$

This can be achieved more easily by substituting $y = x + 2$.

$$\frac{x + 1}{(x + 2)^2} = \frac{y - 1}{y^2} = \frac{y}{y^2} - \frac{1}{y^2} = \frac{1}{y} - \frac{1}{y^2} = \frac{1}{x + 2} - \frac{1}{(x + 2)^2}.$$

Example 13 $\dfrac{2x + 3}{(x + 1)(x + 2)^2} = \dfrac{a}{x + 1} + \dfrac{f(x)}{(x + 2)^2}.$

$$2x + 3 = a(x + 2)^2 + (x + 1)f(x); \quad x = -1, \Rightarrow a = 1.$$

$$2x + 3 - (x + 2)^2 = (x + 1)f(x) = -x^2 - 2x - 1 = -(x + 1)^2,$$

so $$f(x) = -(x + 1).$$

Hence $$\frac{2x + 3}{(x + 1)(x + 2)^2} = \frac{1}{x + 1} - \frac{x + 1}{(x + 2)^2}$$

$$= \frac{1}{x + 1} - \frac{1}{x + 2} + \frac{1}{(x + 2)^2}$$

using example 12 to separate the last fraction.

This result suggests that repeated factors are separated like this:

$$\frac{2x+3}{(x+1)(x+2)^2} = \frac{a}{x+1} + \frac{b}{x+2} + \frac{c}{(x+2)^2}.$$

Hence $2x + 3 = a(x+2)^2 + b(x+1)(x+2) + c(x+1)$.

When $x = -1$, $1 = a$ and $a = 1$ just like the covering-up method

When $x = -2$, $-1 = -1c$ and $c = 1$ the value found by covering $(x+2)^2$.

Equating coefficients of x^2: $0 = a + b$, $b = -a = -1$.

Example 14

$$\frac{x^3+2x-1}{(x-2)(x+1)^3} = \frac{a}{x-2} + \frac{b}{x+1} + \frac{c}{(x+1)^2} + \frac{d}{(x+1)^3}.$$

$$x^3 + 2x - 1 = a(x+1)^3 + b(x-2)(x+1)^2 + \\ + c(x-2)(x+1) + d(x-2).$$

When $x = 2$, $11 = 27a$ and $a = \frac{11}{27}$.
When $x = -1$, $-4 = -3d$ and $d = \frac{4}{3}$.
Coefficients of x^3 give $1 = a + b$, $b = \frac{16}{27}$.

Coefficients of x^2 give $0 = 3a + c$, $c = \dfrac{-11}{9}$.

$$\frac{x^3+2x-1}{(x-2)(x+1)^3} = \frac{11}{27(x-2)} + \frac{16}{27(x+1)} - \frac{11}{9(x+1)^2} + \frac{4}{3(x+1)^3}.$$

Inequalities

In figure 2 three sets of points are shown.

Set 1
Points on the line $y = x + 2$.

Set 2
The shaded area above the line, for which $y > x + 2$.

Set 3
The unshaded area below the line, for which $y < x + 2$.

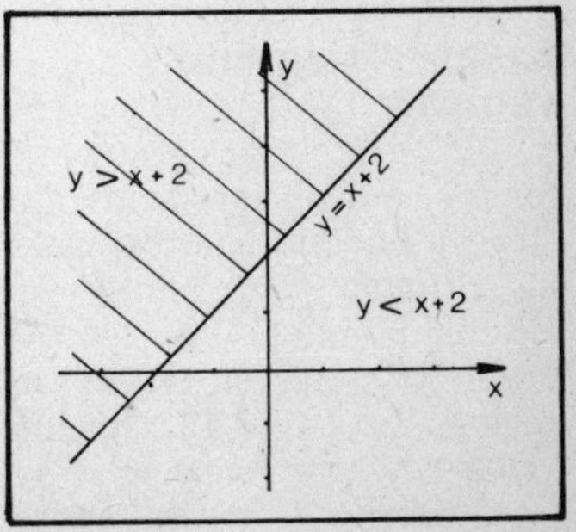

Figure 2

Solving inequalities

Example 15 $3x + 4 > 13,$
$3x > 9,$ subtracting 4 from both sides
$x > 3,$ dividing both sides by 3.

Inequalities can be solved like equations with the following rules.

(1) $p > q \Rightarrow p + r > q + r,$
$6 > 4 \Rightarrow 6 + 3 > 4 + 3$ adding 3 to both sides,
$6 > 4 \Rightarrow 6 - 3 > 4 - 3$ subtracting 3 from both sides.

(2) $p > q \Rightarrow pr > qr$ if $r > 0$; $6 > 4 \Rightarrow 3 \times 6 > 3 \times 4.$

(3) $p > q \Rightarrow pr < qr$ if $r < 0$; $6 > 4 \Rightarrow -3 \times 6 < -3 \times 4.$

Care must be taken when dealing with negative quantities because the negative sign can change the inequality.

Example 16 $3x + 4 < 4x - 3$
$-x + 4 < -3,$ subtracting $4x$
$-x < -7,$ subtracting -4
$x > 7,$ multiplying by -1

or $3x + 4 < 4x - 3$
$3x + 7 < 4x,$ adding 3
$7 < x,$ subtracting $3x$
i.e. $x > 7,$ the same result.

Inequalities and regions

$3x + 4y = 12$ represents a straight line (see figure 3) passing through (4, 0) and (0, 3). For the point (1, 1) $3x + 4y = 3 + 4 = 7 < 12$. (1, 1) lies in region $3x + 4y < 12$. Region below line represents $3x + 4y < 12$ (shaded). Region above line (unshaded) represents $3x + 4y > 12$.

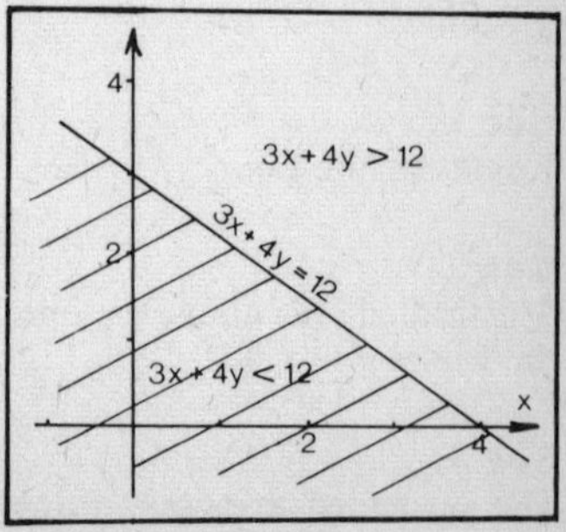

Figure 3

Simultaneous inequalities

Find the region satisfied by the inequalities $x \geq 0$, $y \geq 0$, $x + y \leq 4$, and $y \geq 2x - 1$.

$x \geq 0$ and $y \geq 0$ restricts the solution set to the positive quadrant. It is helpful to shade the region **unwanted** to focus attention on the **wanted** region. $x + y = 4$ is a straight line passing through (0, 4) and (4, 0). $x + y < 4$ is the region below this line {including (0, 0)}. $y = 2x - 1$ passes through (1, 1) and (2, 3).

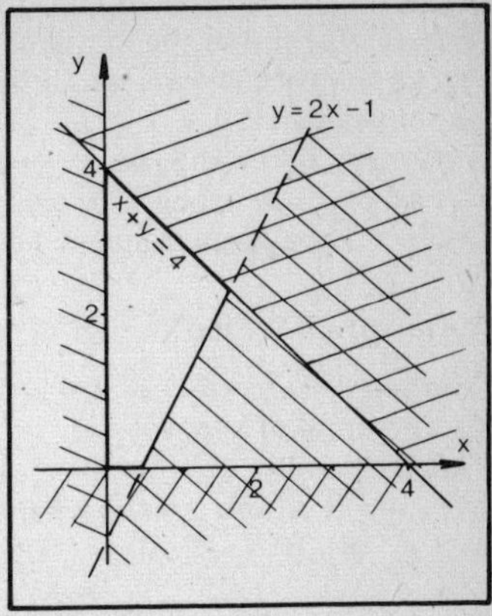

Figure 4

The point (2, 1) satisfies $y < 2x - 1$, since $1 < 4 - 1$, and lies in the unwanted region. The **unshaded** quadrilateral is the solution set of points (figure 4). The boundaries are all included as all the inequalities are 'greater than or equal to' or 'less than or equal to'.

Quadratic inequalities

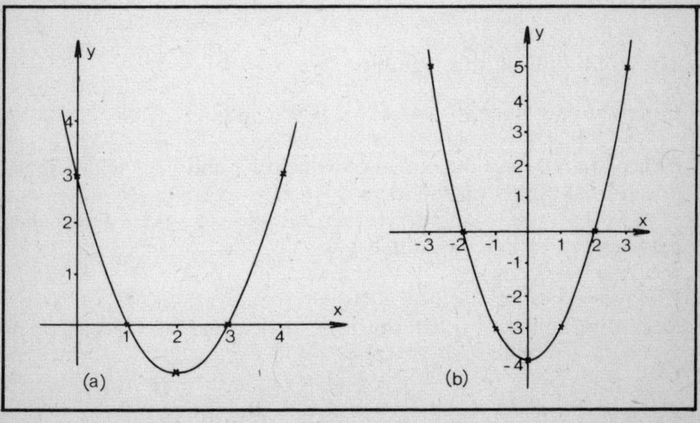

Figure 5

Example 17 Find the values of x which satisfy $x^2 - 4x + 3 < 0$.

Figure 5(a) shows the graph of $y = x^2 - 4x + 3 = (x - 1)(x - 3)$ which cuts the x-axis at $(1, 0)$ and $(3, 0)$. When x lies between these values i.e. $1 < x < 3$, $y = x^2 - 4x + 3 < 0$ and so the solution set is $1 < x < 3$, **or** $x^2 - 4x + 3 = (x - 1)(x - 3) < 0$ is true if one bracket is negative and the other positive. This happens when x is between 1 and 3, i.e. $1 < x < 3$.

Example 18 Find the set of values for $x^2 > 4$.

Method 1 $x^2 > 4 \Leftrightarrow x^2 - 4 > 0$.
Draw graph of $y = x^2 - 4$
(see figure 5b).
$x^2 - 4 > 0$ (above x-axis) when
$x > 2$ or $x < -2$.

Method 2 $x^2 > 4$.
$x^2 = 4 \Leftrightarrow x = \pm 2$.
$x^2 > 4 \Leftrightarrow x > 2$
or $x < -2$.

Algebraic inequalities

Example 19 $(a - b)^2 > 0$ if a and b are real numbers and $a \neq b. \Leftrightarrow a^2 - 2ab + b^2 > 0 \Leftrightarrow a^2 + b^2 > 2ab$.

Writing $a^2 = x$ and $b^2 = y$, $\dfrac{x + y}{2} > \sqrt{x}\sqrt{y} = \sqrt{xy}$.

Thus the arithmetic mean of 2 numbers is greater than their geometric mean.

The equation $ax^2 + bx + c = 0$

The solutions of this equation are $x = \dfrac{-b \pm \sqrt{b^2 - 4ac}}{2a}$ and these are real when $b^2 - 4ac \geq 0 \Rightarrow b^2 \geq 4ac$.

Example 20 Find the ranges of values of x and y for which there are no real points on the locus $y^2 = x(1 - x)$.
$y^2 = x - x^2 > 0$, for real values of $y \Leftrightarrow x(1 - x) > 0 \Rightarrow x$ lies between 0 and 1 (see example 17).
$\Leftrightarrow x > x^2 \Leftrightarrow 0 < x < 1$.
$y^2 = x - x^2 \Leftrightarrow x^2 - x + y^2 = 0$.
Regarding this as a quadratic in x, real values of x occur when $1 > 4y^2 \quad (b^2 > 4ac)$, $\Leftrightarrow y^2 < \frac{1}{4} \Leftrightarrow -\frac{1}{2} < y < \frac{1}{2}$.
$y^2 = x - x^2 \Leftrightarrow y^2 + x^2 - x + \frac{1}{4} = \frac{1}{4} \Leftrightarrow y^2 + (x - \frac{1}{2})^2 = (\frac{1}{2})^2$ which represents a circle centre $(\frac{1}{2}, 0)$ radius $\frac{1}{2}$, and the region inside this circle forms the solution set.

Example 21 Solve for x: $\dfrac{(x-1)(x-3)}{(x+1)(x-2)} > 0.$

$(x+1)(x-2) > 0 \Leftrightarrow x < -1 \quad \text{or} \quad x > 2$

(compare with figure 5a).

$$(x-1)(x-3) > 0 \Leftrightarrow x < 1 \quad \text{or} \quad x > 3.$$

These 4 conditions mean $x < -1$ or $x > 3$.

$(x+1)(x-2) < 0 \Rightarrow -1 < x < 2.$
$(x-1)(x-3) < 0 \Leftrightarrow 1 < x < 3.$ These conditions mean $1 < x < 2$.
Complete solution $x < -1, \quad 1 < x < 2, \quad x > 3.$

Indices and logs

Rules of indices for positive integers

(i) $2^3 \times 2^4 = 2^7$ in general $a^m \times a^n = a^{m+n}$ m,n positive integers.
(ii) $2^5 \div 2^3 = 2^2$ in general $a^m \div a^n = a^{m-n}$ $m > n$.
(iii) $(2^3)^4 = 2^{12}$ in general $(a^m)^n = a^{mn}$.

Fractional and negative indices

(iv) $2^3 \div 2^3 = 2^0$ from rule 2, so $2^0 = 1$. In general $a^0 = 1$.

(v) $2^3 \div 2^5 = 2^{-2}$ from rule 2, so $2^{-2} = \dfrac{1}{2^2}$. In general $a^{-n} = \dfrac{1}{a^n}$.

(vi) $2^{1/2} \times 2^{1/2} = 2^1$ from rule 1, so $2^{1/2} = \sqrt{2}$.
$2^{1/3} \times 2^{1/3} \times 2^{1/3} = 2^1$ from rule 1, so $2^{1/3} = \sqrt[3]{2}$.
(vii) $8^{2/3} = (8^2)^{1/3}$ from rule $3 = \sqrt[3]{8^2} = 4$ from rule 6.
(viii) $8^{2/3} = (8^{1/3})^2 = (\sqrt[3]{8})^2 = 2^2 = 4$. In general $a^{m/n} = \sqrt[n]{a^m} = (\sqrt[n]{a})^m$.

Logarithms

$16 = 2^4, 32 = 2^5 \Leftrightarrow 16 \times 32 = 2^4 \times 2^5 = 2^{4+5} = 2^9 = 512.$

$$\left.\begin{aligned} 16 = 2^4 \Leftrightarrow 4 = \log_2 16 \\ 32 = 2^5 \Leftrightarrow 5 = \log_2 32 \end{aligned}\right\} \begin{aligned} \log_2 16 + \log_2 32 &= \log_2 16 \times 32 \\ &= \log_2 512. \end{aligned}$$

$\log_{10} 16 + \log_{10} 32 \approx 1{\cdot}2041 + 1{\cdot}5051 = 2{\cdot}7092 \approx \log_{10} 512.$

In general $a^x = p \Leftrightarrow x = \log_a p.$
$a^y = q \Leftrightarrow y = \log_a q.$

$pq = a^x a^y = a^{x+y} \Leftrightarrow \log_a pq = x + y = \log_a p + \log_a q.$

$\dfrac{p}{q} = \dfrac{a^x}{a^y} = a^{x-y} \Leftrightarrow \log_a \dfrac{p}{q} = x - y = \log_a p - \log_a q.$

$p^n = (a^x)^n = a^{xn} \Leftrightarrow \log_a p^n = nx = n \log_a p.$

Example 22 Solve for x: $3^x = 8$.

Take logs of both sides: $\log_{10} 3^x = \log_{10} 8$

$$x \log_{10} 3 = \log_{10} 8$$

$$x = \frac{\log_{10} 8}{\log_{10} 3} \approx \frac{0{\cdot}9031}{0{\cdot}4771} = 1{\cdot}893 \text{ to 3 decimal places.}$$

Example 23 $\log_{10} 800 = \log_{10} 100 \times 8 = \log_{10} 100 + \log_{10} 8$

$$\approx 2 + 0{\cdot}9031$$
$$= 2{\cdot}9031.$$

$$\log_{10} 0{\cdot}008 = \log_{10} 0{\cdot}001 \times 8 = \log_{10} 0{\cdot}001 + \log_{10} 8$$
$$= -3 + 0{\cdot}9031$$
$$= \bar{3}{\cdot}9031 \text{ or } -2{\cdot}0969.$$

Change of base of logarithms

From example 22, $3^x = 8 \Leftrightarrow x = \log_3 8 = \dfrac{\log_{10} 8}{\log_{10} 3}$.

In general $a^x = b \Leftrightarrow x = \log_a b$.

Taking logs to base c $\log_c a^x = \log_c b$

$$x \log_c a = \log_c b$$

$$x = \frac{\log_c b}{\log_c a} = \log_a b.$$

This formula gives a method for changing from logs to base a into logs to base c.

Experimental laws

x	1	2	3	4
y	3	5	7	9

This table of values gives a straight line when values of y are plotted against values of x.

The equation of the straight line is $y = 2x + 1$, where 2 is the gradient and 1 the intercept on the y-axis.

$$y = kx^n \Leftrightarrow \log y = \log k + \log x^n$$
$$= \log k + n \log x.$$

If $\log x$ values are plotted against values of $\log y$ (figure 6a), a straight line results in which the gradient is n and the intercept on the y-axis is $\log k$. The values of n and $\log k$ can be calculated from the graph and n and k determined.

$$y = kb^x \Leftrightarrow \log y = \log k + \log b^x$$
$$= \log k + x \log b.$$

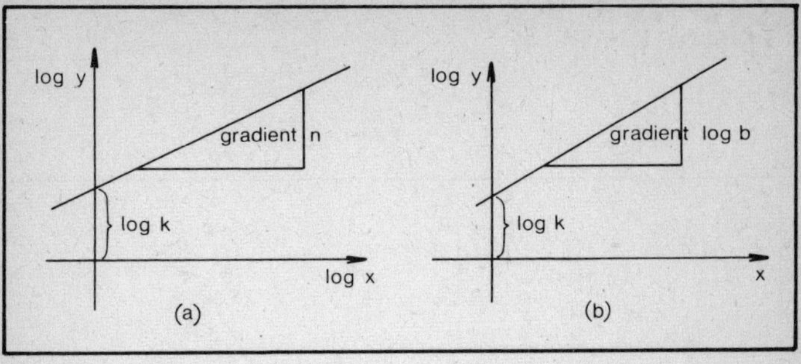

Figure 6

If values of x are plotted against values of $\log y$, a straight line results in which the gradient is $\log b$ and the intercept $\log k$. Hence k and b are determined.

Key terms

Remainder theorem $f(x)$ divided by $(x-a)$ gives remainder $f(a)$.

Factor theorem $f(a)=0 \Rightarrow x-a$ is a factor of $f(x)$.

Special factors $x^2-a^2=(x-a)(x+a)$;
$(x \pm a)^2 = x^2 \pm 2ax + a^2$;
$x^3-a^3=(x-a)(x^2+ax+a^2)$;
$x^3+a^3=(x+a)(x^2-ax+a^2)$.

Sum and product of roots If α and β are the roots of the equation $ax^2+bx+c=0$, $\alpha+\beta=\dfrac{-b}{a}$ and $\alpha\beta=\dfrac{+c}{a}$.

Indices and logarithms If $a^x=b \Leftrightarrow x=\log_a b$.

Change of base $\log_a b = \dfrac{\log_c b}{\log_c a}$.

Chapter 2
Trigonometry

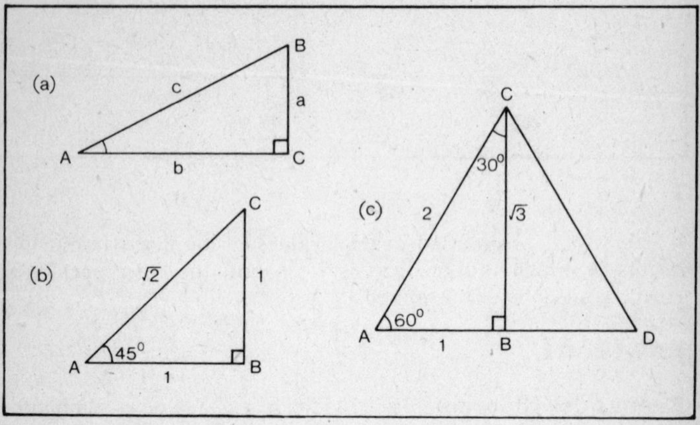

Figure 7

In the right-angled triangle in figure 7(a) the trigonometrical ratios are:

$$\textbf{sine:}\quad \sin A = \frac{a}{c};\quad \textbf{cosine:}\quad \cos A = \frac{b}{c};$$

$$\textbf{tangent:}\quad \tan A = \frac{a}{b}.$$

Pythagoras' theorem

$$a^2 + b^2 = c^2 \Rightarrow \frac{a^2}{c^2} + \frac{b^2}{c^2} = 1 \Rightarrow (\sin A)^2 + (\cos A)^2 = 1.$$

$(\sin A)^2$ is written as $\sin^2 A$, so that $\sin^2 A + \cos^2 A = 1$ (1)

Reciprocal functions

$$\textbf{cosecant:}\quad \operatorname{cosec} A = \frac{c}{a} = \frac{1}{\sin A};$$

$$\textbf{secant:}\quad \sec A = \frac{c}{b} = \frac{1}{\cos A};$$

$$\textbf{cotangent:}\quad \cot A = \frac{b}{a} = \frac{1}{\tan A}.$$

Dividing (1) throughout by $\cos^2 A$ gives $\dfrac{\sin^2 A}{\cos^2 A} + \dfrac{\cos^2 A}{\cos^2 A} = \dfrac{1}{\cos^2 A}$,

which simplifies to $\tan^2 A + 1 = \sec^2 A$.

Dividing (1) throughout by $\sin^2 A$ gives $\dfrac{\sin^2 A}{\sin^2 A} + \dfrac{\cos^2 A}{\sin^2 A} = \dfrac{1}{\sin^2 A}$,

which simplifies to $1 + \cot^2 A = \operatorname{cosec}^2 A$.

Special triangles

In triangle ABC in figure 7(b), AB = 1 = BC and $\angle$ABC = 90°;

$$\text{AC} = \sqrt{2} \simeq 1{\cdot}414 \text{ (Pythagoras)};\quad \sin 45° = \frac{1}{\sqrt{2}} = \cos 45°;$$

$$\tan 45° = 1 = \cot 45°;\quad \sec 45° = \sqrt{2} = \operatorname{cosec} 45°.$$

In figure 7(c) triangle ACD is equilateral with AD = 2, AB = 1, BC = $\sqrt{3} \simeq 1{\cdot}732$ (Pythagoras); $\sin 30° = \frac{1}{2} = \cos 60°$;

$$\cos 30° = \frac{\sqrt{3}}{2} = \sin 60°;\quad \tan 30° = \frac{1}{\sqrt{3}} = \frac{1}{\tan 60°} = \cot 60°.$$

The general angle

Sin A is the length PN (y-coordinate) as OP rotates anticlockwise and is the projection of OP on to the y-axis (see figure 8).

Cos A is the length ON (x-coordinate) as OP rotates anticlockwise and is the projection of OP on to the x-axis.

Tan A is the length XT which becomes infinite when $A = 90°$ and is negative when $90° < A < 180°$.

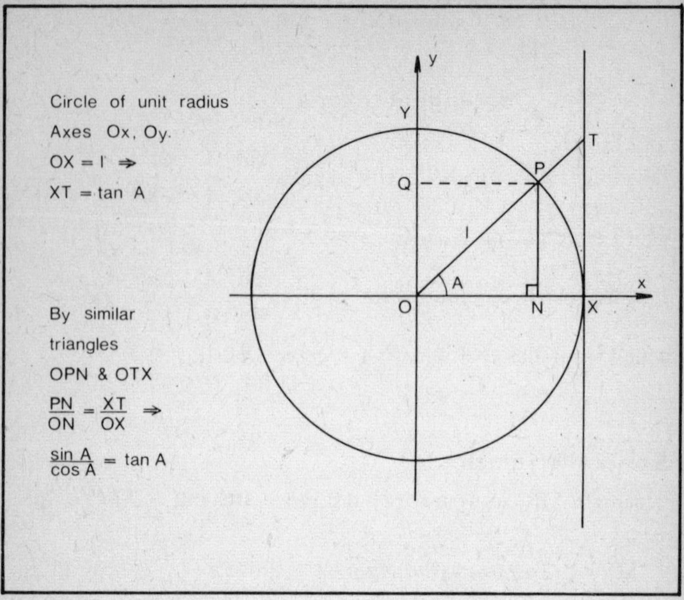

Figure 8

Graph of sin *A*

As A increases from 0° to 90°, sin A increases from 0 to 1 and as A increases from 90° to 180°, sin A decreases from 1 to 0 (see figure 9).

Sin A is positive in the 2nd quadrant (90° to 180°) and negative in the 3rd (180° to 270°) and 4th (270° to 360°) quadrants.

From the symmetry of the graph, $\sin 150° = \sin 30° = +0{\cdot}5$,

$$\sin 210° = \sin 330° = -\sin 30° = -0{\cdot}5.$$

The sine curve repeats its values every 360° and is **periodic**, the **period** being 360°. We have one **oscillation** or **cycle** between 0° and 360°. So $\sin 400° = \sin(360° + 40°) = \sin 40°$. The greatest height is 1 unit and is called the **amplitude**. By extending the graph for negative angles, $\sin(-30°) = -0{\cdot}5 = -\sin 30°$. This is an example of an **odd** function where $\sin(-x) = -\sin x$.

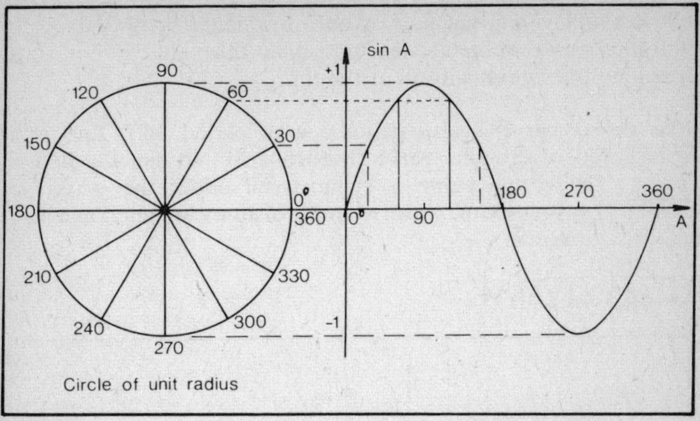

Figure 9

An odd function has rotational symmetry of order 2 about the origin. Such a function is called odd because all the graphs of odd powers of x have this property.

More formally, $f(x)$ is odd if $f(-x) = -f(x)$ for all x.

Graph of cos *A*

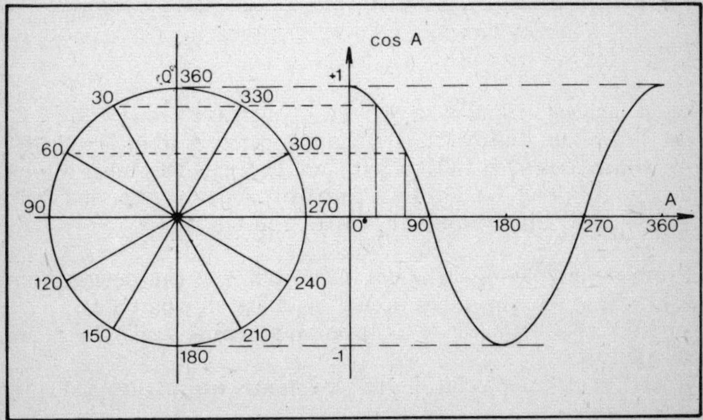

Figure 10

Since $\cos A$ is a projection on to a horizontal axis (x-axis) to draw its curve we rotate the unit circle through $^{+}90°$ or start labelling our angles with $0°$ at the top.

The cosine curve is also periodic with period $360°$, and is a translation of the sine curve through $^{-}90°$, so $\sin(A + 90°) = \cos A$. The cosine curve is symmetrical about the y-axis so $\cos(-A) = \cos A$. Cos A is an example of an **even** function where $\cos(-x) = \cos x$.

Graph of tan *A*

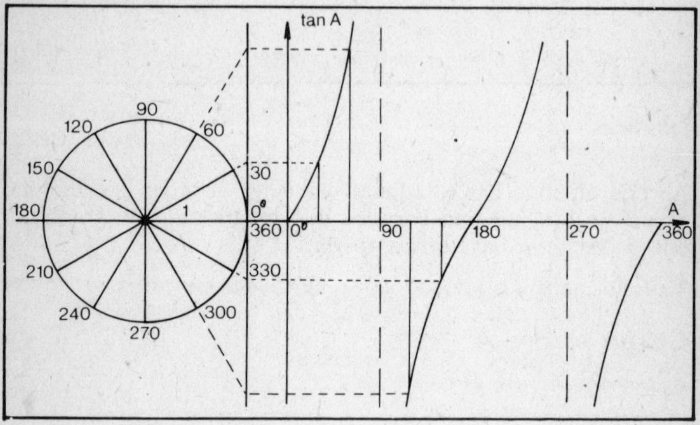

Figure 11

As A increases from $0°$ to $90°$, $\tan A$ increases from 0 to $+\infty$. As A increases from $90°$ to $180°$, $\tan A$ increases from $-\infty$ to $0°$. As A increases from $180°$ to $360°$, $\tan A$ repeats the values it took for $0 \leq A \leq 180°$. So $\tan A$ is a **periodic** function of period $180°$. $\text{Tan}(-A) = -\tan A$ and $\tan A$ is an odd function.

From the graphs of $\sin A$, $\cos A$ and $\tan A$, it can be seen that if $0° < A < 90°$, $\sin A$, $\cos A$ and $\tan A$ are all **positive**.
If $90° < A < 180°$, $\sin A$ is **positive**, $\cos A$ and $\tan A$ are **negative**.
If $180° < A < 270°$, $\sin A$ and $\cos A$ are **negative**, $\tan A$ is **positive**.
If $270 < A < 360°$, $\sin A$ and $\tan A$ are **negative**, $\cos A$ is **positive**.

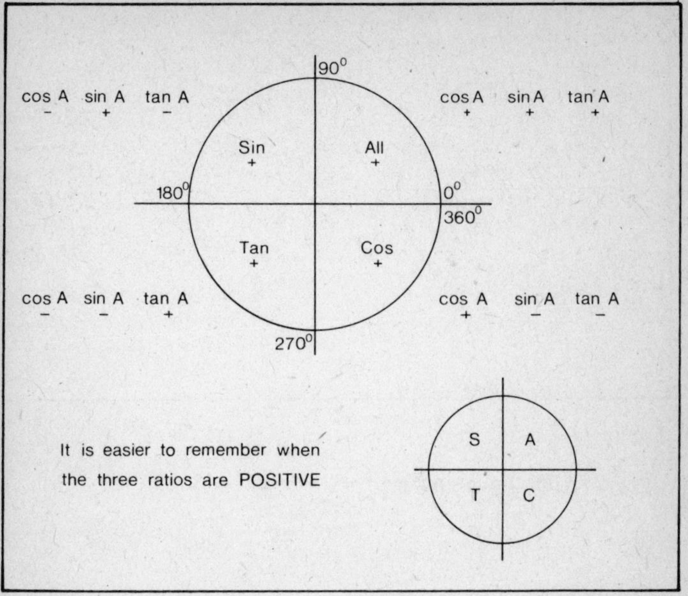

Figure 12

These facts can be summarized with a diagram of the 4 quadrants.

Radians

It is most useful to measure angles in radians rather than degrees. When this is done the gradient of the graph of $\sin x$ when $x = 0$ is equal to 1 and this has significant advantages when differentiating.

Definition of radians

1 radian (1^c) is the angle subtended at the centre of a circle by an arc equal in length to the radius (figure 9a).

In general (figure 9b) θ^c is the ratio of arc length s to radius r.

$$\frac{s}{r} = \theta^c \quad \text{or} \quad s = r\theta^c.$$

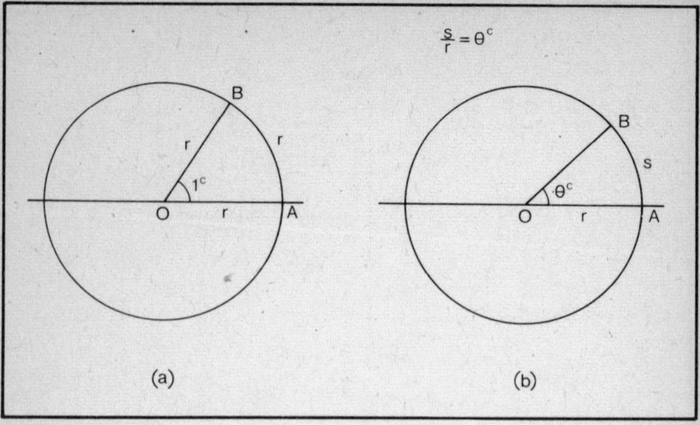

Figure 13

If $s = 2\pi r$ (circumference), $\dfrac{s}{r} = 2\pi^c \Rightarrow 2\pi^c = 360°$.

$$\pi^c = 180°, \quad 1^c = \frac{180°}{\pi} = 57{\cdot}3°.$$

If x is in radians, both axes of $y = \sin x$ are real number lines.

Compound angles

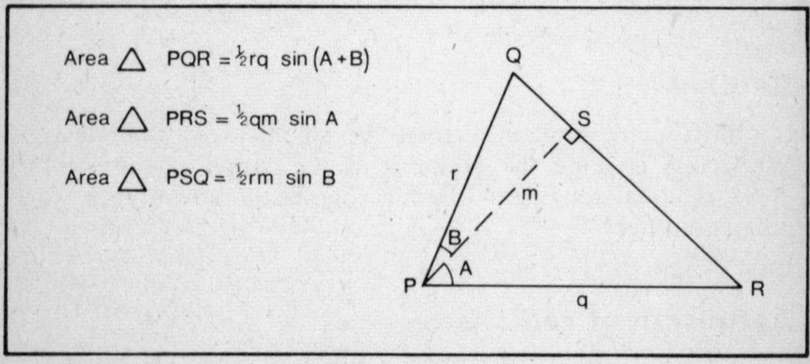

Figure 14

Area $\triangle$PQR = area $\triangle$PRS + area $\triangle$PSQ.

$\frac{1}{2}rq\sin(A + B) = \frac{1}{2}qm\sin A + \frac{1}{2}rm\sin B$,

$rq\sin(A + B) = qr\cos B\sin A + rq\cos A\sin B$.

$\therefore$ **$\sin(A + B) = \sin A\cos B + \cos A\sin B$.**

This is the simplest way of establishing $\sin(A + B)$ but it may be difficult if either A or B exceeds 90°.
Replacing B by $-B$ gives:

$$\sin(A - B) = \sin A \cos(-B) + \cos A \sin(-B),$$
$$\mathbf{\sin(A - B) = \sin A \cos B - \cos A \sin B}$$

since $\cos(-B) = \cos B$ and $\sin(-B) = -\sin B$ (see pages 26 to 28).

Another method uses matrices and vectors.

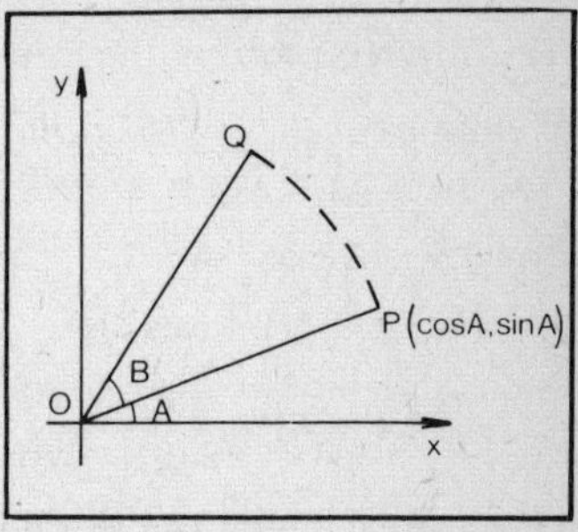

Figure 15

Rotate the vector

$$\mathbf{OP} = \begin{pmatrix} \cos A \\ \sin A \end{pmatrix}$$

through an angle B to OQ, so

$$\mathbf{OQ} = \begin{pmatrix} \cos(A + B) \\ \sin(A + B) \end{pmatrix}$$

The matrix for a rotation of $B°$ about O is

$$\begin{pmatrix} \cos B & -\sin B \\ \sin B & \cos B \end{pmatrix},$$

$$\text{so } \begin{pmatrix} \cos(A + B) \\ \sin(A + B) \end{pmatrix} = \begin{pmatrix} \cos B & -\sin B \\ \sin B & \cos B \end{pmatrix} \begin{pmatrix} \cos A \\ \sin A \end{pmatrix}$$

$$= \begin{pmatrix} \cos B \cos A - \sin B \sin A \\ \sin B \cos A + \cos B \sin A \end{pmatrix}$$

$$\Leftrightarrow \cos(A + B) = \cos A \cos B - \sin A \sin B. \quad (2)$$
$$\sin(A + B) = \sin A \cos B + \cos A \sin B. \quad (3)$$

Substituting $-B$ for B in (2) and (3) and using $\cos(-B) = \cos B$ and $\sin(-B) = -\sin B$ gives

$$\cos(A - B) = \cos A \cos B + \sin A \sin B, \quad (4)$$
$$\text{and } \sin(A - B) = \sin A \cos B - \cos A \sin B. \quad (5)$$

Substituting $B = A$ in (3) gives

$$\sin(A + A) = \sin A \cos A + \cos A \sin A = 2 \sin A \cos A,$$
$$\mathbf{\sin 2A = 2 \sin A \cos A}. \quad (6)$$

Substituting $B = A$ in (2) gives

$\cos(A + A) = \cos A \cos A - \sin A \sin A$,

$\mathbf{\cos 2A = \cos^2 A - \sin^2 A}$. (7)

Using equation (1) i.e. $\cos^2 A = 1 - \sin^2 A$,

$\cos 2A = 1 - \sin^2 A - \sin^2 A$,

$\mathbf{\cos 2A = 1 - 2\sin^2 A}$. (8)

This is often useful in the form $\sin^2 A = \frac{1}{2}(1 - \cos 2A)$.

Using $\sin^2 A = 1 - \cos^2 A$ in (7),

$\cos 2A = \cos^2 A - (1 - \cos^2 A) = \cos^2 A - 1 + \cos^2 A$,

$\mathbf{\cos 2A = 2\cos^2 A - 1}$. (9)

or $\cos^2 A = \frac{1}{2}(1 + \cos 2A)$. (10)

$$\tan(A + B) = \frac{\sin(A + B)}{\cos(A + B)} = \frac{\sin A \cos B + \cos A \sin B}{\cos A \cos B - \sin A \sin B}.$$

Dividing each term top and bottom by $\cos A \cos B$ gives

$$\tan(A + B) = \frac{\tan A + \tan B}{1 - \tan A \tan B}.$$

If $B = A$, $\tan 2A = \dfrac{2\tan A}{1 - \tan^2 A}$.

Example 1 Express $\sin 3A$ in terms of $\sin A$ only.

$$\begin{aligned}\sin 3A = \sin(2A + A) &= \sin 2A \cos A + \cos 2A \sin A \\ &= 2\sin A \cos A \cos A + (1 - 2\sin^2 A)\sin A \\ &= 2\sin A(1 - \sin^2 A) + \sin A - 2\sin^3 A \\ &= 2\sin A - 2\sin^3 A + \sin A - 2\sin^3 A.\end{aligned}$$

$$\mathbf{\sin 3A = 3\sin A - 4\sin^3 A}.$$

A similar method gives

$\mathbf{\cos 3A = 4\cos^3 A - 3\cos A}$, a similar result!

***t* formulae** $\sin 2A = 2\sin A \cos A = \dfrac{2\sin A \cos A}{\sin^2 A + \cos^2 A}$,

using $\sin^2 A + \cos^2 A = 1$ for the denominator.

Dividing top and bottom by $\cos^2 A$ gives $\dfrac{2\tan A}{1 + \tan^2 A}$.

$$\cos 2A = \frac{\cos^2 A - \sin^2 A}{\cos^2 A + \sin^2 A} = \frac{1 - \tan^2 A}{1 + \tan^2 A},$$

dividing top and bottom by $\cos^2 A$

Also $\tan 2A = \dfrac{2 \tan A}{1 - \tan^2 A}$.

Writing $\tan A = t$, we have

$$\sin 2A = \frac{2t}{1+t^2}; \qquad \cos 2A = \frac{1-t^2}{1+t^2}; \qquad \tan 2A = \frac{2t}{1-t^2}.$$

Factor formulae

Adding equations (3) and (5) gives

$$\sin(A+B) + \sin(A-B) = 2 \sin A \cos B. \tag{11}$$

Subtracting (5) from (3)

$$\sin(A+B) - \sin(A-B) = 2 \cos A \sin B. \tag{12}$$

Adding (2) and (4)

$$\cos(A+B) + \cos(A-B) = 2 \cos A \cos B. \tag{13}$$

Subtracting (4) from (2)

$$\cos(A+B) - \cos(A-B) = -2 \sin A \sin B. \tag{14}$$

These formulae are most useful in integration to change products into sums and differences which are generally easier to integrate. By writing $A + B = P$ and $A - B = Q$, i.e.

$$A = \frac{P+Q}{2}, \quad B = \frac{P-Q}{2},$$

equations (11) to (14) become

$$\sin P + \sin Q = 2 \sin \frac{(P+Q)}{2} \cos \frac{(P-Q)}{2}.$$

$$\sin P - \sin Q = 2 \cos \frac{(P+Q)}{2} \sin \frac{(P-Q)}{2}.$$

$$\cos P + \cos Q = 2 \cos \frac{(P+Q)}{2} \cos \frac{(P-Q)}{2}.$$

$$\cos P - \cos Q = -2 \sin \frac{(P+Q)}{2} \sin \frac{(P-Q)}{2}.$$

Equations

In solving trigonometrical equations it is possible to obtain an infinite number of solutions.

Example 2 $\sin x = 0{\cdot}6$.

Four solutions are arrowed in figure 16.
$x = 36{\cdot}9°$ or $180° - 36{\cdot}9° = 143{\cdot}1°$, or $360° + 36{\cdot}9°$, or $360° + 143{\cdot}1°$.

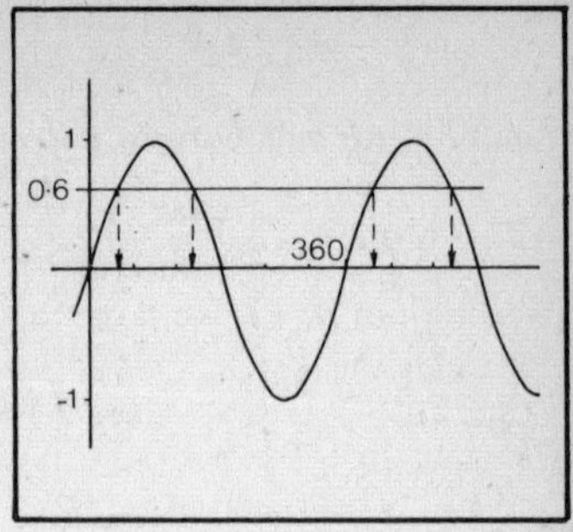

Figure 16

The general solution is

$x = 360k° + 36{\cdot}9°$ or $360k° + 143{\cdot}1°$, where $k = 0, \pm 1, \pm 2$, etc.

Example 3 $\cos x = -0{\cdot}5$; $x = 120°$ or $360° - 120° = 240°$.
General solution $x = 360k° \pm 120°$.

Example 4 $\tan x = 0{\cdot}4$; $x = 21{\cdot}8°$ or $180° + 21{\cdot}8° = 201.8°$.
General solution $x = 180k° + 21{\cdot}8°$.

In the following examples solutions will be given in the range $0° \leq x \leq 360°$.

Example 5 $4 \sin x + 3 \cos x = 1$.

Method 1 Dividing through by $\sqrt{3^2 + 4^2} = 5$, gives

$$\tfrac{4}{5} \sin x + \tfrac{3}{5} \cos x = \tfrac{1}{5}. \qquad (15)$$

The LHS is now a compound angle, if $\frac{4}{5} = \cos A$, $\frac{3}{5} = \sin A$.

$$\cos A \sin x + \sin A \cos x = 0{\cdot}2$$
$$\sin(x + A) = 0{\cdot}2$$
$$x + A = 11{\cdot}5° \quad \text{or} \quad 168{\cdot}5°.$$
$$\sin A = 0{\cdot}6 \Rightarrow A = 36{\cdot}9,$$
$$\text{so} \quad x + 36{\cdot}9° = 11{\cdot}5° \quad \text{or} \quad 168{\cdot}5° \quad \text{or} \quad 371{\cdot}5.$$
$$x = 11{\cdot}5° - 36{\cdot}9 = -25{\cdot}9°$$
$$\text{or} \quad x = 168{\cdot}5 - 36{\cdot}9 = 131{\cdot}9°$$
$$\text{or} \quad x = 371{\cdot}5 - 36{\cdot}9 = 334{\cdot}6°.$$

If $\frac{4}{5} = \sin B$ and $\frac{3}{5} = \cos B$, equation (15) becomes

$$\sin B \sin X + \cos B \cos x = 0{\cdot}2$$
$$\cos(x - B) = 0{\cdot}2, \text{ leading to the same result.}$$

Method 2 From page 33,

$$t = \tan \tfrac{1}{2}x; \quad \sin x = \frac{2t}{1+t^2}; \quad \cos x = \frac{1-t^2}{1+t^2}.$$

Substituting in $4 \sin x + 3 \cos x = 1$,

$$\frac{8t}{1+t^2} + \frac{3(1-t^2)}{1+t^2} = 1$$

$$8t + 3 - 3t^2 = 1 + t^2$$
$$2t^2 - 4t - 1 = 0.$$

Thus

$$t = \frac{+4 \pm \sqrt{16 + 4 \times 2 \times 1}}{4} = \frac{4 \pm \sqrt{24}}{4} = 2{\cdot}225 \quad \text{or} \quad -0{\cdot}225.$$

$$t = \tan \tfrac{1}{2}x = 2{\cdot}225 \quad \text{or} \quad -0{\cdot}225,$$
$$\tfrac{1}{2}x = 65{\cdot}8^\circ \quad \text{or} \quad 180^\circ - 12{\cdot}7^\circ = 167{\cdot}3^\circ,$$
$$x = 131{\cdot}6^\circ \quad \text{or} \quad 334{\cdot}6^\circ.$$

Example 6 $\sin 3x + 3 + 4 \sin x = 0$.

Use $\sin 3x = 3 \sin x - 4 \sin^3 x$.
Then $3 \sin x - 4 \sin^3 x + 3 + 4 \sin x = 0$,
$4 \sin^3 x - 7 \sin x - 3 = 0$, a cubic in $\sin x$.

It is not easy to solve a cubic equation. If one root can be found, the cubic expression can then be expressed as the product of a linear factor and a quadratic factor.

Consider $f(y) = 4y^3 - 7y - 3 = 0$ and use the **factor theorem**.

$$f(1) = 4 - 7 - 3 = -6,$$
$$f(2) = 32 - 14 - 3 = 15.$$

There is a root between 1 and 2. This will not give us a solution since $\sin x$ would be > 1.

$$f(0) = -3$$
$$f(-1) = -4 + 7 - 3 = 0$$

Hence $y + 1$ is a factor of $f(y)$.

$f(y) = (y+1)(4y^2 - 4y - 3) = 0$, by long division or inspection.

$(y+1)(2y-3)(2y+1) = 0$.

$y + 1 = 0$	or $2y - 3 = 0$	or $2y + 1 = 0$,
$y = -1$	or $y = 1{\cdot}5$	or $y = -\frac{1}{2}$.
$\sin x = -1$	no solution,	$\sin x = -0{\cdot}5$,
$x = 270^\circ$		$x = 210^\circ$ or 330°.

Example 7 $\sec 2x - 2\tan 2x + 2\tan x = 0.$ (16)

If $t = \tan x$; $\sec 2x = \dfrac{1}{\cos 2x} = \dfrac{1+t^2}{1-t^2}$ and $\tan 2x = \dfrac{2t}{1-t^2}$.

Equation (16) becomes $\dfrac{1+t^2}{1-t^2} - \dfrac{4t}{1-t^2} + 2t = 0.$

$$1 + t^2 - 4t + 2t - 2t^3 = 0 \Leftrightarrow 2t^3 - t^2 + 2t - 1 = 0.$$

Consider $f(t) = 2t^3 - t^2 + 2t - 1 = 0.$ (17)

$f(1) = 2 - 1 + 2 - 1 = 1$; $f(0) = -1 \Leftrightarrow$ a root exists between 0 and 1.
$f(-1) = -2 - 1 - 2 - 1 = -6.$
$f(\frac{1}{2}) = \frac{1}{4} - \frac{1}{4} + 1 - 1 = 0 \Leftrightarrow (2t - 1)$ is a factor.
$f(t) = (2t - 1)(t^2 + 1) = 0.$ This can be deduced directly from (17).
$(2t - 1) = 0$ or $t^2 + 1 = 0.$
$t = \frac{1}{2}$ or no real solution.
$\tan x = \frac{1}{2} \Rightarrow x = 26{\cdot}6^\circ$ or $206{\cdot}6^\circ$.

Example 8 $\sin x + \sin 2x - \sin 3x = 0.$ (18)

Using $\sin P - \sin Q = 2\cos\frac{1}{2}(P+Q)\sin\frac{1}{2}(P-Q)$,

$$\begin{aligned}\sin x - \sin 3x &= 2\cos\tfrac{1}{2}(x+3x)\sin\tfrac{1}{2}(x-3x)\\ &= 2\cos 2x\sin(-x) = -2\cos 2x\sin x.\end{aligned}$$

Equation (18) becomes

$$\begin{aligned}\sin 2x - 2\cos 2x\sin x &= 0,\\ 2\sin x\cos x - 2\cos 2x\sin x &= 0,\\ 2\sin x(\cos x - \cos 2x) &= 0,\\ \sin x = 0 \Rightarrow x &= 0^\circ, 180^\circ, 360^\circ.\end{aligned}$$

$$\begin{aligned}\text{Or}\quad \cos x - \cos 2x = 0 \Rightarrow\ & \cos x - 2\cos^2 x + 1 = 0,\\ & 2\cos^2 x - \cos x - 1 = 0,\\ & (2\cos x + 1)(\cos x - 1) = 0,\\ & \cos x = -\tfrac{1}{2} \quad\text{or}\quad \cos x = 1,\\ & x = 120^\circ, 240^\circ \quad\text{or}\quad 0^\circ, 360^\circ.\end{aligned}$$

Solutions are: $0^\circ, 120^\circ, 180^\circ, 240^\circ$ or 360°.

Inverse trigonometric functions

$f(x) = \sin x = 0{\cdot}5 \Rightarrow x = \sin^{-1}(0{\cdot}5) = 30^\circ = \dfrac{\pi}{6}$, the angle whose sine is $0{\cdot}5$.

$$f(30^\circ) = 0{\cdot}5 \Rightarrow f^{-1}(0{\cdot}5) = 30^\circ.$$

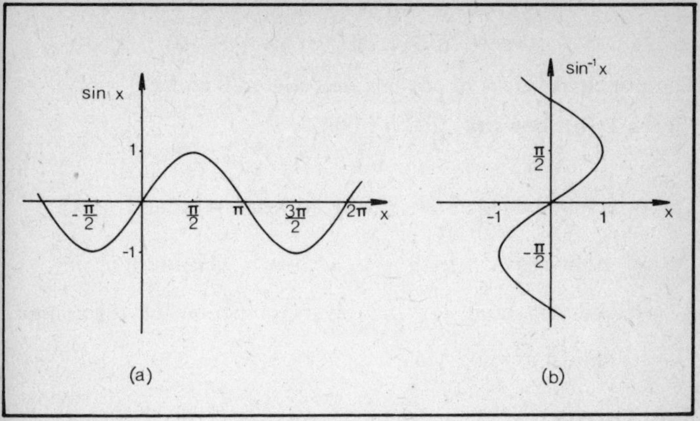

Figure 17

Sin^{-1} x or arc sin x is the inverse function for sin x, i.e. the angle whose sine is x. Sin$^{-1}(0{\cdot}5) = \dfrac{\pi^c}{6}$ (30°) or $\dfrac{5\pi^c}{6}$ (150°) and has two values between 0° and 360°.

To make sure that sin^{-1} x is a one-valued function we must restrict the range of values of sin x to be between $-\frac{1}{2}\pi^c$ and $+\frac{1}{2}\pi^c$, and the values obtained are called the **principal values** of the function.

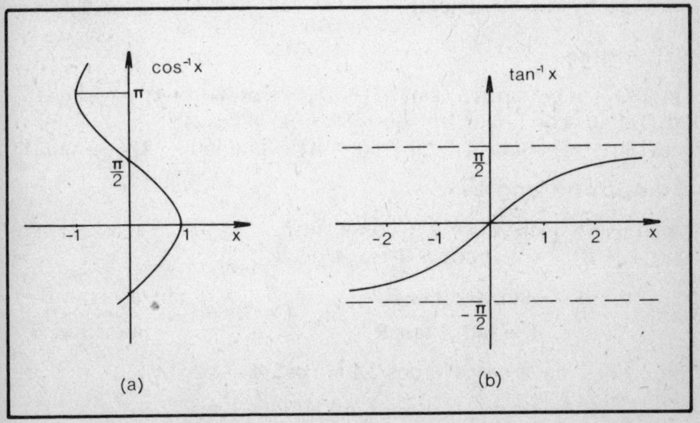

Figure 18

Figure 18(a) shows the graph of $\cos^{-1} x$ (or arc cos x).

$\cos x = 0{\cdot}5 \Rightarrow x = \cos^{-1}(0{\cdot}5) = \frac{1}{3}\pi^c$ or $-\frac{1}{3}\pi^c$ ($60°$ or $-60°$).

The principal values of $\cos^{-1} x$ lie between 0 and π^c.

Figure 18(b) shows the graph of $\tan^{-1} x$.

$$\tan x = 1 \Rightarrow x = \tan^{-1}(1) = \tfrac{1}{4}\pi^c \quad (45°).$$

The principal values of $\tan^{-1} x$ lie between $-\frac{1}{2}\pi^c$ and $+\frac{1}{2}\pi^c$.

Do not confuse $\sin^{-1} x$ with $\dfrac{1}{\sin x} = \operatorname{cosec} x$. The notation $\sin^{-1} x$ is reserved to stand for the inverse function of $\sin x$, and $\dfrac{1}{\sin x}$ is shown as $(\sin x)^{-1}$.

Key terms

$$\operatorname{cosec} A = \frac{1}{\sin A}; \quad \sec A = \frac{1}{\cos A}; \quad \cot A = \frac{1}{\tan A}.$$

$$\sin^2 A + \cos^2 A = 1; \quad 1 + \tan^2 A = \sec^2 A;$$
$$1 + \cot^2 A = \operatorname{cosec}^2 A.$$

Special angles $\sin 45° = \cos 45° = \dfrac{1}{\sqrt{2}}$; $\tan 45° = 1$;

$$\sin 30° = \cos 60° = 0{\cdot}5; \quad \sin 60° = \cos 30° = \frac{\sqrt{3}}{2};$$

$$\tan 60° = \frac{1}{\tan 30°} = \sqrt{3}.$$

All angles

$\sin(180 - A) = \sin A$; $\sin(180 + A) = \sin(360 - A) = -\sin A$;
$\cos(180 \pm A) = -\cos A$; $\cos(360 \pm A) = \cos A$;
$\tan(180 + A) = \tan A$; $\tan(180 - A) = \tan(360 - A) = -\tan A$.

Compound angles

$\sin(A \pm B) = \sin A \cos B \pm \cos A \sin B$;
$\cos(A \pm B) = \cos A \cos B \mp \sin A \sin B$;

$$\tan(A + B) = \frac{\tan A + \tan B}{1 - \tan A \tan B}; \quad \tan(A - B) = \frac{\tan A - \tan B}{1 + \tan A \tan B};$$

$$\sin 2A = 2 \sin A \cos A; \quad \cos 2A = \cos^2 A - \sin^2 A$$

$$\tan 2A = \frac{2 \tan A}{1 - \tan^2 A}; \qquad = 2\cos^2 A - 1$$
$$= 1 - 2\sin^2 A.$$

Chapter 3
Differentiation

Differentiation is the name given to a process which determines the rate of change of one variable with respect to another. It is a section of that branch of mathematics called calculus. In it, we consider the rates of change rather than the amount of change and do so by investigating the effect of very small increases or decreases in the variables concerned.

Before proceeding further we need to examine the idea of a limit of a function, since this is required for our discussion of the process of differentiation.

Limits

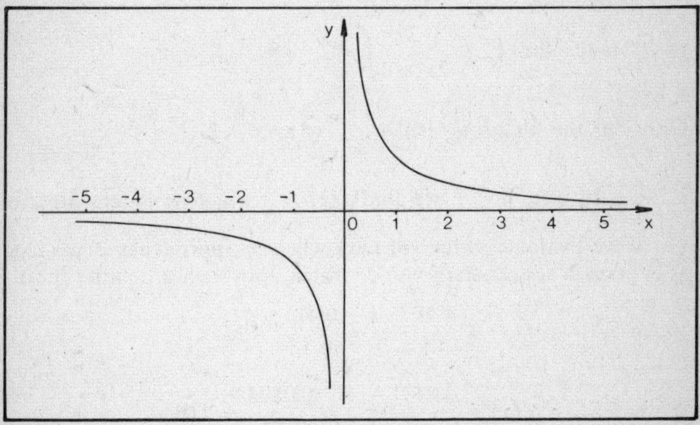

Figure 19

Consider the function given by $y = f(x) = \frac{2}{x}$. The graph of this function is shown in figure 19. It can be clearly seen that
(a) as $x \to \infty, f(x) \to 0$,
(b) as $x \to -\infty, f(x) \to 0$,
but $f(x) \neq 0$ for any value of x and we say that $f(x)$ tends to a limit of zero.

Statements (a) and (b) above are written

$$\lim_{x\to\infty} f(x) = 0 \quad \text{and} \quad \lim_{x\to-\infty} f(x) = 0.$$

Note that as $x \to \pm\infty, f(x) \to 0$, but in one case the values of $f(x)$ are just positive and in the other they are just negative. It is usual to say that $f(x)$ approaches zero from above or below as the case may be.

The curve is discontinuous at $x = 0$ and we say that at $x = 0$ there is a point of discontinuity.

It is not always necessary to sketch the function to determine any limits. If it can be expressed in a convenient form, the limit is easily recognizable.

Thus in the form $y = 2 - \dfrac{3}{2x+1}$ it is clear that as x gets larger the value of $\dfrac{3}{2x+1}$ decreases and hence the value of y becomes nearer to 2. Hence $\lim\limits_{x\to\infty}\left(2 - \dfrac{3}{2x+1}\right) = 2.$

Consider the function $f(x) = \dfrac{x^2-4}{x-2}$ as $x \to 2$.

If we substitute $x = 2$ we find $f(2) = \dfrac{0}{0}$ which is indeterminate, but if we evaluate values of $f(x)$ when x approaches 2 we find $f(x)$ takes a sequence of values which approach a definite limit.

If $x = 2{\cdot}1$, $\quad f(x) = \dfrac{4{\cdot}41 - 4}{2{\cdot}1 - 2} = \dfrac{0{\cdot}41}{0{\cdot}1} = 4{\cdot}1.$

If $x = 2{\cdot}01$, $\quad f(x) = \dfrac{4{\cdot}0401 - 4}{2{\cdot}01 - 2} = \dfrac{0{\cdot}0401}{0{\cdot}01} = 4{\cdot}01.$

If $x = 2{\cdot}001$, $\quad f(x) = \dfrac{4{\cdot}004001 - 4}{2{\cdot}001 - 2} = \dfrac{0{\cdot}004001}{0{\cdot}001} = 4{\cdot}001.$

Hence as $x \to 2$ the value of $f(x) \to 4$ but will never actually equal 4.

We write $\lim\limits_{x\to 2}\left(\dfrac{x^2-4}{x-2}\right) = 4.$

We usually effect the above process by adapting to an algebraic method.

Hence if $f(x) = \dfrac{x^2 - 4}{x - 2}$ consider the value of $f(2 + h)$.

We have $f(2 + h) = \dfrac{(2 + h)^2 - 4}{(2 + h) - 2} = \dfrac{h^2 + 4h}{h} = 4 + h$.

Now as h tends to 0, the value of x tends to 2 and hence $f(2) = 4$.

To show that $\lim\limits_{\theta \to 0} \left(\dfrac{\sin \theta}{\theta}\right) = 1$.

Sometimes we are required to find limits that involve the trigonometrical functions. One important result is that $\sin \theta \simeq \theta$ when θ is small. Remember that θ must be measured in radians.

Consider figure 20 in which two radii OA and OC enclose a sector of a circle centre O. Draw a tangent at A and produce OC to meet the tangent at B. Join AC.
Let the radius of the circle be r and angle AOC $= \theta$ (radians).
In ΔOAB, AB $= r \tan \theta$.
$\therefore$ area ΔOAB $= \frac{1}{2}r^2 \tan \theta$,
and area sector OAC $= \frac{1}{2}r^2\theta$
and area ΔOAC $= \frac{1}{2}r^2 \sin \theta$.

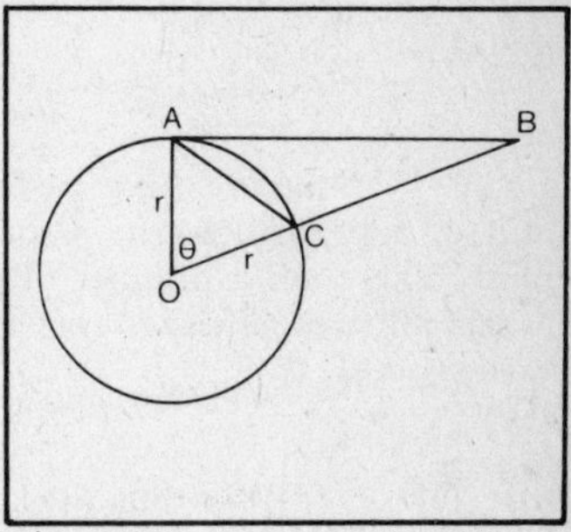

Figure 20

Clearly area ΔOAB > area sector OAC > area ΔOAC,
i.e. $\frac{1}{2}r^2 \tan \theta > \frac{1}{2}r^2\theta > \frac{1}{2}r^2 \sin \theta$.
Since $r^2 > 0$, we have $\tan \theta > \theta > \sin \theta$.
Dividing by $\sin \theta$, which is positive since θ is acute

$$\frac{1}{\cos \theta} > \frac{\theta}{\sin \theta} > 1.$$

Hence as $\theta \to 0$, $\dfrac{1}{\cos \theta} \to 1$.

$\therefore \dfrac{\theta}{\sin \theta} \to 1$ or $\theta \simeq \sin \theta$ and we write $\lim\limits_{\theta \to 0} \left(\dfrac{\sin \theta}{\theta}\right) = 1$.

This also gives $\cos\theta \to 1$ as $\theta \to 0$, but by using the identity $\cos\theta = 1 - 2\sin^2\left(\frac{\theta}{2}\right)$ we can obtain a better approximation. Namely $\cos\theta \simeq 1 - 2\left(\frac{\theta}{2}\right)^2 = 1 - \frac{1}{2}\theta^2$.

Example 1 Find the $\lim\limits_{\theta\to 0}\left(\frac{3\theta\sin\theta}{1-\cos 2\theta}\right)$.

$$\text{Now } \lim_{\theta\to 0}\left(\frac{3\theta\sin\theta}{1-\cos 2\theta}\right) = \lim_{\theta\to 0}\left(\frac{3\theta\sin\theta}{2\sin^2\frac{1}{2}\theta}\right)$$

$$= \lim_{\theta\to 0}\left(\frac{6\theta\sin\frac{1}{2}\theta\cos\frac{1}{2}\theta}{2\sin^2\frac{1}{2}\theta}\right)$$

$$= \lim_{\theta\to 0}\left(\frac{3\theta\cos\frac{1}{2}\theta}{\sin\frac{1}{2}\theta}\right)$$

$$= 6, \quad \text{since } \cos\tfrac{1}{2}\theta \to 1 \quad \text{and} \quad \frac{\frac{1}{2}\theta}{\sin\frac{1}{2}\theta} \to 1.$$

In this example we have assumed in the final stages that the limit of a product is the same as the product of the separate limits. There are a number of properties of limits which should be remembered. These are given below

(i) $\lim\limits_{x\to a}\{f(x)+g(x)\} = \lim\limits_{x\to a} f(x) + \lim\limits_{x\to a} g(x)$.

(ii) $\lim\limits_{x\to a}\{kf(x)\} = k\lim\limits_{x\to a} f(x)$, where k is a constant.

(iii) $\lim\limits_{x\to a}\{f(x)g(x)\} = \left\{\lim\limits_{x\to a} f(x)\right\}\left\{\lim\limits_{x\to a} g(x)\right\}$.

The derived function

The gradient of a linear function is constant, but when the graph is a curve **the gradient is different at different points** on the curve. A general function which represents the gradient of the curve is called the **derived function** and the process of finding this function is called **differentiation**.

First consider a geometrical approach to a limit. If we take a point P on a curve and a second point Q at a different point, then the gradient of the secant PQ can be found as this is a straight line (see page 108). If we now reduce the increment from P to Q so that Q moves to a new position nearer P, i.e. Q′, the gradient of PQ′ can now be found.

Repeating this process we can find the gradient of PQ″ and so on until Q becomes a point adjacent to P. In this case the gradient of the secant passing through P and a neighbouring point Q is very close to the gradient of the tangent to the curve drawn at P.

We say that the gradient of a curve at a given point is the gradient of the tangent to the curve at that point. It can be found by finding the limit of the gradient of a secant PQ as $Q \to P$. This is shown in figure 21.

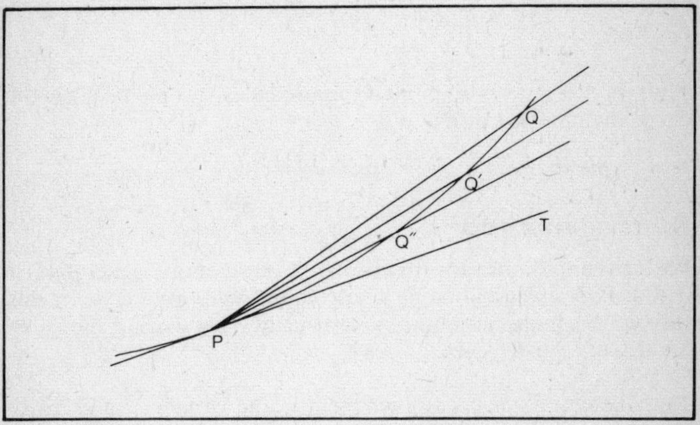

Figure 21

Example 2 Find the gradient of the tangent to the curve $y = 3x^2$ at the point P(1, 3).

Consider figure 22. Let Q be a point on the curve near to P whose x-coordinate is given by $x = 1 + h$. Hence the y-coordinate of Q will be $3(1 + h)^2$.
By considering ΔQPN,

$$\text{the gradient of PQ} = \frac{3(1+h)^2 - 3}{(1+h) - 1} = \frac{6h + 3h^2}{h} = 6 + 3h.$$

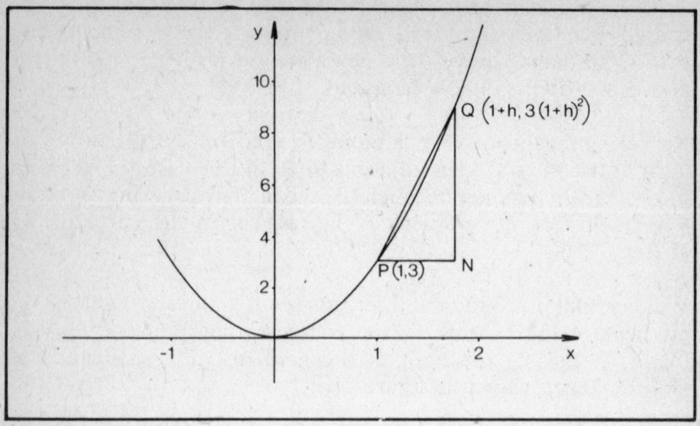

Figure 22

Thus as $h \to 0$, i.e. the point Q approaches the point P on the curve the gradient of PQ $\to 6$.

$\therefore$ The gradient of the tangent at (1, 3) is 6.

General method

We can apply the above process a little more generally by letting P be a general point whose coordinates are (x, y). In this case we use a special notation instead of h by writing the coordinates of Q as $(x + \delta x, y + \delta y)$.

Note δx simply means an increment in the variable x and δy an increment in the variable y. It does not mean the product of δ and x.

Consider the curve $y = x^3$, where P is a fixed point (x, y) and Q is a neighbouring point whose coordinates are given by $(x + \delta x, y + \delta y)$.

Since Q lies on the curve, $y + \delta y = (x + \delta x)^3$.
Since P lies on the curve, $y = x^3$.
Subtracting: $\delta y = (x + \delta x)^3 - x^3$.
Dividing by δx we have

the gradient of the chord $= \dfrac{\delta y}{\delta x} = \dfrac{(x + \delta x)^3 - x^3}{\delta x}$

$$= \frac{x^3 + 3x^2(\delta x) + 3x(\delta x)^2 + (\delta x)^3 - x^3}{\delta x}$$

$$= \frac{3x^2(\delta x) + 3x(\delta x)^2 + (\delta x)^3}{\delta x}.$$

$$\therefore \quad \frac{\delta y}{\delta x} = 3x^2 + 3x(\delta x) + (\delta x)^2.$$

Now the gradient of the tangent is the limit as $\delta x \to 0$ and is denoted by $\frac{dy}{dx}$.

$$\therefore \quad \frac{dy}{dx} = \lim_{\delta x \to 0} \left(\frac{\delta y}{\delta x}\right) = \lim_{\delta x \to 0} \{3x^2 + 3x(\delta x) + (\delta x)^2\}.$$

Hence the derived function is given by $\frac{dy}{dx} = 3x^2$.

Note that when $x = 2$, the gradient of the tangent is $3(2)^2 = 12$, and when $x = 0$, the gradient of the tangent $= 0$ and is thus parallel to the x-axis.
The above process is known as **differentiation from first principles**.

By inspection, from a number of results we have a general rule for the differentiation of a function of the form ax^n. It can be proved by induction.

If $y = ax^n$, $\quad \frac{dy}{dx} = nax^{n-1}$.

We can apply this rule to functions of this form to obtain their gradients without working from first principles. It applies even if n is negative or fractional.

Example 3 Find the gradients of the following functions at the points stated: (i) $y = 6x^3$ at $(2, 48)$; (ii) $y = \frac{1}{x}$ at $(1, 1)$; (iii) $y = \sqrt{x}$ at $(4, 2)$.

(i) Since $y = 6x^3$, $\frac{dy}{dx} = 18x^2 \Rightarrow$ gradient at $(2, 48) = 72$.

(ii) Since $y = \frac{1}{x} = x^{-1}$, $\frac{dy}{dx} = -1x^{-2} = \frac{-1}{x^2}$

$\Rightarrow$ gradient at (1, 1) = −1.

(iii) Since $y = \sqrt{x} = x^{\frac{1}{2}}$, $\frac{dy}{dx} = \frac{1}{2}x^{-\frac{1}{2}} = \frac{1}{2\sqrt{x}}$

$\Rightarrow$ gradient at (4, 2) = $\frac{1}{4}$.

By using an alternative approach we can derive the gradient function in a different form. Consider a mapping diagram for $x \rightarrow 2x$ for the interval $\{x: 0 \leq x \leq 4\}$ as shown in figure 23(a).

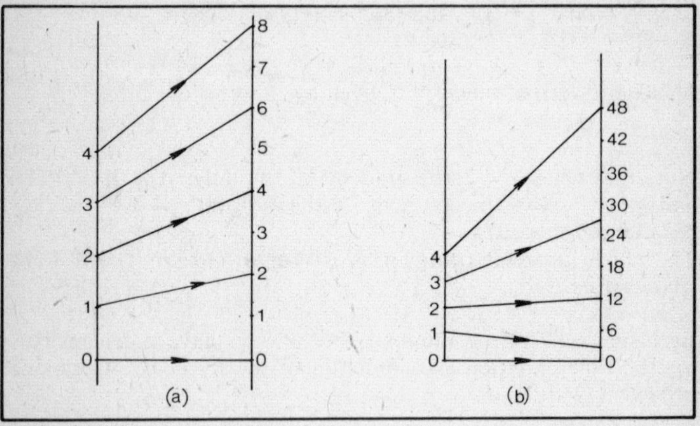

Figure 23

It is clear that in figure 23(a) we are considering a doubling function and the mapping is a uniform stretch of scale factor 2. Whereas in figure 23(b), which shows the mapping $x \rightarrow 3x^2$ over the same interval, no constant stretching is taking place. In this case we consider an average scale factor. For a function $f(x)$ over an interval $a \leq x \leq b$ the average scale factor is given by $\frac{\{f(b) - f(a)\}}{(b - a)}$. For $f(x) = 3x^2$, this factor over an interval $0 \leq x \leq 4$ would give

$$\frac{f(4) - f(0)}{4 - 0} = \frac{48 - 0}{4 - 0} = 12.$$

Now we can find the stretch experienced by a particular point by introducing a limiting process.
Consider again $x \to 3x^2$ as we approach $x = 1$ from above, i.e. the interval will be $1 \leq x \leq 1 + h$.

$$\therefore \quad \text{The average scale factor is } \frac{3(1+h)^2 - 3(1)^2}{(1+h) - 1} = \frac{6h + h^2}{h} = 6 + h.$$

Now as $h \to 0$ the factor $\to 6$. Compare our method and result in example 2 on page 43 and clearly we have formed the derivative of the function $3x^2$ at $(1, 3)$.

The derivative of $f(x)$ is written as $f'(x)$ and the derivative at a given point $x = a$ is given as $f'(a)$.

Hence over an interval $a \leq x \leq b$ the derivative of $f(x)$ at $x = a$ is defined as $f'(a) = \lim_{b \to a} \left\{ \frac{f(b) - f(a)}{b - a} \right\}$,

or, by replacing b with $a + h$, we have

$$f'(a) = \lim_{h \to 0} \left\{ \frac{f(a+h) - f(a)}{h} \right\}.$$

For $f(x) = \frac{1}{x}$ we can apply this method to find the differential at $(2, \frac{1}{2})$.

$$f'(2) = \lim_{h \to 0} \left\{ \frac{f(2+h) - f(2)}{h} \right\} = \lim_{h \to 0} \left\{ \frac{1/(2+h) - \frac{1}{2}}{h} \right\}$$

$$= \lim_{h \to 0} \left\{ \frac{2 - (2+h)}{2(2+h)h} \right\} = \lim_{h \to 0} \left\{ \frac{-h}{2(2+h)h} \right\}$$

$$= \lim_{h \to 0} \left\{ \frac{-1}{2(2+h)} \right\} = -\frac{1}{4}.$$

Polynomials

For a polynomial in x, say $y = 3x^2 + 7x - 6$, we can differentiate term by term using the general rule. Let y be a function which is the sum of two terms u and v, both of which are functions of x.

Then $y = u + v$. (1)

Now an increment of δx in x would cause increments of δu, δv and δy in u, v and y respectively.
Hence $\quad y + \delta y = (u + \delta u) + (v + \delta v)$. (2)
Subtracting (1) from (2) we have

$$\delta y = \delta u + \delta v.$$

Dividing by δx gives $\quad \dfrac{\delta y}{\delta x} = \dfrac{\delta u}{\delta x} + \dfrac{\delta v}{\delta x}.$

As $\delta x \to 0$; δu, δv and δy all approach 0.

$$\frac{dy}{dx} = \lim_{\delta x \to 0} \left(\frac{\delta y}{\delta x}\right) = \lim_{\delta x \to 0} \left(\frac{\delta u}{\delta x} + \frac{\delta v}{\delta x}\right).$$

Thus $\quad \dfrac{dy}{dx} = \dfrac{du}{dx} + \dfrac{dv}{dx}.$

This can be extended to any number of terms in the sum. Hence for $y = 3x^2 + 7x - 6$, we have $\dfrac{dy}{dx} = 6x + 7$.

Products

Consider the product $y = uv$, where u and v are both functions of x. An increment of δx in x will cause increments of δu, δv and δy in u, v and y respectively.

Hence $\quad y + \delta y = (u + \delta u)(v + \delta v).$
$\therefore \quad y + \delta y = uv + u(\delta v) + v(\delta u) + (\delta u)(\delta v).$

But $\quad y = uv.$

$\therefore$ Subtracting gives $\quad \delta y = u(\delta v) + v(\delta u) + (\delta u)(\delta v).$
Dividing by δx we have

$$\frac{\delta y}{\delta x} = u\frac{\delta v}{\delta x} + v\frac{\delta u}{\delta x} + \frac{\delta u}{\delta x}(\delta v).$$

In the limit as $\delta x \to 0$; $\delta u, \delta v, \delta y \to 0 \quad \therefore \quad \boldsymbol{\dfrac{dy}{dx} = u\dfrac{dv}{dx} + v\dfrac{du}{dx}}.$

Example 4 Differentiate $y = (x^2 - 3x + 7)(2x^3 - 5x)$.

Let $u = x^2 - 3x + 7. \quad \therefore \quad \dfrac{du}{dx} = 2x - 3.$

Let $v = 2x^3 - 5x$. $\quad\therefore\quad \dfrac{dv}{dx} = 6x^2 - 5.$

Using the result for a product

$$\frac{dy}{dx} = u\frac{dv}{dx} + v\frac{du}{dx}$$

$$= (x^2 - 3x + 7)(6x^2 - 5) + (2x^3 - 5x)(2x - 3).$$

Quotients

Consider the quotient $y = \dfrac{u}{v}$, where u and v are both functions of x. A process similar to that used for products will give an equivalent formula for quotients.

$$\therefore \frac{dy}{dx} = \frac{v(du/dx) - u(dv/dx)}{v^2}.$$

Example 5 Differentiate $y = \dfrac{x}{x^2 + 2}$.

Let $u = x$, thus $\dfrac{du}{dx} = 1$; let $v = x^2 + 2$, thus $\dfrac{dv}{dx} = 2x$.

Using the quotient result $\dfrac{dy}{dx} = \dfrac{(x^2 + 2)1 - x(2x)}{(x^2 + 2)^2} = \dfrac{2 - x^2}{(x^2 + 2)^2}$.

Function of a function

Consider $(3x^2 - 7)^4$. This is a function (i.e. the fourth power) of $3x^2 - 7$ which is itself a function of x. Hence $(3x^2 - 7)^4$ is called a function of a function of x.

If we let $y = (3x^2 - 7)^4$ and $u = 3x^2 - 7$, then $y = u^4$. Consider an increment of δx in x which will cause an increment of δu and δy in u and y respectively.
Since δu, δy, and δx are finite quantities (although small),

$$\frac{\delta y}{\delta x} = \frac{\delta y}{\delta u} \times \frac{\delta u}{\delta x}.$$

Clearly, in the limit, as $\delta x \to 0$, δu, $\delta y \to 0$, and $\dfrac{dy}{dx} = \dfrac{dy}{du} \times \dfrac{du}{dx}$.

Hence in the above example, $y = u^4$ and $\dfrac{dy}{du} = 4u^3$.

$u = 3x^2 - 7$ and $\dfrac{du}{dx} = 6x$.

$\therefore$ Using the function of a function result,

$$\frac{dy}{dx} = 4u^3 \times 6x = 24x(3x^2 - 7)^3.$$

Example 6 Differentiate $\sqrt{\dfrac{(1+x)}{(2+x)}}$ with respect to x.

This problem can be solved in two ways (i) by using the quotient rule and (ii) by using the product rule.

(i) Let $y = \dfrac{(1+x)^{1/2}}{(2+x)^{1/2}}$.

$$\left.\begin{aligned}\text{Then}\quad u &= (1+x)^{1/2}. \quad \therefore \frac{du}{dx} = \tfrac{1}{2}(1+x)^{-1/2} \times 1\\ v &= (2+x)^{1/2}. \quad \therefore \frac{dv}{dx} = \tfrac{1}{2}(2+x)^{-1/2} \times 1\end{aligned}\right\}\text{using function of a function.}$$

Using the quotient rule

$$\frac{dy}{dx} = \frac{\{(2+x)^{1/2}\}\frac{1}{2}(1+x)^{-1/2} - \{(1+x)^{1/2}\}\frac{1}{2}(2+x)^{-1/2}}{(2+x)}.$$

Multiplying numerator and denominator by $(1+x)^{1/2}(2+x)^{1/2}$.

$$\frac{dy}{dx} = \frac{\frac{1}{2}(2+x) - \frac{1}{2}(1+x)}{(2+x)^{3/2}(1+x)^{1/2}} = \frac{1}{2(2+x)^{3/2}(1+x)^{1/2}}$$

$$= \frac{1}{2\sqrt{(1+x)(2+x)^3}}.$$

(ii) By writing the function in the form $y = (1+x)^{1/2}(2+x)^{-1/2}$ it is possible to use the product formula.

Now $u = (1+x)^{1/2}$, $\dfrac{du}{dx} = \frac{1}{2}(1+x)^{-1/2} \times 1 = \frac{1}{2}(1+x)^{-1/2}$.

$v = (2+x)^{-1/2}$, $\dfrac{dv}{dx} = -\frac{1}{2}(2+x)^{-3/2} \times 1 = -\frac{1}{2}(2+x)^{-3/2}$.

$$\therefore \frac{dy}{dx} = u\frac{dv}{dx} + v\frac{du}{dx}$$

$$= (1+x)^{1/2}\{-\tfrac{1}{2}(2+x)^{-3/2}\} + (2+x)^{-1/2}\{\tfrac{1}{2}(1+x)^{-1/2}\}.$$

$$= -\tfrac{1}{2}(1+x)^{1/2}(2+x)^{-3/2} + \tfrac{1}{2}(1+x)^{-1/2}(2+x)^{-1/2}.$$

Extracting a factor of $\frac{1}{2}(1+x)^{-1/2}(2+x)^{-3/2}$ we obtain

$$\frac{dy}{dx} = \frac{-(1+x)+(2+x)}{2(1+x)^{1/2}(2+x)^{3/2}} = \frac{1}{2\sqrt{(1+x)(2+x)^3}}.$$

Implicit functions

An equation such as $3x^2 - xy + 3y = 7$ contains y not defined directly in terms of x, i.e. y is an implicit function of x. In this case the equation can be rearranged to give y as an explicit function of x, i.e. $y(3-x) = 7 - 3x^2$. $\therefore y = \dfrac{(7-3x^2)}{(3-x)}$. However some equations cannot be dealt with in this way, for example, $x^2 + xy - y^2 = 8$. In such examples we differentiate with respect to x, term by term, i.e.

$$\frac{d}{dx}(x^2) + \frac{d}{dx}(xy) - \frac{d}{dx}(y^2) = \frac{d}{dx}(8).$$

$$\therefore 2x + \left(x\frac{dy}{dx} + y\right) - 2y\frac{dy}{dx} = 0.$$

Remember that (i) $\dfrac{d}{dx}(xy) = x\dfrac{dy}{dx} + y$, since xy is a product;

(ii) $\dfrac{d}{dx}(y^2) = 2y\dfrac{dy}{dx}$, since y^2 is a function of a function

Hence $(x-2y)\dfrac{dy}{dx} = -y - 2x \Rightarrow \dfrac{dy}{dx} = -\dfrac{2x+y}{x-2y}$.

If y is a function of x, an increment of δx in x will cause an increment of δy in y. Since δy and δx are finite, small quantities

$$\frac{\delta y}{\delta x} = \frac{1}{\delta x/\delta y}.$$

As $\delta x \to 0$, $\delta y \to 0$ so, taking the limit,

$$\boldsymbol{\frac{dy}{dx} = \frac{1}{dx/dy}}.$$

Example 7 Differentiate the equation of the parabola $y^2 = 4ax$.

We can tackle this in a number of ways producing an answer in different forms.

(i) Since $y^2 = 4ax$, $x = \dfrac{y^2}{4a}$.

$\therefore$ Differentiating with respect to y, $\dfrac{dx}{dy} = \dfrac{2y}{4a} = \dfrac{y}{2a}$.

But $\dfrac{dy}{dx} = \dfrac{1}{dx/dy} = \dfrac{2a}{y}$.

(ii) Since $y^2 = 4ax$, $y = 2a^{\frac{1}{2}}x^{\frac{1}{2}}$.

Differentiating with respect to x, $\dfrac{dy}{dx} = \dfrac{2}{2}a^{\frac{1}{2}}x^{-\frac{1}{2}} = \sqrt{\dfrac{a}{x}}$.

(iii) By differentiating implicitly with respect to x.

$$\frac{d}{dx}(y^2) = \frac{d}{dx}(4ax).$$

$$2y\frac{dy}{dx} = 4a \quad \therefore \quad \frac{dy}{dx} = \frac{2a}{y} = \frac{2a}{2a^{\frac{1}{2}}x^{\frac{1}{2}}} = \sqrt{\frac{a}{x}}.$$

Parametric form

Sometimes, instead of writing an implicit equation relating two variables x and y, each is given in terms of a third variable, called a **parameter**. For example, in the equation $y^2 = 4ax$ above we could write two equations, $x = at^2$, $y = 2at$ which give the variables x and y in terms of t, the parameter. Eliminating the parameter, if possible, will give the usual equation, i.e.

$$x = a\left(\frac{y}{2a}\right)^2 \quad \therefore \ x = \frac{y^2}{4a} \Rightarrow y^2 = 4ax.$$

To find the gradient of curves given in parametric form we use the function of a function rule.

Example 8 Find the gradient of the parabola whose equations are given parametrically as $x = at^2$, $y = 2at$.

$\therefore \dfrac{dx}{dt} = 2at$ and $\dfrac{dy}{dt} = 2a$.

$$\therefore \frac{dy}{dx} = \frac{dy}{dt} \times \frac{dt}{dx} = \frac{2a}{2at} = \frac{1}{t}.$$

This gives yet another form for the gradient of the parabola $y^2 = 4ax$. This time in terms of a parameter t.

Trigonometrical functions

Consider the function $y = \sin x$. We can find the gradient from first principles by applying the process for algebraic functions. Let $P(x, y)$ and $Q(x + \delta x, y + \delta y)$ be two neighbouring points on the curve where δx is an increment in x, causing an increment δy in y.

Therefore $\quad y + \delta y = \sin(x + \delta x) \quad$ and $\quad y = \sin x$.

Subtracting $\quad \delta y = \sin(x + \delta x) - \sin x$

$\qquad\qquad = 2\cos(x + \tfrac{1}{2}\delta x)\sin(\tfrac{1}{2}\delta x)$ by factor formula.

Hence dividing by δx, the gradient of the chord PQ is given by

$$\frac{\delta y}{\delta x} = \frac{2\cos(x + \tfrac{1}{2}\delta x)\sin(\tfrac{1}{2}\delta x)}{\delta x} = \cos(x + \tfrac{1}{2}\delta x) \times \frac{\sin \tfrac{1}{2}\delta x}{\tfrac{1}{2}\delta x}.$$

In the limit as $\delta x \to 0$, $\delta y \to 0$,

$$\begin{aligned} \frac{dy}{dx} = \lim_{\delta x \to 0} \left(\frac{\delta y}{\delta x}\right) &= \lim_{\delta x \to 0} \left\{\cos(x + \tfrac{1}{2}\delta x) \times \frac{\sin \tfrac{1}{2}\delta x}{\tfrac{1}{2}\delta x}\right\} \\ &= \left\{\lim_{\delta x \to 0} \cos(x + \tfrac{1}{2}\delta x)\right\} \times \left\{\lim_{\delta x \to 0} \frac{\sin \tfrac{1}{2}\delta x}{\tfrac{1}{2}\delta x}\right\} \\ &= \cos x, \quad \text{since } \lim_{\theta \to 0} \left(\frac{\sin \theta}{\theta}\right) = 1. \end{aligned}$$

Hence $\qquad \dfrac{d}{dx}(\sin x) = \cos x.$

Remember that in the derivation of the gradient above, x must be measured in radians in order that the limit of $\dfrac{\sin \theta}{\theta}$ as $\theta \to 0$ may be applied. A similar method can be used to show that $\dfrac{d}{dx}(\cos x) = -\sin x.$

Composite functions

To find the differentials of composite functions we can use the methods discussed for products, quotients and the function of a function.

Example 9 Find the differential coefficient of tan x.

Let $y = \tan x = \dfrac{\sin x}{\cos x}$.

Using the quotient rule, $u = \sin x, \dfrac{du}{dx} = \cos x;$

$$v = \cos x, \frac{dv}{dx} = -\sin x.$$

$$\therefore \frac{dy}{dx} = \frac{v(du/dx) - u(dv/dx)}{v^2} = \frac{\cos x(\cos x) - \sin x(-\sin x)}{\cos^2 x}$$

$$= \frac{\cos^2 x + \sin^2 x}{\cos^2 x} = \frac{1}{\cos^2 x}.$$

$$\therefore \frac{d}{dx}(\tan x) = \sec^2 x.$$

Example 10 Differentiate sin $4x$ with respect to x.

Let $y = \sin 4x$ and $u = 4x$. $\therefore\ y = \sin u$.

Since $\dfrac{dy}{du} = \cos u$ and $\dfrac{du}{dx} = 4$, we have:

$$\frac{dy}{dx} = \frac{dy}{du} \times \frac{du}{dx} = 4 \cos u \Rightarrow \frac{dy}{dx} = 4 \cos 4x.$$

Example 11 Differentiate $\sin^3 x$ with respect to x.

Let $y = \sin^3 x$. If $u = \sin x$, $y = u^3$.

$$\therefore \frac{du}{dx} = \cos x, \quad \frac{dy}{du} = 3u^2.$$

Hence $\dfrac{dy}{dx} = \dfrac{dy}{du} \times \dfrac{du}{dx} = 3u^2 \cos x = 3 \sin^2 x \cos x.$

$$\therefore \frac{d}{dx}(\sin^3 x) = 3 \sin^2 x \cos x.$$

Example 12 Differentiate with respect to x the following functions: (i) $x^2 \sin x$, (ii) $\dfrac{x^2}{\cos 2x}$.

(i) Let $y = x^2 \sin x$. Since this is a product where $u = x^2$ and $v = \sin x$, we can use the product rule.

Now $\dfrac{du}{dx} = 2x$ and $\dfrac{dv}{dx} = \cos x$.

$$\therefore \frac{dy}{dx} = u\frac{dv}{dx} + v\frac{du}{dx} = x^2 \cos x + 2x \sin x.$$

Hence $\dfrac{d}{dx}(x^2 \sin x) = x(x \cos x + 2 \sin x)$.

(ii) Let $y = \dfrac{x^2}{\cos 2x}$. This is a quotient with $u = x^2$ and $v = \cos 2x$.

Now $\dfrac{du}{dx} = 2x$ and $\dfrac{dv}{dx} = -2 \sin 2x$.

$$\therefore \frac{dy}{dx} = \frac{v(du/dx) - u(dv/dx)}{v^2} = \frac{\cos 2x(2x) - x^2(-2 \sin 2x)}{\cos^2 2x}$$

$$= \frac{2x \cos 2x + 2x^2 \sin 2x}{\cos^2 2x}.$$

Hence $\dfrac{d}{dx}\left(\dfrac{x^2}{\cos 2x}\right) = \dfrac{2x(\cos 2x + x \sin 2x)}{\cos^2 2x}$.

It is wise to remember a number of standard differentials so that they may be used in more complicated problems without difficulty. They are listed below for reference and you should check the derivation of each.

$\dfrac{d}{dx}(\sin x) = \cos x$; $\quad \dfrac{d}{dx}(\cos x) = -\sin x$;

$\dfrac{d}{dx}(\tan x) = \sec^2 x$; $\quad \dfrac{d}{dx}(\sec x) = \sec x \tan x$;

$\dfrac{d}{dx}(\cot x) = -\operatorname{cosec}^2 x$; $\quad \dfrac{d}{dx}(\operatorname{cosec} x) = -\operatorname{cosec} x \cot x$.

Inverse circular functions

If $y = \sin^{-1} x$, then y is the angle whose sine is x. The function $\sin^{-1} x$ is called the inverse sine and similar functions exist for $\cos^{-1} x$, $\tan^{-1} x$ and so on. These have been defined and discussed on page 36–7 and the graphs drawn. They can be written as arc sin x, arc cos x, etc. To differentiate $\sin^{-1} x$ we proceed as follows.

Let $y = \sin^{-1} x$, in which case $x = \sin y$.

Differentiating with respect to y we have $\dfrac{dx}{dy} = \cos y$.

$$\frac{dy}{dx} = \frac{1}{\cos y} = \frac{1}{\sqrt{1 - \sin^2 y}} = \frac{1}{\sqrt{1 - x^2}}.$$

Hence $\quad \dfrac{d}{dx}(\sin^{-1} x) = \dfrac{1}{\sqrt{1 - x^2}}.$

Note that the sign of the differential coefficient may be positive or negative for a given value of x. This is clear from the graph on page 37 as $\sin^{-1} x$ is a many valued function, although if we restrict the range by taking only the **principal values** the gradient will always be positive. The gradients of $\boldsymbol{y = \cos^{-1} x}$ and $\boldsymbol{y = \tan^{-1} x}$ can be derived in a similar way. The results are:

$$\frac{d}{dx}(\cos^{-1} x) = \frac{-1}{\sqrt{1 - x^2}} \quad \text{and} \quad \frac{d}{dx}(\tan^{-1} x) = \frac{1}{1 + x^2}.$$

For $\boldsymbol{y = \sec^{-1} x}$ we apply the same technique.

Since $y = \sec^{-1} x$, $x = \sec y$.

Differentiating with respect to y, $\dfrac{dx}{dy} = \sec y \tan y$.

$$\therefore \frac{dy}{dx} = \frac{1}{\sec y \tan y} = \frac{1}{\sec y\sqrt{\sec^2 y - 1}}.$$

$$\therefore \frac{dy}{dx} = \frac{1}{x\sqrt{x^2 - 1}} \quad \text{and we write} \quad \frac{d}{dx}(\sec^{-1} x) = \frac{1}{x\sqrt{x^2 - 1}}.$$

Example 13 Differentiate $\sin^{-1}\left(\dfrac{x}{a}\right)$ with respect to x.

Let $y = \sin^{-1}\left(\dfrac{x}{a}\right)$. $\quad \therefore \dfrac{x}{a} = \sin y \Rightarrow x = a \sin y$.

Differentiate with respect to y to give $\frac{dx}{dy} = a \cos y$.

$$\frac{dy}{dx} = \frac{1}{a \cos y} = \frac{1}{a\sqrt{1 - \sin^2 y}} = \frac{1}{a\sqrt{1 - (x/a)^2}} = \frac{1}{\sqrt{a^2 - x^2}}.$$

Hence $\frac{d}{dx}\left\{\sin^{-1}\left(\frac{x}{a}\right)\right\} = \frac{1}{\sqrt{a^2 - x^2}}$.

A list of the differential coefficients of the standard inverse circular functions is listed below for reference. They should be learnt thoroughly so that they can be applied accurately.

$$\frac{d}{dx}(\sin^{-1} x) = \frac{1}{\sqrt{1 - x^2}}; \qquad \frac{d}{dx}\left\{\sin^{-1}\left(\frac{x}{a}\right)\right\} = \frac{1}{\sqrt{a^2 - x^2}};$$

$$\frac{d}{dx}(\cos^{-1} x) = \frac{-1}{\sqrt{1 - x^2}}; \qquad \frac{d}{dx}\left\{\cos^{-1}\left(\frac{x}{a}\right)\right\} = \frac{-1}{\sqrt{a^2 - x^2}};$$

$$\frac{d}{dx}(\tan^{-1} x) = \frac{1}{1 + x^2}; \qquad \frac{d}{dx}\left\{\tan^{-1}\left(\frac{x}{a}\right)\right\} = \frac{a}{a^2 + x^2};$$

$$\frac{d}{dx}(\cot^{-1} x) = \frac{-1}{1 + x^2}; \qquad \frac{d}{dx}\left\{\cot^{-1}\left(\frac{x}{a}\right)\right\} = \frac{-a}{a^2 + x^2};$$

$$\frac{d}{dx}(\sec^{-1} x) = \frac{1}{x\sqrt{x^2 - 1}}; \qquad \frac{d}{dx}\left\{\sec^{-1}\left(\frac{x}{a}\right)\right\} = \frac{a}{x\sqrt{x^2 - a^2}};$$

$$\frac{d}{dx}(\operatorname{cosec}^{-1} x) = \frac{-1}{x\sqrt{x^2 - 1}}; \qquad \frac{d}{dx}\left\{\operatorname{cosec}^{-1}\left(\frac{x}{a}\right)\right\} = \frac{-a}{x\sqrt{x^2 - a^2}}.$$

Successive differentiation

Given y as a function of x we have seen how to form the gradient or differential coefficient. Unless y is a linear function, the gradient function will also be a function of x and therefore we could differentiate this as well. We would form the gradient of the gradient function which would give the rate of change of the gradient. This is called the **second differential** of y and is written $\frac{d}{dx}\left(\frac{dy}{dx}\right)$ or more usually $\frac{d^2y}{dx^2}$.

If we use the notation of $f(x)$ for the function of x, the first differential is $f'(x)$ and the second $f''(x)$. This notation can be extended to cover the **third differential** of the original function by writing $\frac{d^3y}{dx^3}$ or $f'''(x)$.

Note that for higher derivatives, i.e. $\frac{d^6y}{dx^6}$ or $\frac{d^ny}{dx^n}$ we would write $f^{(6)}(x)$ and $f^{(n)}(x)$ or sometimes y_6 and y_n respectively.

Example 14 Find $\frac{dy}{dx}, \frac{d^2y}{dx^2}$ for the functions given by:

(i) $y = x^3 + 3x^2 + 4x$, (ii) $y = x^2 \tan x$.

(i) Since $y = x^3 + 3x^2 + 4x$,

$$\frac{dy}{dx} = 3x^2 + 6x + 4 \quad \text{and} \quad \frac{d^2y}{dx^2} = 6x + 6.$$

(ii) Since $y = x^2 \tan x$

$$\frac{dy}{dx} = x^2 \sec^2 x + 2x \tan x \text{ (using product rule).}$$

$$\begin{aligned}\frac{d^2y}{dx^2} &= x^2(2 \sec x \sec x \tan x) + 2x \sec^2 x \\ &\qquad + 2x \sec^2 x + 2 \tan x. \\ &= 2x^2 \sec^2 x \tan x + 4x \sec^2 x + 2 \tan x.\end{aligned}$$

Care must be taken if the given equations are expressed in terms of a parameter. If $x = f(t)$ and $y = g(t)$ then $\frac{dy}{dx}$ will be found in terms of t.

$$\text{Hence} \quad \frac{d^2y}{dx^2} = \frac{d}{dx}\left(\frac{dy}{dx}\right) = \frac{d}{dt}\left(\frac{dy}{dx}\right) \times \frac{dt}{dx}.$$

Example 15 Find $\frac{dy}{dx}$ and $\frac{d^2y}{dx^2}$ if $x = (t^2 - 1)^2$ and $y = t^3$.

$$\frac{dx}{dt} = 4t(t^2 - 1) \quad \text{and} \quad \frac{dy}{dt} = 3t^2 \quad \text{giving} \quad \frac{dy}{dx} = \frac{3t}{4(t^2 - 1)}.$$

$$\text{Hence} \quad \frac{d^2y}{dx^2} = \frac{d}{dx}\left(\frac{dy}{dx}\right) = \frac{d}{dt}\left\{\frac{3t}{4(t^2 - 1)}\right\}\frac{dt}{dx}$$

$$= \left\{\frac{4(t^2-1)3 - 3t \times 8t}{16(t^2-1)^2}\right\} \times \frac{1}{4t(t^2-1)}.$$

$$\therefore \frac{d^2y}{dx^2} = \frac{-12(1+t^2)}{64t(t^2-1)^3} = \frac{-3(t^2+1)}{16t(t^2-1)^3}.$$

Maximum and minimum points

Points on a graph where the gradient is zero are called **stationary points**. They are maximum or minimum points considered in a localized sense or a point of inflexion. Consider the curve shown in figure 24 on which have been drawn the tangents at various points.

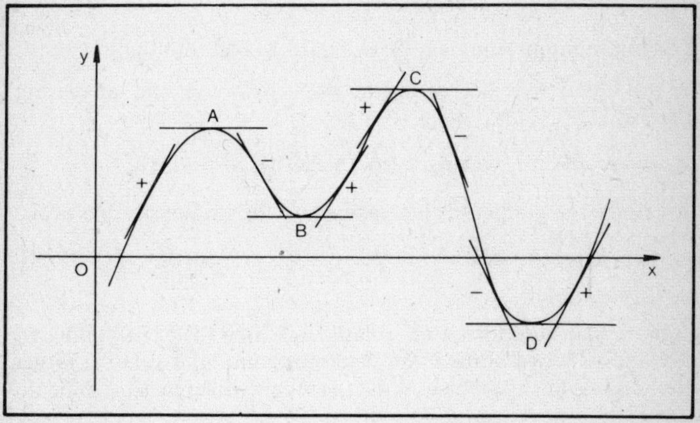

Figure 24

By considering the gradients of these tangents, we observe that at each of the points A, B, C and D the gradient is zero, since the tangent is parallel to the x-axis. Also, we note that the gradient changes sign as x increases through each of the points.

For A and C the gradient changes from positive to negative. For B and D the gradient changes from negative to positive. Also for A and C, the curve has its greatest value in that region, so that if points are chosen on the curve close to and on either side of A or C, the value of the ordinate will be less than the ordinate of A or C.

Similarly, B and D provide the lowest points of the curve in that region and any points chosen close to and on either side of B or D will have ordinates greater than the ordinates of B or D. We say that A and C are **maximum points** of the curve and that B and D are **minimum points**.

Now at a maximum point, say A, in figure 24 we can see that:

(i) the function is increasing in value before A and decreasing after A;

(ii) the gradient $\frac{dy}{dx}$ changes from positive to negative;

(iii) thus from (ii) the gradient is decreasing and hence the rate of change of the gradient is negative, i.e. $\frac{d^2y}{dx^2} < 0$.

For a minimum point, say B, in figure 24 we note that:

(i) the function is decreasing in value before B and increasing after B;

(ii) the gradient $\frac{dy}{dx}$ changes from negative to positive;

(iii) hence the gradient is increasing and the rate of change of the gradient is positive i.e. $\frac{d^2y}{dx^2} > 0$.

In particular questions we normally use either (ii) or (iii) above to distinguish between maximum and minimum points. Don't forget method (ii) which can be useful if the second differential is difficult.

Example 16 Find the maximum and minimum values of the function $2x^3 - 9x^2 + 12x$.

Let $y = 2x^3 - 9x^2 + 12x$. $\therefore \quad \frac{dy}{dx} = 6x^2 - 18x + 12$. (1)

The curve will have a stationary point when $\frac{dy}{dx} = 0$, i.e.

$$\begin{aligned} 6x^2 - 18x + 12 &= 0, \\ x^2 - 3x + 2 &= 0, \\ (x-1)(x-2) &= 0. \\ \therefore x = 1 \quad \text{or} \quad x &= 2. \end{aligned}$$

$\therefore$ when $x = 1$, $y = 2(1)^3 - 9(1)^2 + 12(1) = 5$,
when $x = 2$, $y = 2(2)^3 - 9(2)^2 + 12(2) = 4$.

The curve has stationary points at (1, 5) and (2, 4).

To distinguish between them, consider the second differential.

From (1): $\dfrac{d^2y}{dx^2} = 12x - 18.$

When $x = 1$, $\dfrac{d^2y}{dx^2} < 0$, $\therefore$ a maximum point.

When $x = 2$, $\dfrac{d^2y}{dx^2} > 0$, $\therefore$ a minimum point.

$\therefore$ The curve has a maximum at (1, 5) and a minimum at (2, 4).

If we use the method of considering the gradient on each side of $x = 1$ and $x = 2$, it is best to draw up a table. From equation (1):

$$\frac{dy}{dx} = 6(x^2 - 3x + 2) = 6(x - 1)(x - 2).$$

x	$1 - h$	1	$1 + h$
$\dfrac{dy}{dx}$	+	0	−

x	$2 - h$	2	$2 + h$
$\dfrac{dy}{dx}$	−	0	+

Hence at $x = 1$ the gradient changes from positive to negative and gives a maximum. At $x = 2$ the gradient changes from negative to positive and gives a minimum.

Points of inflexion

Consider the graph of $y = 2x^3 - 9x^2 + 12x$. We have already found a maximum point at (1, 5) and a minimum at (2, 4) so, knowing that it passes through the origin, its sketch can be drawn. This is shown in figure 25(a) on page 62.

We notice that at point A in figure 25(a) the curvature of the curve is changing from concave downwards to concave upwards. Now when the curve is concave downwards, $\dfrac{dy}{dx}$ is decreasing and hence

$$\frac{d^2y}{dx^2} < 0.$$

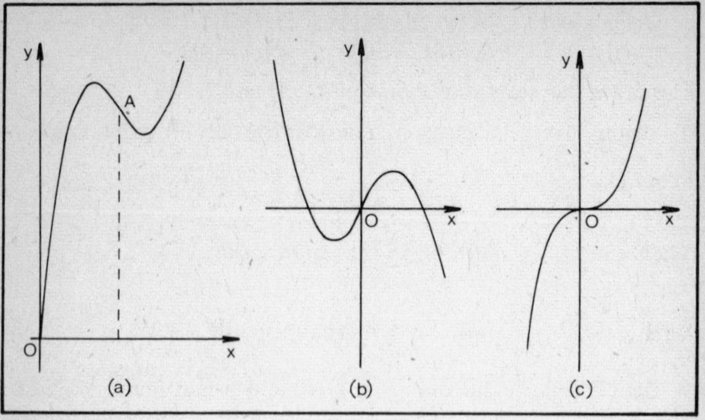

Figure 25

When the curve is concave upwards $\frac{dy}{dx}$ is increasing and hence $\frac{d^2y}{dx^2} > 0.$

$\therefore$ At the point of change, called the **point of inflexion**, $\frac{d^2y}{dx^2} = 0$ and $\frac{d^2y}{dx^2}$ changes sign at this point.

Hence at a point of inflexion:
(i) the curvature changes from concave downwards to concave upwards, figure 25(a), or vice versa, figure 25(b) at point O;
(ii) $\frac{dy}{dx}$ is increasing before and decreasing after, or vice versa;
(iii) this means that $\frac{d^2y}{dx^2}$ changes sign on either side of the point;
(iv) $\frac{dy}{dx}$ will be a maximum or a minimum, $\therefore \frac{d^2y}{dx^2} = 0.$

In general it is not necessary for $\frac{dy}{dx} = 0$, although this is often the case as in figure 25(c) which shows the curve $y = x^3$.

$\therefore$ If $y = 2x^3 - 9x^2 + 12x,$

$$\frac{dy}{dx} = 6x^2 - 18x + 12,$$

$$\frac{d^2y}{dx^2} = 12x - 18.$$

Now $\frac{d^2y}{dx^2} = 0$ when $12x - 18 = 0$, i.e. $x = 1\frac{1}{2}$.

Since on either side of $x = 1\frac{1}{2}$, $\frac{d^2y}{dx^2}$ changes sign, the point where $x = 1\frac{1}{2}$ defines a point of inflexion. Remember this can be checked by noting that the gradient of $\frac{d^2y}{dx^2}$ is non zero!

The following example will illustrate how we can apply the methods for finding maximum and minimum points to other problems.

Example 17 Find the least area of metal required to make a cylindrical container from thin sheet metal in order that it might have a capacity of 2000π cm^3.

Let the radius be r cm, the height h cm, and the total surface area S cm^2.

$\therefore$ The area of two ends $= 2\pi r^2$ and the area of the curved surface $= 2\pi rh$. Thus the total surface area, $S = 2\pi r^2 + 2\pi rh$.

But vol $= \pi r^2 h = 2000\pi$.

$$\therefore\ h = \frac{2000}{r^2}. \qquad (1)$$

Hence total surface area $S = 2\pi r^2 + (2\pi r)\frac{2000}{r^2}$,

or

$$S = 2\pi r^2 + \frac{4000\pi}{r}. \qquad (2)$$

To find the minimum, form $\frac{dS}{dr}$ and equate to zero:

$$\frac{dS}{dr} = 4\pi r - \frac{4000\pi}{r^2}.$$

$$\frac{dS}{dr} = 0 \quad \text{when} \quad 4\pi r - \frac{4000\pi}{r^2} = 0, \text{ i.e. when } r^3 = 1000 \Rightarrow r = 10.$$

Now $\dfrac{d^2S}{dr^2} = 4\pi + \dfrac{8000\pi}{r^3}$, which is positive for $r = 10$.

$\therefore$ When $r = 10$, the surface area is a minimum. In this case $h = 20$ from equation (1).

Using equation (2):

$$\text{Total surface area} = 200\pi + 400\pi = 600\pi \text{ cm}^3.$$

Small changes

Since we defined $\dfrac{dy}{dx}$ as the limit of $\dfrac{\delta y}{\delta x}$ as $\delta x \to 0$, it is true that

$\dfrac{\delta y}{\delta x} \simeq \dfrac{dy}{dx}$ when δx is small.

Hence $\delta y \simeq \dfrac{dy}{dx}\,\delta x.$

We can use this approximation to find the effect on one variable of a small change in the other.

Example 18 An error of 2% is made in measuring the radius of a circle. What is the error in the area?

$$\text{Now} \quad \frac{dA}{dr} \simeq \frac{\delta A}{\delta r} \quad \text{or} \quad \delta A \simeq \frac{dA}{dr}\,\delta r. \qquad (1)$$

Now $\delta r = \dfrac{2r}{100}$.

Also $A = \pi r^2$, $\therefore \dfrac{dA}{dr} = 2\pi r$.

Thus in equation (1) we have $\delta A \simeq 2\pi r \left(\dfrac{2r}{100}\right) = \dfrac{4\pi r^2}{100}$.

$$\%\text{ error in the area} = \frac{\delta A}{A} \times 100 = \frac{4\pi r^2 \times 100}{100\pi r^2}\% = 4\%.$$

Key terms

If $y = ax^n$, then the gradient function $\dfrac{dy}{dx} = nax^{n-1}$.

If $y = f(x)$, then $\dfrac{dy}{dx} = f(x) = \lim\limits_{b \to a} \left\{ \dfrac{f(b) - f(a)}{b - a} \right\}$

$$= \lim_{h \to 0} \left\{ \frac{f(a+h) - f(a)}{h} \right\}.$$

The product rule If $y = u(x) \times v(x)$, then $\dfrac{dy}{dx} = u\dfrac{dv}{dx} + v\dfrac{du}{dx}$.

The quotient rule If $y = \dfrac{u(x)}{v(x)}$,

then $\dfrac{dy}{dx} = \dfrac{v(du/dx) - u(dv/dx)}{v^2}$.

Composite functions $y = y(u)$ and $u = u(x)$ gives

$$\frac{dy}{dx} = \frac{dy}{du} \times \frac{du}{dx}.$$

A **maximum point** is defined by $\dfrac{dy}{dx} = 0$ and $\dfrac{d^2y}{dx^2} < 0$.

A **minimum point** is defined by $\dfrac{dy}{dx} = 0$ and $\dfrac{d^2y}{dx^2} > 0$.

A **point of inflexion** occurs when $\dfrac{d^2y}{dx^2} = 0$ and $\dfrac{d^2y}{dx^2}$ changes sign on either side of the point.

$\dfrac{d}{dx}(\sin x) = \cos x$; $\quad \dfrac{d}{dx}(\cos x) = -\sin x$;

$\dfrac{d}{dx}(\tan x) = \sec^2 x$; $\quad \dfrac{d}{dx}(\cot x) = -\operatorname{cosec}^2 x$;

$\dfrac{d}{dx}\left(\sin^{-1}\dfrac{x}{a}\right) = \dfrac{1}{\sqrt{a^2 - x^2}}$; $\quad \dfrac{d}{dx}\left(\tan^{-1}\dfrac{x}{a}\right) = \dfrac{a}{a^2 + x^2}$.

Chapter 4
Integration

This branch of calculus is concerned, in one of its aspects, about the inverse of differentiation. That is, if the gradient of a function is known, is it possible to find the function from which that gradient was derived?

Now we know from chapter 3 that

$$\frac{d}{dx}(x^2 + 3) = 2x \text{ and } \frac{d}{dx}(x^2 - 7) = 2x.$$

Indeed any function of the form $x^2 \pm k$ will have a derived function of $2x$. Clearly, then, if we are given the gradient of a function as $2x$, the integral function must be of the form $x^2 + c$, where c is some positive or negative constant. In general, it is not possible to determine the value of this constant and the resulting function is called an **indefinite integral**.

Example 1 If $f'(x) = x^4$ find $f(x)$.

If $f'(x) = x^4$, then $f(x) = \frac{1}{5}x^5 + c$, where c is an arbitrary constant. Figure 26 gives a graphical interpretation of the above process.

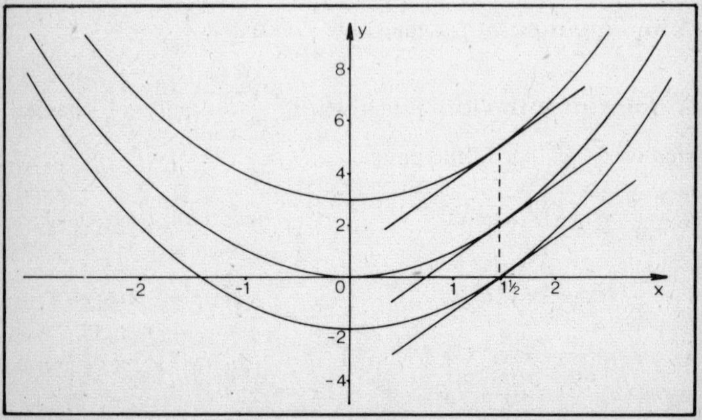

Figure 26

Each of the curves $y = x^2$, $y = x^2 + 3$, $y = x^2 - 2$ is shown with its gradient at the point where $x = 1\frac{1}{2}$. All of these curves have the same gradient function $2x$ represented by parallel tangents. To determine the value of the arbitrary constant, a set of conditions must be given. This may be the coördinates of one point on the curve.

Example 2 A curve has a gradient function of $3x^2$ and passes through the point (2, 10). Find the equation of the curve.

If $\frac{dy}{dx} = 3x^2$, then $y = x^3 + c$.

As the curve passes through the point (2, 10) these values satisfy the equation, i.e. $10 = (2)^3 + c$ and $c = 2$.
$\therefore$ the equation is $y = x^3 + 2$.

A special notation is used to denote the operation of integration. This is $\int$ which is derived from the old English form of the letter s, being the first letter of the word sum. The idea of a sum as an aspect of integration will be discussed later.

We write $\int 3x^2\,dx = x^3 + c$.

The meaning of the dx will also be made clear from the summation process, but it does indicate with respect to which variable we are integrating. That is $\int f(x)\,dx$ indicates that the function $f(x)$ is to be integrated with respect to x. In general, using the result $\frac{d}{dx}(ax^n) = nax^{n-1}$ we can deduce that:

$$\int ax^n\,dx = \frac{ax^{n+1}}{(n+1)} + c.$$

This will hold for all positive, negative, integral or fractional values of n except $n = -1$. The integral of the sum of any number of terms is equal to the sum of the separate integrals.

Example 3 Find $\int (x^3 - 3x^2 + 8x + 7)\,dx$.

Integrating with respect to x we obtain:

$$\int (x^3 - 3x^2 + 8x + 7)\,dx = \frac{x^4}{4} - x^3 + 4x^2 + 7x + c.$$

When $n = -1$ the general rule breaks down. It will be shown on page 96 that $\frac{d}{dx}(\ln x) = \frac{1}{x}$.

Hence $\int \frac{1}{x}\,dx = \ln x + c = \ln(kx)$ by writing $c = \ln k$.

Area under a curve

Another view of integration can be obtained by considering the area under a given curve. The technique involves dividing the area into a number of elementary strips and summing over the complete range. This process will involve some error, but by increasing the number of elements a better approximation to the actual area can be found. Indeed, if the number of elements is allowed to tend to infinity, the calculated area will, in the limit, be the actual area.

Consider $f(x) = 2x + 3$. If we wish to find the area bounded by the graph and the x-axis between $x = 0$ and $x = 10$ we could divide the area into n strips of equal width as shown in figure 27.

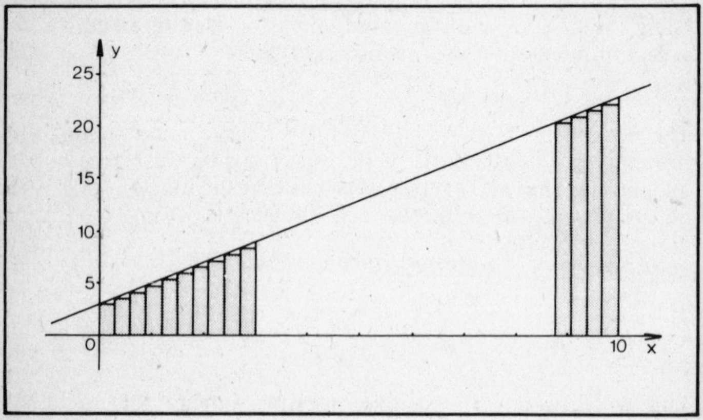

Figure 27

If the area is divided in n strips, the width of each is $\frac{10}{n}$.

We calculate the ordinate for each value of x so that the area of each rectangle can be found.

When $x = 0$, $\quad f(x) = 3$, i.e. $f(0) = 3$.

When $x = \frac{10}{n}$, $\quad f\left(\frac{10}{n}\right) = 2 \times \frac{10}{n} + 3$.

$$\text{When } x = 2 \times \frac{10}{n}, \quad f\left(2 \times \frac{10}{n}\right) = 2 \times \frac{20}{n} + 3.$$

$$\vdots$$

$$\text{When } x = 10 - \frac{10}{n}, \quad f\left(10 - \frac{10}{n}\right) = 2\left(10 - \frac{10}{n}\right) + 3.$$

Hence the sum of all the rectangles is:

$$3 \times \frac{10}{n} + \left(\frac{20}{n} + 3\right)\frac{10}{n} + \left(\frac{40}{n} + 3\right)\frac{10}{n} + \left(\frac{60}{n} + 3\right)\frac{10}{n} +$$
$$\cdots + \left(20 - \frac{20}{n} + 3\right)\frac{10}{n}$$
$$= \frac{10}{n}\left[3 + \left(\frac{20}{n} + 3\right) + \left(\frac{40}{n} + 3\right) + \left(\frac{60}{n} + 3\right) + \right.$$
$$\left. \cdots + \left\{\frac{20(n-1)}{n} + 3\right\}\right].$$

The terms inside the square bracket form an arithmetic progression with a common difference of $\frac{20}{n}$ and with a first term of 3. Since there are n terms, the sum is given by $\frac{n}{2}\left\{6 + (n-1)\frac{20}{n}\right\}$. For work on arithmetic progressions, see page 147.

$\therefore$ The sum of areas of the rectangles is

$$\frac{10}{n} \times \frac{n}{2}\left\{6 + (n-1)\frac{20}{n}\right\} = 5\left(26 - \frac{20}{n}\right) = 130 - \frac{100}{n}.$$

From this form of our answer we can conclude that as $n \to \infty$, the area $\to$ 130. Hence the area under this straight line = 130. This could, of course, be found from the area of the trapezium, i.e. $\frac{1}{2}(3 + 23) \times 10 = 26 \times 5 = 130$.

We can also apply this process to a non-linear function. Consider $f(x) = x^2$ as shown in figure 28. Again divide the required area into n elementary rectangles of width h. A typical element PQRS will have an area given by $(rh)^2 \times h = r^2h^3$.

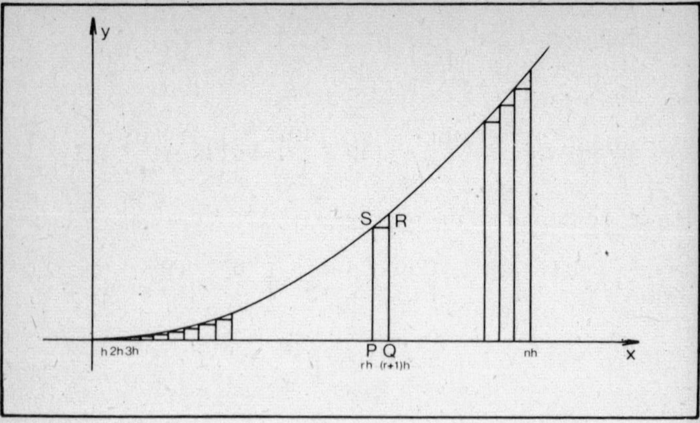

Figure 28

Hence the total area is $\sum_{r=1}^{n-1} r^2h^3 = h^3 \sum_{r=1}^{n-1} r^2$.

$\therefore$ Area $= h^3 \times \dfrac{(n-1)}{6}(n)(2n-1)$. See the result for $\sum r^2$ on page 153 with n replaced by $n - 1$.

$$\therefore \text{Area} = \frac{h^3n(n-1)(2n-1)}{6}.$$

If the maximum value of $x = b$, then $h = \dfrac{b}{n}$,

$$\text{thus}\quad \text{area} = \frac{b^3n}{6n^3}(n-1)(2n-1) = \frac{b^3}{6}\left(1-\frac{1}{n}\right)\left(2-\frac{1}{n}\right).$$

As $n \to \infty$, area $\to \frac{1}{3}b^3$. We say that the area is given by the limit as $n \to \infty$ and is equal to $\frac{1}{3}b^3$.

To generalize this method we consider one typical element obtained by dividing an area under a given curve into a large number of rectangles of width δx. Let the curve be given by the function $y = f(x)$ and let one typical element be formed between the ordinates x and $x + \delta x$.

Clearly the chosen element is formed by selecting two points $P(x, y)$ and $Q(x + \delta x, y + \delta y)$ on the curve and completing the lower rectangle as in figure 28. The area of the rectangle is $y\,\delta x$ and thus the sum of all such rectangles is given by

$$\sum_{x=a}^{x=b} y\,\delta x.$$

If the number of rectangles is increased by letting $\delta x \to 0$, the total area is

$$\lim_{\delta x \to 0} \sum_{x=a}^{x=b} y\,\delta x$$

For this we use the integral notation and the final form is given by the following.

The area under a curve $y = f(x)$ over the range $x = a$ to $x = b$ is

$$\int_a^b f(x)\,dx.$$

Since the evaluation of this result will give a definite numerical answer, the integral is called a **definite integral** and no arbitrary constant is necessary.

If $F(x)$ is the integral function of $f(x)$, then

$$\int_a^b f(x)\,dx = F(b) - F(a).$$

A special notation is used to represent the value of $F(b) - F(a)$. We write

$$F(b) - F(a) = \Big[F(x)\Big]_a^b.$$

Example 4 $\displaystyle\int_1^3 (x^2 + 3x)\,dx = \left[\frac{x^3}{3} + \frac{3x^2}{2}\right]_1^3$

$$= \left(\frac{27}{3} + \frac{27}{2}\right) - \left(\frac{1}{3} + \frac{3}{2}\right) = 20\tfrac{2}{3}.$$

Example 5 Evaluate (i) $\displaystyle\int_0^2 f(x)\,dx$, (ii) $\displaystyle\int_2^3 f(x)\,dx$ and (iii) $\displaystyle\int_0^3 f(x)\,dx$, where $f(x) = x^2 - 2x$.

(i) $\displaystyle\int_0^2 f(x)\,dx = \int_0^2 (x^2 - 2x)\,dx = \left[\frac{x^3}{3} - x^2\right]_0^2$

$$= (\tfrac{8}{3} - 4) - (0 - 0) = -1\tfrac{1}{3}.$$

(ii) $\int_2^3 f(x)\,dx = \int_2^3 (x^2 - 2x)\,dx = \left[\frac{x^3}{3} - x^2\right]_2^3$

$$= (\tfrac{27}{3} - 9) - (\tfrac{8}{3} - 4) = 1\tfrac{1}{3}.$$

(iii) $\int_0^3 f(x)\,dx = \int_0^3 (x^2 - 2x)\,dx = \left[\frac{x^3}{3} - x^2\right]_0^3$

$$= (\tfrac{27}{3} - 9) - (0 - 0) = 0.$$

The solution to part (i) is negative, which indicates that the area is below the axis, not that the area is negative. Since the results to part (i) and (ii) are numerically equal, the areas are equal. Notice that when the integral over the whole range from 0 to 3 is evaluated, the result is the algebraic sum of the first two answers, that is, not the total actual area of $2\tfrac{2}{3}$.

This becomes clear if we sketch the graph of $y = x^2 - 2x$. The curve and the relevant areas are shown in figure 29.

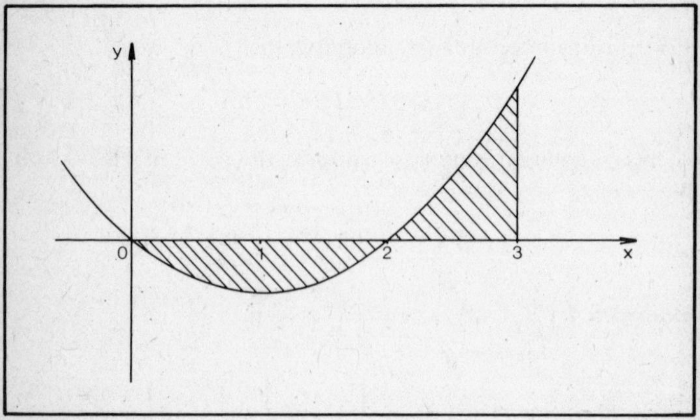

Figure 29

In general if $y = f(x)$ is a function taking only negative values over an interval $x = a$ to $x = b$, then $y = -f(x)$ is reflection of the curve in the x-axis possessing the same area, i.e.

an area given by $\int_a^b -f(x)\,dx = \left[-F(x)\right]_a^b$

$$= -\left[F(x)\right]_a^b = -\int_a^b f(x)\,dx.$$

Thus the integral evaluates the negative of the area bounded by $y = f(x)$, $x = a$, $x = b$ and the x-axis.

We can extend our discussion on areas to the problem of evaluating an area not bounded by the x-axis. The two examples which follow illustrate the methods.

Example 6 Find the area enclosed by the curve $x = 2y^2$, the y-axis and the line $y = 3$.

We choose an elementary strip parallel to the x-axis, PQRS as shown in figure 30(a) where P(x, y) and Q$(x + \delta x, y + \delta x)$ are two neighbouring points on the curve.

Hence the area of the element is approximately given by the area of the inner rectangle RSPT. The area is $x\ \delta y$.

By summing and taking the limit as $\delta y \to 0$ we obtain a value for the total area, i.e.

$$\lim_{\delta y \to 0} \sum_{y=0}^{y=3} x\ \delta y = \int_0^3 x\ dy = \int_0^3 2y^2\ dy.$$

The required area is $\left[\dfrac{2y^3}{3}\right]_0^3 = (2 \times \frac{27}{3} - 0) = \frac{54}{3}$.

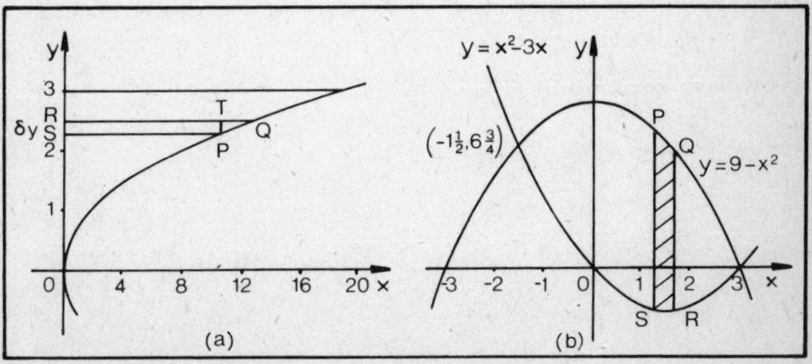

Figure 30

Example 7 Find the area enclosed by the curves $y = x^2 - 3x$, and $y = 9 - x^2$.

We must first find the points of intersection of the curves so that we know the range for the summation. The curves intersect when $x^2 - 3x = 9 - x^2$.

$$\Leftrightarrow 2x^2 - 3x - 9 = 0 \Leftrightarrow (2x + 3)(x - 3) = 0,$$
$$\Rightarrow x = -1\tfrac{1}{2} \quad \text{or} \quad x = 3.$$

$\therefore$ the curves intersect at $(-1\frac{1}{2}, 6\frac{3}{4})$ and $(3, 0)$.

By noting where the curves cut the axes, a sketch can be drawn.

When $y = 0$, $x^2 - 3x = 0 \Rightarrow x = 0$ or 3.
When $y = 0$, $9 - x^2 = 0 \Rightarrow x = \pm 3$.

By letting $x = 0$ it is clear that the curves cut the y-axis at the origin and $y = 9$ respectively. The completed sketch is shown in figure 30(b).

In considering an element of area we must ensure that it is typical and hence a strip parallel to the y-axis must be chosen, for example, PQRS.

Now the height of PS is $Y - y$, where Y and y are the ordinates of P and S respectively. Note that it is the difference $Y - y$, since in the position shown y will be negative and will effectively produce the sum of Y and y as required. Since the width of the element is δx, the area of PQRS $\simeq (Y - y)\,\delta x$.

Hence by summing and taking the limit as $\delta x \to 0$ the total area is

$$\lim_{\delta x \to 0} \sum_{x=-1\frac{1}{2}}^{x=3} (Y - y)\,\delta x = \int_{-3/2}^{3} (Y - y)\,dx.$$

Now $Y = 9 - x^2$ and $y = x^2 - 3x$

$$\therefore \text{ area required} = \int_{-3/2}^{3} \{(9 - x^2) - (x^2 - 3x)\}\,dx$$

$$= \int_{-3/2}^{3} (9 + 3x - 2x^2)\,dx = \left[9x + \frac{3x^2}{2} - \frac{2x^3}{3}\right]_{-3/2}^{3}$$

$$= (27 + \tfrac{27}{2} - 18) - (-\tfrac{27}{2} + \tfrac{27}{8} + \tfrac{9}{4})$$

$$= \tfrac{45}{2} + \tfrac{63}{8} = 30\tfrac{3}{8}.$$

Remember that in the calculation of area the choice of element is not necessarily unique, but it must be typical of the whole region and must be such that it will produce a function which can be integrated, preferably easily.

Standard integrals

We have considered the meaning of integration and now turn our attention to the integrals of specific functions. In chapter 3 we derived the differentials of a number of different functions. By considering integrations as the inverse operation to differentiation, we can list some standard integrals which should be learnt thoroughly.

$$\frac{d}{dx}(ax^n) = nax^{n-1}; \qquad \int ax^n\,dx = \frac{ax^{n+1}}{n+1} + c, \quad n \neq -1;$$

$$\frac{d}{dx}(\sin x) = \cos x; \qquad \int \cos x\,dx = \sin x + c;$$

$$\frac{d}{dx}(\cos x) = -\sin x; \qquad \int \sin x\,dx = -\cos x + c;$$

$$\frac{d}{dx}(\tan x) = \sec^2 x; \qquad \int \sec^2 x\,dx = \tan x + c;$$

$$\frac{d}{dx}(\cot x) = -\operatorname{cosec}^2 x; \qquad \int \operatorname{cosec}^2 x\,dx = -\cot x + c;$$

$$\frac{d}{dx}(\sec x) = \sec x \tan x; \qquad \int \sec x \tan x\,dx = \sec x + c;$$

$$\frac{d}{dx}(\operatorname{cosec} x) = -\operatorname{cosec} x \cot x;$$

$$\int \operatorname{cosec} x \cot x\,dx = -\operatorname{cosec} x + c.$$

The inverse trigonometrical functions were also discussed in chapter 3 and produced the following results.

$$\frac{d}{dx}\left\{\sin^{-1}\left(\frac{x}{a}\right)\right\} = \frac{1}{\sqrt{a^2 - x^2}}; \qquad \int \frac{dx}{\sqrt{a^2 - x^2}} = \sin^{-1}\left(\frac{x}{a}\right) + c;$$

$$\frac{d}{dx}\left\{\cos^{-1}\left(\frac{x}{a}\right)\right\} = \frac{-1}{\sqrt{a^2 - x^2}}; \qquad \int \frac{dx}{\sqrt{a^2 - x^2}} = -\cos^{-1}\left(\frac{x}{a}\right) + c;$$

$$\frac{d}{dx}\left\{\tan^{-1}\left(\frac{x}{a}\right)\right\} = \frac{a}{a^2 + x^2}; \qquad \int \frac{a}{a^2 + x^2}\,dx = \tan^{-1}\left(\frac{x}{a}\right) + c;$$

The exponential and logarithmic functions are discussed fully in chapter 5. The differentials and integrals of these functions will be quoted here and used later in the chapter.

$$\frac{d}{dx}(e^{ax}) = ae^{ax}; \qquad \int e^{ax}\,dx = \frac{e^{ax}}{a} + c;$$

$$\frac{d}{dx}(\ln x) = \frac{1}{x}; \qquad \int \frac{dx}{x} = \ln x + c = \ln kx;$$

$$\frac{d}{dx}(a^x) = a^x \ln a \qquad \int a^x\,dx = \frac{a^x}{\ln a} + c.$$

Methods of integration

The above list of integrals enables the student to integrate any function given in one of these forms. However, many functions which we are required to integrate are more complex and may well be made up of a combination of these functions. We now consider various special methods for dealing with specific types of functions.

Inspection

This method demands a sound knowledge of the function of a function technique used in differentiation (see page 49). To use this method, you must be able to recognize the presence of a function and its differential. The following examples will illustrate the method.

Example 8 Find $\int x(3x^2 + 7)^4\,dx$.

Consider $\frac{d}{dx}(3x^2 + 7)^5 = 5(3x^2 + 7)^4 \times 6x = 30x(3x^2 + 7)^4$.

Since our result differs from the function we require to integrate only by a constant it follows that:

$$\int x(3x^2 + 7)^4\, dx = \tfrac{1}{30}(3x^2 + 7)^5 + c.$$

Example 9 Find $\int 8 \sin^4 x \cos x\, dx$.

Since $\frac{d}{dx}(\sin^5 x) = 5 \sin^4 x \cos x$, it follows that

$$\int 8 \sin^4 x \cos x\, dx = \tfrac{8}{5} \sin^5 x + c.$$

Powers of sin *x* and cos *x*

Odd powers In these integrals we split the function into an even power multiplied by a single sine or cosine. By using Pythagoras' theorem, $\cos^2 x + \sin^2 x = 1$, or the equivalent forms, we can rearrange the integral to give functions that can be integrated by inspection.

Example 10 Find $\int \cos^5 x\, dx$.

$$\int \cos^5 x\, dx = \int \cos^4 x \cos x\, dx = \int (1 - \sin^2 x)^2 \cos x\, dx$$

$$= \int (1 - 2 \sin^2 x + \sin^4 x)\cos x\, dx$$

$$= \int (\cos x - 2 \sin^2 x \cos x + \sin^4 x \cos x)\, dx.$$

$$\therefore \quad \int \cos^5 x\, dx = \sin x - \tfrac{2}{3} \sin^3 x + \tfrac{1}{5} \sin^5 x + c.$$

Even powers Functions of this type can be integrated by rearranging the integrand by using the double-angle formulae, i.e. $\cos^2 x = \frac{1}{2}(1 + \cos 2x)$ and $\sin^2 x = \frac{1}{2}(1 - \cos 2x)$ (see page 32).

Example 11 Find $\int \cos^4 x\, dx$.

$$\int \cos^4 x\, dx = \int (\cos^2 x)^2\, dx = \frac{1}{4}\int (1 + \cos 2x)^2\, dx$$

$$= \frac{1}{4}\int (1 + 2 \cos 2x + \cos^2 2x)\, dx$$

$$= \frac{1}{4}\int \{1 + 2\cos 2x + \tfrac{1}{2}(1 + \cos 4x)\}\, dx$$

$$= \frac{1}{4}\left(\frac{3x}{2} + \sin 2x + \frac{\sin 4x}{8}\right) + c$$

$$= \frac{3x}{8} + \frac{\sin 2x}{4} + \frac{\sin 4x}{32} + c.$$

Functions of the type $f'(x)/f(x)$

Consider a function of the form $y = \ln f(x)$. In this case y is a function of a function of x and hence

$$\frac{dy}{dx} = \frac{1}{f(x)} \times f'(x) = \frac{f'(x)}{f(x)}.$$

Hence $\displaystyle\int \frac{f'(x)}{f(x)}\, dx = \ln[f(x)] + c = \ln[kf(x)]$, where $\ln k = c$.

This particular type is extremely useful.

$$\int \tan x\, dx = \int \frac{\sin x}{\cos x}\, dx = -\int \frac{-\sin x}{\cos x}\, dx = -\ln \cos x + c.$$

Another useful integral which can be evaluated by this method is

$$\int \sec x\, dx = \int \frac{\sec x(\sec x + \tan x)}{\sec x + \tan x}\, dx = \int \frac{\sec^2 x + \sec x \tan x}{\sec x + \tan x}\, dx.$$

The numerator is now the differential of the denominator and hence

$$\int \sec x\, dx = \ln(\sec x + \tan x) + c.$$

Substitution

Suppose we wish to evaluate $\int f(x)\, dx$. We can sometimes effect this by making a substitution and evaluating a different integral. There are two basic ways in which we can make the substitution:

(i) by putting $u = h(x)$,
(ii) by putting $x = g(u)$.

(i) If $f(x)$ is a function of x, and $y = \int f(x)\, dx$, then $\dfrac{dy}{dx} = f(x)$.

$$\therefore \quad \frac{dy}{du} = \frac{dy}{dx} \times \frac{dx}{du} = f(x)\frac{dx}{du} \quad \text{as} \quad \frac{dy}{dx} = f(x).$$

$$y = \int f(x)\frac{dx}{du}\, du.$$

Example 12 Consider again the integral in example 8, i.e. $\int x(3x^2 + 7)^4\, dx$. Now if we make a substitution $u = 3x^2 + 7$, then $\frac{du}{dx} = 6x$, which gives $\frac{dx}{du} = \frac{1}{6x}$.

$$\text{Hence} \quad \int x(3x^2 + 7)^4\, dx = \int x(3x^2 + 7)^4 \frac{dx}{du}\, du$$

$$= \int \tfrac{1}{6}u^4\, du = \frac{u^5}{30} + c.$$

$$\therefore \quad \int x(3x^2 + 7)^4\, dx = \tfrac{1}{30}(3x^2 + 7)^5 + c \text{ as before.}$$

Example 13 Find $\int \frac{x}{\sqrt{x-2}}\, dx$.

Let $u = \sqrt{x-2} = (x-2)^{1/2}$. Thus $u^2 = x - 2 \Leftrightarrow x = u^2 + 2$.

Now $\frac{du}{dx} = \tfrac{1}{2}(x-2)^{-1/2} \Rightarrow \frac{dx}{du} = 2(x-2)^{1/2}$.

$$\text{Hence} \quad \int \frac{x}{\sqrt{x-2}}\, dx = \int \frac{x}{\sqrt{x-2}} \frac{dx}{du}\, du = \int (u^2 + 2)2\, du$$

$$= \frac{2u^3}{3} + 4u + c = \frac{2u}{3}(u^2 + 6) + c.$$

Replacing u by $(x-2)^{1/2}$

$$\int \frac{x}{\sqrt{x-2}}\, dx = \tfrac{2}{3}(x-2)^{1/2}(x - 2 + 6) + c = \tfrac{2}{3}(x+4)\sqrt{x-2} + c.$$

Example 14 Find $\int \sin^2 x \cos^3 x\, dx$.

Let $u = \sin x \Rightarrow \frac{du}{dx} = \cos x$, i.e. $\frac{dx}{du} = \frac{1}{\cos x}$.

$$\int \sin^2 x \cos^3 x\, dx = \int \sin^2 x \cos^3 x \frac{dx}{du}\, du$$

$$= \int \sin^2 x(1 - \sin^2 x)\cos x \frac{dx}{du}\, du$$

$$= \int u^2(1 - u^2)\, du = \int (u^2 - u^4)\, du$$

$$= \frac{u^3}{3} - \frac{u^5}{5} + c = \frac{u^3}{15}(5 - 3u^2) + c.$$

Replacing u by $\sin x$ we obtain

$$\int \sin^2 x \cos^3 x\, dx = \tfrac{1}{15} \sin^3 x(5 - 3\sin^2 x) + c.$$

This method is also applicable to definite integrals in which case it is often more convenient to change the limits to those of the new variable.

Example 15 Evaluate $\displaystyle\int_1^2 \frac{8x}{(2x+1)^3}\, dx.$

Let $u = 2x + 1 \Rightarrow x = \tfrac{1}{2}(u - 1)$

$$\therefore \quad \frac{du}{dx} = 2 \Rightarrow \frac{dx}{du} = \tfrac{1}{2}.$$

Under this transformation the interval $1 \leq x \leq 2$ maps on to $3 \leq u \leq 5$.

$$\int_1^2 \frac{8x}{(2x+1)^3}\, dx = \int_{x=1}^{x=2} \frac{8x}{(2x+1)^3} \frac{dx}{du}\, du = \int_3^5 \frac{4 \times \tfrac{1}{2}(u-1)}{u^3}\, du$$

$$= \int_3^5 2(u^{-2} - u^{-3})\, du = 2\Big[-u^{-1} + \tfrac{1}{2}u^{-2}\Big]_3^5$$

$$= 2\left[\left(\frac{-1}{5} + \frac{1}{50}\right) - \left(\frac{-1}{3} + \frac{1}{18}\right)\right] = 2\left(\frac{-9}{50} + \frac{5}{18}\right)$$

$$= 2\left(\frac{-81 + 125}{450}\right) = \frac{88}{450} = \frac{44}{225}.$$

(ii) When we use the substitution $x = f(u)$ a very similar process evolves.

Consider $y = \int g(x)\, dx$ and let $x = f(u)$.

Now $\frac{dy}{dx} = g(x) = g[f(u)]$.

Hence $\quad \frac{dy}{du} = \frac{dy}{dx} \times \frac{dx}{du} = g[f(u)]\,\frac{dx}{du}$.

$\therefore\ y = \int g[f(u)]\,\frac{dx}{du}\,du.$

Example 16 Find $\int_0^3 \frac{3}{\sqrt{9 - x^2}}\,dx$.

Let $x = 3\sin u \Rightarrow \frac{dx}{du} = 3\cos u$.

The interval $0 \le x \le 3$ becomes $0 \le u \le \frac{\pi}{2}$ under this transformation.

$$\int_0^3 \frac{3}{\sqrt{9 - x^2}}\,\frac{dx}{du}\,du = \int_0^{\pi/2} \left(\frac{3}{\sqrt{9 - 9\sin^2 u}}\right) 3\cos u\,du = \int_0^{\pi/2} 3\,du$$

$$= \Big[3u\Big]_0^{\pi/2} = \frac{3\pi}{2}.$$

The use of partial fractions

We can often rearrange a function by using partial fractions before attempting the integration. The methods of partial fractions were discussed in chapter 1 and we shall use the results in this section.

Example 17 Find $\int \frac{x - 2}{x^2 - 4x - 5}\,dx$.

Now $x^2 - 4x - 5 \equiv (x + 1)(x - 5)$

$$\therefore \frac{x - 2}{(x + 1)(x - 5)} \equiv \frac{A}{x + 1} + \frac{B}{(x - 5)} \equiv \frac{A(x - 5) + B(x + 1)}{(x + 1)(x - 5)}.$$

Equating coefficients gives $A = \frac{1}{2}$, $B = \frac{1}{2}$.

$$\therefore \int \frac{x - 2}{(x + 1)(x - 5)}\,dx = \int \frac{1}{2(x + 1)}\,dx + \int \frac{1}{2(x - 5)}\,dx$$

$$= \tfrac{1}{2}\ln(x + 1) + \tfrac{1}{2}\ln(x - 5) + c$$

$$= \tfrac{1}{2}\ln(x+1)(x-5)+c$$
$$= \log k\sqrt{x^2-4x-5}.$$

Notice that this result could have been obtained by writing the integral in the form $\dfrac{f'(x)}{f(x)}$, i.e.

$$\int \frac{x-2}{x^2-4x-5}\,dx = \frac{1}{2}\int \frac{2x-4}{x^2-4x-5}\,dx = \tfrac{1}{2}\ln(x^2-4x-5)+c.$$

Remember that if the integrand is a rational function in which the numerator has equal or higher degree than the denominator, it is best to divide out the fraction first.

Example 18 Find $\displaystyle\int \frac{x^2}{x+1}\,dx$.

Now $\dfrac{x^2}{x+1} \equiv x-1+\dfrac{1}{x+1}$ by division (see page 11).

$$\int \frac{x^2}{x+1}\,dx = \int \left(x-1+\frac{1}{x+1}\right)dx = \tfrac{1}{2}x^2-x+\ln(x+1)+c.$$

Integration by parts

The product rule for differentiation was given on page 48, that is, if u and v are functions of x, $\dfrac{d}{dx}(uv) = u\dfrac{dv}{dx}+v\dfrac{du}{dx}$.

Integrating both sides with respect to x:

$$uv = \int u\frac{dv}{dx}\,dx + \int v\frac{du}{dx}\,dx,$$

or

$$\int u\frac{dv}{dx}\,dx = uv - \int v\frac{du}{dx}\,dx. \tag{1}$$

This result provides an equivalent of the product rule for differentiation, in integration. For a definite integral between values of a and b, the result becomes

$$\int_a^b u\frac{dv}{dx}\,dx = \Big[uv\Big]_a^b - \int_a^b v\frac{du}{dx}\,dx. \tag{2}$$

Example 19 Find $\displaystyle\int x\sin x\,dx$.

Let $u = x \Rightarrow \dfrac{du}{dx} = 1,$

and $\dfrac{dv}{dx} = \sin x \Rightarrow v = -\cos x.$

At this stage we are choosing a particular integral and thus we let the constant of integration be zero. Using (1)

$$\int x \sin x \, dx = -x \cos x - \int -\cos x \, dx$$

$$= -x \cos x + \int \cos x \, dx$$

$$= -x \cos x + \sin x + c.$$

It may be necessary to repeat the process to completely evaluate the integral.

Example 20 Evaluate $\int_0^1 x^2 e^x \, dx.$

Let $u = x^2 \Rightarrow \dfrac{du}{dx} = 2x$ and $\dfrac{dv}{dx} = e^x \Rightarrow v = e^x.$

(Integration and differentiation of the exponential function e^x is discussed later on page 99.)

Integrating by parts:

$$\int_0^1 x^2 e^x \, dx = \Big[x^2 e^x\Big]_0^1 - \int_0^1 2xe^x \, dx.$$

The second integral can be evaluated using parts again.

$$u = 2x \Rightarrow \frac{du}{dx} = 2 \quad \text{and} \quad \frac{dv}{du} = e^x \Rightarrow v = e^x.$$

$$\therefore \int_0^1 x^2 e^x \, dx = \Big[x^2 e^x\Big]_0^1 - \left\{\Big[2xe^x\Big]_0^1 - \int_0^1 2e^x \, dx\right\}$$

$$= \Big[x^2 e^x - 2xe^x\Big]_0^1 + \Big[2e^x\Big]_0^1$$

$$= e^1 - 2e^1 + 2e^1 - 2$$

$$= e - 2 = 0{\cdot}718.$$

Some functions do not appear to be products but can be integrated by parts using 1 as one of the terms.

Example 21 Find $\int \tan^{-1} x \, dx$.

Let $u = \tan^{-1} x \Rightarrow \frac{du}{dx} = \frac{1}{1+x^2}$ and $\frac{dv}{dx} = 1 \Rightarrow v = x$.

Integrating by parts:

$$\int \tan^{-1} x \, dx = x \tan^{-1} x - \int \frac{x}{1+x^2} \, dx$$

$$= x \tan^{-1} x - \tfrac{1}{2} \ln(1+x^2) + c.$$

Example 22 Find $\int \ln x \, dx$.

Let $u = \ln x \Rightarrow \frac{du}{dx} = \frac{1}{x}$ and $\frac{dv}{dx} = 1 \Rightarrow v = x$.

Integrating by parts:

$$\int \ln x \, dx = x \ln x - \int x \frac{1}{x} \, dx = x \ln x - x + c.$$

Special methods

In integration there are many special techniques which help in the evaluation of given integrals. The following examples illustrate a few of the more common ones.

(i) Integrals of the type $\int \frac{dx}{ax^2 + bx + c}$ can be evaluated by completing the square in the denominator.

Example 23 Find $\int \frac{dx}{x^2 + 6x + 17}$.

Now $x^2 + 6x + 17 \equiv (x+3)^2 + 8$.

Thus $$\int \frac{dx}{x^2 + 6x + 17} = \int \frac{dx}{(x+3)^2 + (\sqrt{8})^2}.$$

Since $x + 3$ is linear, this integral is of the form $\tan^{-1}\left(\frac{x}{a}\right)$.

$$\int \frac{dx}{x^2 + 6x + 17} = \frac{1}{\sqrt{8}} \tan^{-1}\left(\frac{x+3}{\sqrt{8}}\right) + c.$$

(ii) Integrals of the type $\int \frac{Ax + B}{ax^2 + bx + c}\,dx$ can often be solved by rearranging the numerator to include the differential of the denominator.

Example 24 Find $\int \frac{2x + 5}{x^2 + 4x + 5}\,dx.$

Now $\frac{d}{dx}(x^2 + 4x + 5) = 2x + 4.$

Hence

$$\int \frac{2x+5}{x^2+4x+5}\,dx = \int \frac{2x+4}{x^2+4x+5}\,dx + \int \frac{1}{x^2+4x+5}\,dx$$

$$= \ln(x^2 + 4x + 5) + \int \frac{1}{(x+2)^2 + 1^2}\,dx$$

$$= \ln(x^2 + 4x + 5) + \tan^{-1}(x + 2) + c.$$

(iii) Integrals of the form $\int \frac{dx}{a + b\cos x}$ can usually be solved by using the substitution $t = \tan\left(\frac{x}{2}\right)$.

Now if $t = \tan\left(\frac{x}{2}\right)$, $\tan x = \frac{2t}{1 - t^2}$ (from the double-angle formula for $\tan 2A$).

Hence $\sin x = \frac{2t}{1 + t^2}$, $\cos x = \frac{1 - t^2}{1 + t^2}$

and $\frac{dt}{dx} = \frac{1}{2}\sec^2\left(\frac{x}{2}\right) \Rightarrow \frac{dx}{dt} = \frac{2}{1 + t^2}.$

Example 25 Evaluate $\int_0^{\pi/2} \frac{dx}{3 + 5\cos x}.$

Using the substitution $t = \tan\left(\frac{x}{2}\right)$,

$$\int_0^{\pi/2} \frac{dx}{3 + 5\cos x} = \int_{x=0}^{x=\pi/2} \frac{1}{3 + 5\cos x} \frac{dx}{dt}\, dt$$

$$= \int_0^1 \frac{1}{3 + 5\{(1 - t^2)/(1 + t^2)\}} \left(\frac{2}{1 + t^2}\right) dt$$

$$= \int_0^1 \frac{2\, dt}{3(1 + t^2) + 5(1 - t^2)} = \int_0^1 \frac{2\, dt}{8 - 2t^2}$$

$$= \int_0^1 \frac{dt}{4 - t^2} = \frac{1}{4}\int_0^1 \left(\frac{1}{2 - t} + \frac{1}{2 + t}\right) dt$$

$$= \frac{1}{4}\Big[\ln(2 + t) - \ln(2 - t)\Big]_0^1$$

$$= \left[\tfrac{1}{4} \ln\left(\frac{2 + t}{2 - t}\right)\right]_0^1 = \tfrac{1}{4} \ln 3.$$

Volumes of revolution

We can use integration to calculate the volumes of solids by a technique similar to that developed for finding areas.

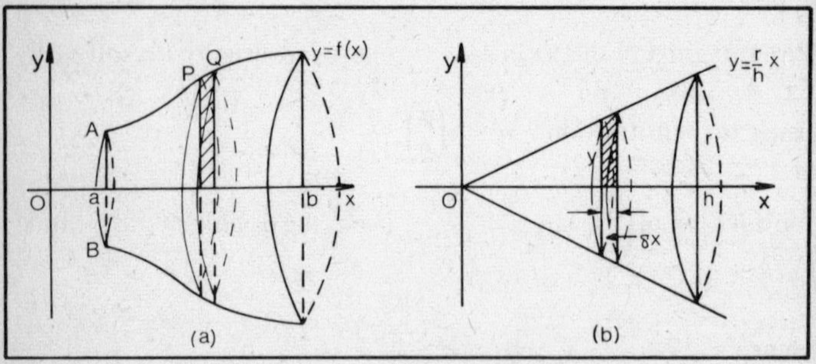

Figure 31

Consider a function $y = f(x)$. If the curve is rotated about the x-axis it forms a surface and encloses a volume. For the shape shown in figure 31(a) the rotation would form a bowl on a circular base with diameter AB. Clearly then, if we choose an element of area, shown shaded, and rotate it about the x-axis, an element of volume is formed.

If the points $P(x, y)$ and $Q(x + \delta x, y + \delta y)$ lie on the curve, then the volume of this approximate cylinder is $\pi y^2 \delta x$. By summing all such elements in the range $x = a$ to $x = b$ and taking the limit as $\delta x \to 0$, we obtain an expression for the volume, i.e.

$$\text{Volume} = \lim_{\delta x \to 0} \left[\sum_{x=a}^{x=b} \pi y^2 \delta x \right] = \int_a^b \pi y^2 \, dx.$$

Example 26 Find the volume of a cone of height h and base radius r.

Let $f(x)$ in this case be given by $y = \left(\frac{r}{h}\right)x$. This will give a cone on rotation of radius r and height h (see figure 31b). Consider an element of volume formed by rotating the shaded area about the x-axis.

Volume of the element $\simeq \pi y^2 \delta x$.

Summing all such elements between $x = 0$ and $x = h$ gives the total volume as $\sum_{x=0}^{x=h} \pi y^2 \delta x$.

In the limit as $\delta x \to 0$,

$$\text{Volume} = \pi \int_0^h y^2 \, dx = \pi \int_0^h \frac{r^2 x^2}{h^2} \, dx = \pi \left[\frac{r^2 x^3}{3h^2} \right]_0^h = \tfrac{1}{3}\pi r^2 h.$$

This method can be used even if the rotation is not about the x-axis.

Example 27 Find the volume generated when the curve $y = x^2 - 2x + 4$ is rotated about the line $y = 7$.

To find the points of intersection solve simultaneously.

$$\therefore x^2 - 2x + 4 = 7$$
$$\Leftrightarrow x^2 - 2x - 3 = 0$$
$$\Leftrightarrow (x - 3)(x + 1) = 0$$
$$\Rightarrow x = -1 \quad \text{or} \quad 3.$$

An element of volume is now formed by rotating the shaded element of area about $y = 7$.

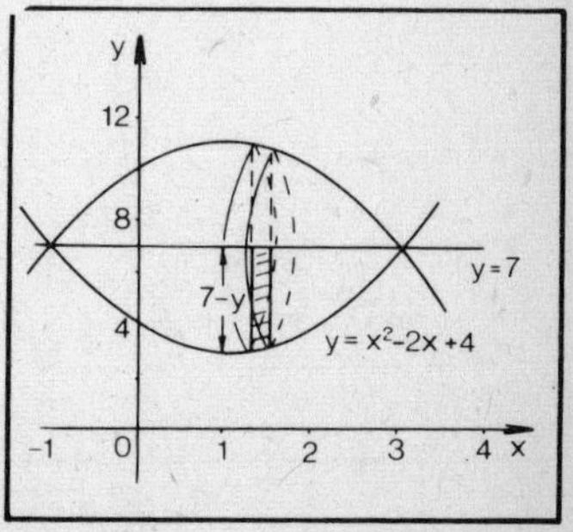

Figure 32

The volume of element

$$\simeq \pi(7 - y)^2\,\delta x$$
$$= \pi(3 + 2x - x^2)^2\,\delta x.$$

$$\text{Total volume} = \pi \int_{-1}^{3} (3 + 2x - x^2)^2\,dx$$

$$= \pi \int_{-1}^{3} (9 + 12x - 2x^2 - 4x^3 + x^4)\,dx$$

$$= \pi \left[9x + 6x^2 - \frac{2x^3}{3} - x^4 + \frac{x^5}{5}\right]_{-1}^{3}$$

$$= \pi\{(27 + 54 - 18 - 81 + \tfrac{243}{5}) - (-9 + 6 + \tfrac{2}{3} - 1 - \tfrac{1}{5})\}$$

$$= \pi\{30\tfrac{3}{5} - (-3\tfrac{8}{15})\} = \frac{512\pi}{15}.$$

Note that, unlike the equivalent problem for finding areas, the calculation for volume cannot result in a negative answer.

Area of a sector

Areas can be found using polar coordinates. If we require the area of the sector AOB between two radii OA and OB which make angles of α and β with the fixed line Ox, we select an element of area OPQ as shown in figure 33(a).

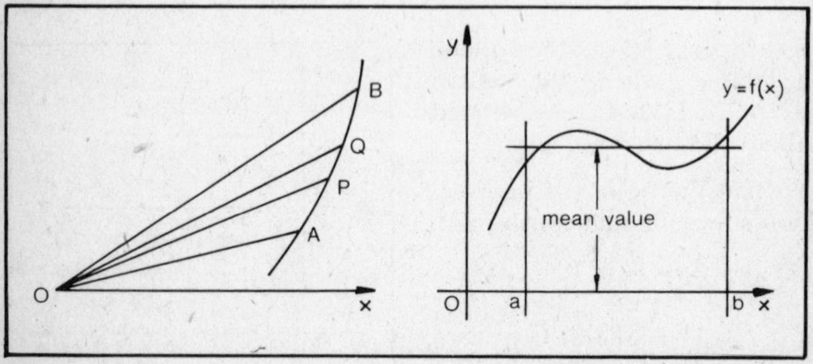

Figure 33

If angle POQ = $\delta\theta$ and angle POX = θ, the area of the sector

POQ $\simeq \frac{1}{2}r^2\theta$. By summing between $\theta = \alpha$ and $\theta = \beta$ and taking the limit as $\delta\theta \to 0$, we have:

$$\text{Total area of sector AOB} = \int_\alpha^\beta \tfrac{1}{2}r^2\, d\theta.$$

Example 28 Find the area of the loop of the curve $r = 3\sin 2\theta$ between $\theta = 0$ and $\theta = \dfrac{\pi}{2}$.

$$\text{Now}\quad \text{Area} = \int_0^{\pi/2} \tfrac{1}{2}r^2\, d\theta = \int_0^{\pi/2} \tfrac{9}{2}\sin^2 2\theta\, d\theta$$

$$= \int_0^{\pi/2} \tfrac{9}{2} \times \tfrac{1}{2}(1 - \cos 4\theta)\, d\theta = \frac{9}{4}\left[\theta - \frac{\sin 4\theta}{4}\right]_0^{\pi/2}$$

$$= \frac{9}{4}\left(\frac{\pi}{2} - 0\right) = \frac{9\pi}{8}.$$

Mean values

If $y = f(x)$ then the mean or average value of $f(x)$ over a given range $a \le x \le b$ is defined by

$$\frac{1}{b-a}\int_a^b f(x)\, dx.$$

Consider $y = f(x)$ as shown in figure 33(b) and let $y_1, y_2, y_3, y_4, \ldots, y_n$ be the ordinates corresponding to $x = a$, $x = a + \delta x$, $x = a + 2\,\delta x, \ldots, x = a + (n-1)\,\delta x$ where $n\,\delta x = b - a$. The mean of these values is:

$$\frac{1}{n}(y_1 + y_2 + \cdots + y_n) = \frac{y_1\,\delta x + y_2\,\delta x + \cdots + y_n\,\delta x}{n\,\delta x}$$

$$= \frac{1}{b-a}\sum_{x=a}^{x=b} y\,\delta x.$$

If this result tends to a limit as $n \to \infty$, i.e. $\delta x \to 0$, then the limit is $\dfrac{1}{b-a}\displaystyle\int_a^b y\, dx$.

Geometrically, this can be seen as the height of the rectangle having the same area as the area under the graph over the interval $a \le x \le b$ (see figure 33b).

Example 29 Find the mean value of the function $f(x) = \sin x$ over the interval $0 \leq x \leq \pi$.

$$\text{The required mean} = \frac{1}{\pi - 0}\int_0^\pi \sin x \, dx = \frac{1}{\pi}\Big[-\cos x\Big]_0^\pi$$

$$= \frac{1}{\pi}\{1 - (-1)\} = \frac{2}{\pi} \simeq 0{\cdot}637.$$

Numerical integration

It may not always be possible to find the integral $\int_a^b f(x)\, dx$ by the methods described in the chapter. Indeed, it may not be possible to integrate a function by an analytical process at all and in this case we can use an approximate method.

The two common methods are:
(i) the trapezium rule,
(ii) Simpson's rule.

The trapezium rule

In this method the area is divided into a number of strips and the points where the ordinates meet the curve joined by straight lines forming trapezia (see figure 34).

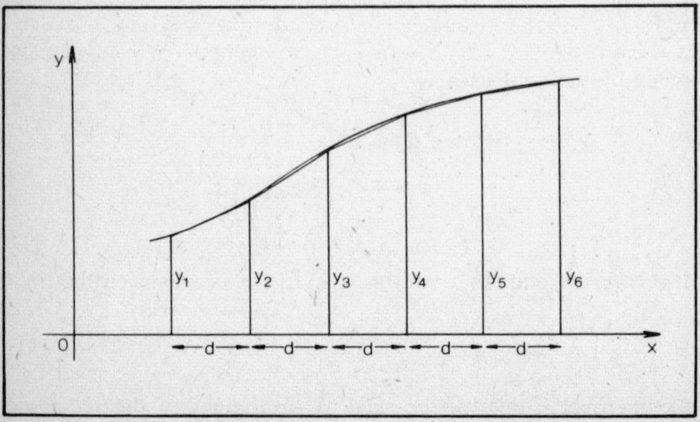

Figure 34

In this case we have chosen six ordinates $y_1, y_2, \ldots, y_6$ spaced at equal intervals d.

$$\text{The total area is } \left(\frac{y_1 + y_2}{2}\right)d + \left(\frac{y_2 + y_3}{2}\right)d + \left(\frac{y_3 + y_4}{2}\right)d + \left(\frac{y_4 + y_5}{2}\right)d + \left(\frac{y_5 + y_6}{2}\right)d$$

$$= \tfrac{1}{2}d(y_1 + 2y_2 + 2y_3 + 2y_4 + 2y_5 + y_6).$$

This is the trapezium rule for six ordinates. If n ordinates are used, the trapezium rule becomes

$$\text{Area} = \tfrac{1}{2}d(y_1 + 2y_2 + 2y_3 + \cdots + 2y_{n-1} + y_n).$$

Example 30 Use the trapezium rule to estimate the value of

$$\int_1^2 \frac{2}{x}\,dx.$$

Using six ordinates:

$$\text{Area} = \tfrac{1}{2}d(y_1 + 2y_2 + 2y_3 + 2y_4 + 2y_5 + y_6)$$
$$= \tfrac{1}{2}d\{(y_1 + y_6) + 2(y_2 + y_3 + y_4 + y_5)\}$$

With six ordinates, $d = 0{\cdot}2$

x	1	1·2	1·4	1·6	1·8	2·0
y	2	1·6667	1·4286	1·2500	1·1111	1·0

Now

$$\begin{array}{rlrl} y_1 = & 2{\cdot}0000 & y_2 = & 1{\cdot}6667 \\ y_6 = & 1{\cdot}0000 & y_3 = & 1{\cdot}4286 \\ y_1 + y_6 = & 3{\cdot}0000 & y_4 = & 1{\cdot}2500 \\ & & y_5 = & 1{\cdot}1111 \\ & & & 5{\cdot}4564 \times 2 \\ & 10{\cdot}9128 & = & 10{\cdot}9128 \\ & 13{\cdot}9128 & & \end{array}$$

$$\therefore \quad \text{Area} = \frac{0{\cdot}2}{2} \times 13{\cdot}9128 = 1{\cdot}3913 \text{ to 4 decimal places.}$$

$$\text{Note that } \int_1^2 \frac{2}{x}\,dx = 2\Big[\ln x\Big]_1^2 = 2 \ln 2 = 1{\cdot}3862.$$

Simpson's rule

This method improves on the accuracy obtained by the trapezium rule by joining the tops of the ordinates with a smooth curve instead of straight lines. We assume a quadratic passes through three consecutive points on the curve. Let the quadratic be of the form $y = ax^2 + bx + c$ after translating, so that the y-axis lies along the centre ordinate.

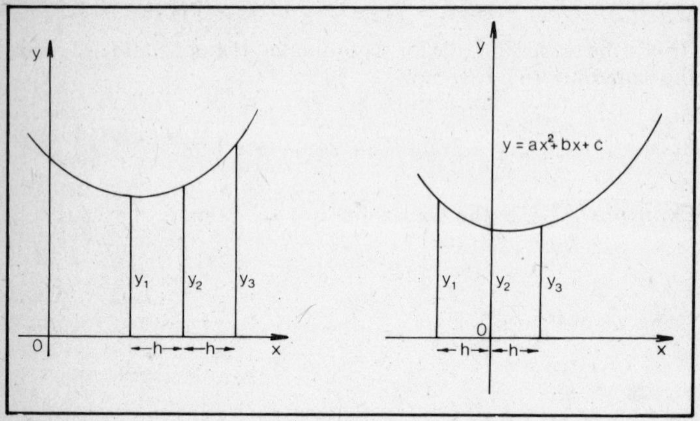

Figure 35

Let the distance between successive ordinates be h. The coefficients a, b and c can be found since $(-h, y_1)$, $(0, y_2)$, (h, y_3) lie on the curve.

Thus

$$y_1 = ah^2 - bh + c, \quad (1)$$
$$y_2 = c, \quad (2)$$
$$y_3 = ah^2 + bh + c. \quad (3)$$

From equations (1) and (3)

$$y_1 + y_3 = 2ah^2 + 2c. \quad (4)$$

Now the area under the quadratic is:

$$A = \int_{-h}^{h} (ax^2 + bx + c)\,dx = \left[\tfrac{1}{3}ax^3 + \tfrac{1}{2}bx^2 + cx\right]_{-h}^{h}$$

$$= \tfrac{2}{3}ah^3 + 2ch = \tfrac{1}{3}h(2ah + 6c) = \tfrac{1}{3}h(y_1 + y_3 + 4y_2).$$

This is usually written as: Area $= \tfrac{1}{3}h(y_1 + 4y_2 + y_3)$.

The result holds for a quadratic arc in any position. This uses only three ordinates but, in practice, we normally require more than three ordinates. So, provided there is an odd number of ordinates, we can apply the rule a number of times. In figure 36, seven ordinates are shown.

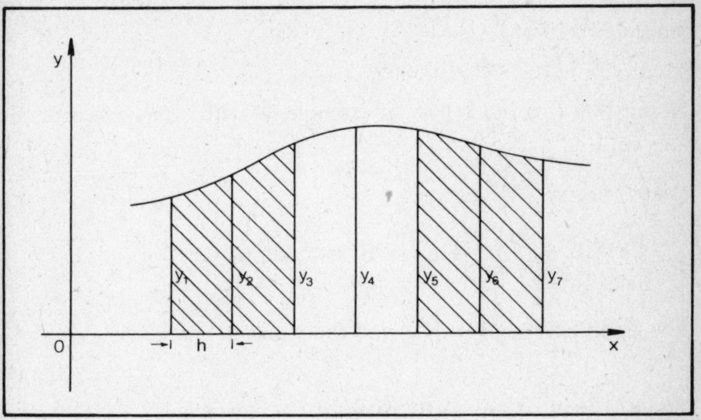

Figure 36

The total area is given by applying Simpson's rule three times.

$$\text{Area} = \tfrac{1}{3}h(y_1 + 4y_2 + y_3) + \tfrac{1}{3}h(y_3 + 4y_4 + y_5) + \tfrac{1}{3}h(y_5 + 4y_6 + y_7),$$

$$\therefore \quad \text{Area} = \tfrac{1}{3}h(y_1 + 4y_2 + 2y_3 + 4y_4 + 2y_5 + 4y_6 + y_7).$$

Learn the pattern of the result. The first and last terms have a coefficient of 1 and each of the other terms has a coefficient of 4 and 2 alternately.

For example, for 5 ordinates:

$$\text{Area} = \tfrac{1}{3}h(y_1 + 4y_2 + 2y_3 + 4y_4 + y_5).$$

Example 31 Use Simpson's rule with five ordinates to evaluate

$$\int_1^2 \frac{2}{x}\,dx.$$

x	1	1·25	1·50	1·75	2
y	2	1·6000	1·3333	1·1428	1

$y_1 = 2.0000$	$y_3 = 1.3333$	$y_2 = 1.6000$
$y_5 = 1.0000$	$\times 2$	$y_4 = 1.1428$
3.0000	2.6666	2.7428
2.6666		$\times 4$
10.9712		10.9712
16.6378		

Since $d = 0.25 = \frac{1}{4}$, $\frac{1}{3}d = \frac{1}{12}$.

Area $= \frac{1}{12} \times 16.6378 = 1.3865$.

Notice that this is a repeat of example 30 with a result nearer to 2 ln 2.

Key terms

$$\int ax^n\,dx = \frac{ax^{n+1}}{n+1} + c, \text{ for all rational } n \text{ except } n = -1.$$

The **area** between $y = f(x)$ and the x-axis is $\int_a^b f(x)\,dx$.

The **volume of revolution** about the x-axis is $\pi \int_a^b \{f(x)\}^2\,dx$.

Integration by parts $\int u \frac{dv}{dx}\,dx = uv - \int v \frac{du}{dx}\,dx$.

The **mean value** of $f(x)$ over $a \leq x \leq b$ is $\frac{1}{b-a}\int_a^b f(x)\,dx$.

The **area of a sector** is $\int \frac{1}{2}r^2\,d\theta$.

The **trapezium rule** is Area $= \frac{1}{2}d(y_1 + 2y_2 + 2y_3 + \cdots + y_n)$.

Simpson's rule is Area $= \frac{1}{3}h(y_1 + 4y_2 + y_3)$,

or for five ordinates Area $= \frac{1}{3}h(y_1 + 4y_2 + 2y_3 + 4y_4 + y_5)$,

or for $2n + 1$ ordinates

Area $= \frac{1}{3}h(y_1 + 4y_2 + 2y_3 + 4y_4 + \cdots + 2y_{2n-1} + 4y_{2n} + y_{2n+1})$.

Chapter 5
Exponential and Logarithmic Functions

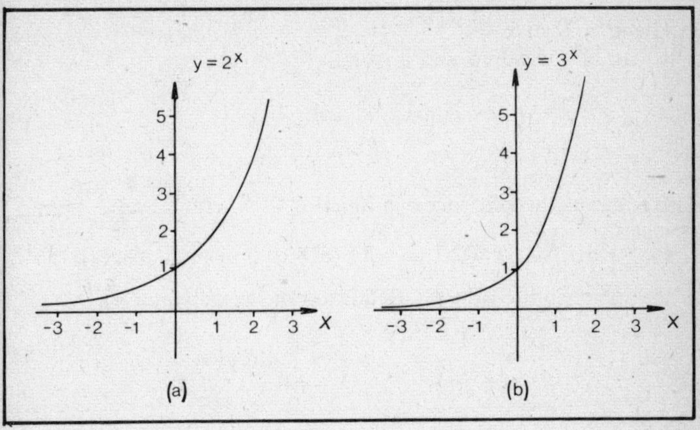

Figure 37

Figure 37(a) shows the graph of the function $y = 2^x$. From chapter 1 we know that $2^{-1} = \frac{1}{2}$, $2^0 = 1$ and $2^{1/2} = \sqrt{2} \simeq 1{\cdot}414$, and these and other intermediate values enable a smooth curve to be drawn. (2^k, where k is irrational is defined to lie on the curve).

The gradient at $x = 0$ can be measured and found to be $\simeq 0{\cdot}7$. We shall call this value M_2 (the gradient of 2^x when $x = 0$). If $y = 2^x$, $y + \delta y = 2^{x+\delta x}$, if a small change δx in x corresponds to a small change δy in y.

$$\delta y = 2^{x+\delta x} - y = 2^{x+\delta x} - 2^x = 2^x(2^{\delta x} - 1).$$

$$\frac{dy}{dx} = \lim_{\delta x \to 0} \left(\frac{\delta y}{\delta x}\right) = \lim_{\delta x \to 0} \left(\frac{2^x(2^{\delta x} - 1)}{\delta x}\right)$$

$$x = 0 \Rightarrow \left(\frac{dy}{dx}\right)_{x=0} = M_2 = \lim_{\delta x \to 0} \left(\frac{2^0(2^{\delta x} - 1)}{\delta x}\right) = \lim_{\delta x \to 0} \left(\frac{2^{\delta x} - 1}{\delta x}\right).$$

In general $\dfrac{dy}{dx} = \lim_{\delta x \to 0} \left(\dfrac{\delta y}{\delta x}\right) = \lim_{\delta x \to 0} \left(\dfrac{2^x(2^{\delta x} - 1)}{\delta x}\right) = 2^x M_2,$

i.e. $$\frac{dy}{dx} \simeq 0{\cdot}7 \times 2^x \quad \text{if} \quad y = 2^x.$$

Similarly (in figure 37b) $y = 3^x \Rightarrow \frac{dy}{dx} = 3^x M_3$, where M_3 is the gradient of 3^x at $x = 0$.
This can be measured and is $\simeq 1{\cdot}1$.

$$y = 3^x \Rightarrow \frac{dy}{dx} \simeq 1{\cdot}1 \times 3^x.$$

This suggests the existence of a function e^x where $\frac{dy}{dx} = 1 \times e^x$ and $2 < e < 3$. In fact $e \simeq 2{\cdot}7$ ($e = 2{\cdot}71828$ to 5 decimal places).

$$y = e^x \Leftrightarrow x = \ln y \quad \text{(definition of logs, chapter 1).}$$

$$\frac{dx}{dy} = \frac{1}{dy/dx} = \frac{1}{e^x} = \frac{1}{y} \Leftrightarrow \frac{d}{dy}(\ln y) = \frac{1}{y},$$

or $$\frac{d}{dx}(\ln x) = \frac{1}{x} \Leftrightarrow \int \frac{1}{x}\,dx = \ln x + k$$

$$= \ln cx \quad \text{if} \quad \ln c = k.$$

Differentiating $\log_{10} x$ gives:

$$\begin{aligned} y = \log_{10} x \Leftrightarrow y + \delta y &= \log_{10}(x + \delta x) \\ \Leftrightarrow \qquad \delta y &= \log_{10}(x + \delta x) - y \\ &= \log_{10}(x + \delta x) - \log_{10} x \\ &= \log_{10}\left(1 + \frac{\delta x}{x}\right). \end{aligned}$$

So $$\frac{\delta y}{\delta x} = \frac{1}{\delta x}\log_{10}\left(1 + \frac{\delta x}{x}\right).$$

$$\frac{dy}{dx} = \lim_{\delta x \to 0}\left(\frac{\delta y}{\delta x}\right) = \lim_{\delta x \to 0}\left\{\log_{10}\left(1 + \frac{\delta x}{x}\right)^{1/\delta x}\right\}$$

Let $\frac{\delta x}{x} = t$ then $\delta x \to 0 \Leftrightarrow t \to 0$ and $\frac{1}{\delta x} = \frac{1}{xt}$.

$$\frac{dy}{dx} = \lim_{t \to 0}\{\log_{10}(1 + t)^{1/xt}\} = \lim_{t \to 0}\left\{\frac{1}{x}\log_{10}(1 + t)^{1/t}\right\}. \qquad (1)$$

We need to evaluate the limit of $(1 + t)^{1/t}$ as $t \to 0$. We do this by taking progressively smaller values of t, i.e.

$$
\begin{array}{lll}
t = 0{\cdot}1 & (1+t)^{1/t} = (1{\cdot}1)^{10} & = 2{\cdot}59374246, \\
t = 0{\cdot}01 & (1+t)^{1/t} = (1{\cdot}01)^{100} & = 2{\cdot}70481386, \\
t = 0{\cdot}001 & (1+t)^{1/t} = (1{\cdot}001)^{1000} & = 2{\cdot}71692302, \\
t = 0{\cdot}0001 & (1+t)^{1/t} = (1{\cdot}0001)^{10000} & = 2{\cdot}71814591, \\
t = 0{\cdot}00001 & (1+t)^{1/t} = (1{\cdot}00001)^{1000000} & = 2{\cdot}71828182.
\end{array}
$$

It is clear that the limit is approaching a number which can be calculated to any degree of accuracy. This number we call e and correct to 9 decimal places $e = 2{\cdot}718281828$.
From the binomial theorem (see chapter 8),

$$
\begin{aligned}
&(1+t)^{1/t} \\
&= 1 + t \times \frac{1}{t} + \frac{1}{t}\left(\frac{1}{t} - 1\right)\frac{t^2}{2} + \frac{1}{t}\left(\frac{1}{t} - 1\right)\left(\frac{1}{t} - 2\right)\frac{t^3}{2 \times 3} + \cdots \\
&= 2 + \frac{1-t}{2} + \frac{(1-t)(1-2t)}{2 \times 3} + \frac{(1-t)(1-2t)(1-3t)}{2 \times 3 \times 4} + \cdots
\end{aligned}
$$

Thus $\lim_{t \to 0} (1+t)^{1/t} = 2 + \frac{1}{2} + \frac{1}{6} + \frac{1}{24} + \cdots$

$$= 2{\cdot}5 + 0{\cdot}1\dot{6} + 0{\cdot}041\dot{6} + 0{\cdot}008\dot{3} + 0{\cdot}0013\dot{8} + \cdots$$

The sum of the first five terms $= 2{\cdot}7180\dot{5}$ and this will approach the value of e.

From equation (1): $\dfrac{dy}{dx} = \dfrac{1}{x}\log_{10} e.$

Consequently $y = \ln x \Leftrightarrow \dfrac{dy}{dx} = \dfrac{1}{x}\ln e = \dfrac{1}{x}$, since $\ln e = 1$,

or $\displaystyle\int \frac{1}{x}\,dx = \ln x + k = \ln cx.$

From chapter 4 $\displaystyle\int x^m\,dx = \frac{x^{m+1}}{m+1}$,

and thus integrating $f(x) = \dfrac{1}{x}$ gives

$$\int \frac{1}{x}\,dx = \frac{x^0}{0} \quad \text{which is infinite.}$$

Define $F(a) = \int_1^a \frac{1}{x}\,dx$, so $F(1) = \int_1^1 \frac{1}{x}\,dx = 0.$

$F(a)$ = area under graph from $x = 1$ to $x = a$ (horizontal shaded area in figure 38b).

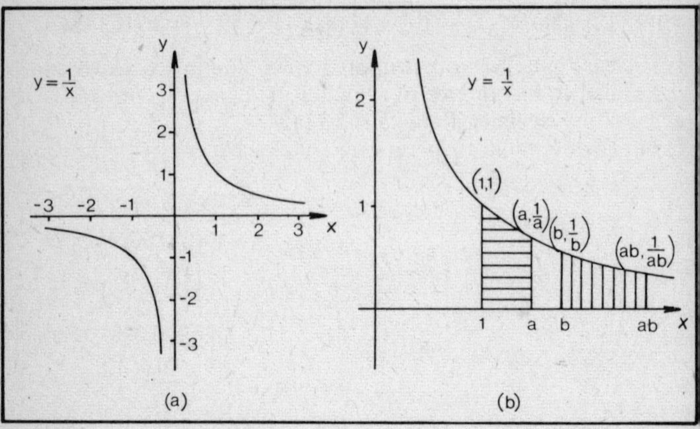

Figure 38

Consider the transformation represented by the matrix $\begin{pmatrix} b & 0 \\ 0 & \frac{1}{b} \end{pmatrix}$

Under this transformation $(1, 0) \to (b, 0)$; $(1, 1) \to \left(b, \frac{1}{b}\right)$;

$(a, 0) \to (ab, 0)$; $\left(a, \frac{1}{a}\right) \to \left(ab, \frac{1}{ab}\right)$.

The horizontal shaded area is transformed into the vertical shaded area.

The determinant of $\begin{pmatrix} b & 0 \\ 0 & \frac{1}{b} \end{pmatrix} = 1 \Rightarrow$ area is preserved under the transformation.

So $\int_1^a \frac{1}{x}\,dx = \int_b^{ab} \frac{1}{x}\,dx = \int_1^{ab} \frac{1}{x}\,dx - \int_1^b \frac{1}{x}\,dx,$

i.e. $$F(a) = F(ab) - F(b),$$

or $F(a) + F(b) = F(ab)$ a fundamental law of logarithms.

Replacing b by $\frac{b}{a}$ gives $F(a) + F\left(\frac{b}{a}\right) = F(b),$

so $F(b) - F(a) = F\left(\frac{b}{a}\right)$ another logarithm law.

Writing $b = a$ gives $\int_1^a \frac{1}{x}\,dx = \int_{a^3}^{a^2} \frac{1}{x}\,dx.$

$b = a^2$ gives $\int_1^a \frac{1}{x}\,dx = \int_{a^2}^{a^3} \frac{1}{x}\,dx$ and so on.

$$F(a^n) = \int_1^{a^n} \frac{1}{x}\,dx = \int_1^a \frac{1}{x}\,dx + \int_a^{a^2} \frac{1}{x}\,dx + \int_{a^2}^{a^3} \frac{1}{x}\,dx + \cdots$$

$$+ \int_{a^{n-1}}^{a^n} \frac{1}{x}\,dx = n\int_1^a \frac{1}{x}\,dx = n\,F(a)$$ another logarithm law.

These 3 properties of $F(a)$ suggest F is a logarithmic function. Since $F(a)$ is continuously increasing there exists a number e such that

$$\int_1^e \frac{1}{x}\,dx = 1$$

If $F(e) = 1$ and $F(a)$ is a logarithmic function, then $F(e) = \log e = 1 \Leftrightarrow$ the base of the logarithms is e.

We now have two important functions:

$$y = e^x \Leftrightarrow \frac{dy}{dx} = e^x \quad \text{and} \quad x = \ln y = \ln e^x. \tag{2}$$

$$y = \ln x \Leftrightarrow \frac{dy}{dx} = \frac{1}{x} \quad \text{and} \quad x = e^y = e^{\ln x} \therefore \tag{3}$$

$\text{Log}_e\, x$ is the natural logarithm of x and is also denoted by $\ln x$. Values of $\ln x$ are printed in books of mathematical tables. e^x is an exponential function (the exponent is x) and is also denoted by $\exp x$.

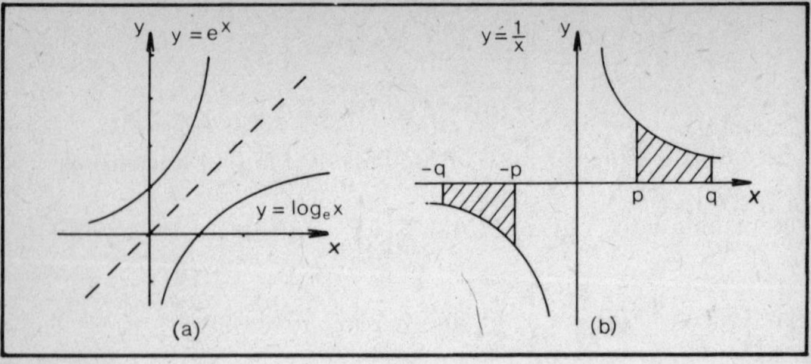

Figure 39

$y = e^x$ and $y = \ln x$ are inverse functions of each other as is shown by relations (2) and (3). Their graphs are related by a reflection in the line $y = x$ (figure 39a).

Since $\log(-x)$ is not defined, care must be taken when integrating $\frac{1}{x}$ if the limits are negative. From the graph of $y = \frac{1}{x}$ in figure 39(b), the shaded areas are equal in size since the graph is symmetrical.

$$\int_{-q}^{-p} \frac{1}{x}\, dx = -\int_{p}^{q} \frac{1}{x}\, dx, \quad \text{since the area } A \text{ is below the } x\text{-axis.}$$

$$= -(\ln q - \ln p)$$

$$= \ln p - \ln q. \quad \text{This is in fact } \Big[\ln |x|\Big]_{-q}^{-p}$$

Consequently $\int_{q}^{p} \frac{1}{x}\, dx = \ln |x|$ if p, q are both positive or both negative.

$\int_{-2}^{+1} \frac{1}{x}\, dx$ consists of two infinite areas and cannot be evaluated.

Derivative of $e^{f(x)}$

$$y = e^{f(x)} \Leftrightarrow \ln y = \ln e^{f(x)} = f(x) \ln e = f(x).$$

Differentiating $\frac{1}{y}\frac{dy}{dx} = \frac{df(x)}{dx} \Leftrightarrow \frac{dy}{dx} = \frac{df(x)}{dx}\, y = \frac{df(x)}{dx}\, e^{f(x)}.$

Example 1 $y = e^{x^2} \Rightarrow \dfrac{dy}{dx} = 2xe^{x^2}$.

Example 2 $y = e^{x^3+x} \Rightarrow \dfrac{dy}{dx} = (3x^2 + 1)e^{x^3+x}$.

Example 3 $y = e^{\sin x} \Rightarrow \dfrac{dy}{dx} = \cos x \, e^{\sin x}$.

Example 4 $\displaystyle\int_0^1 e^{2x}\,dx = \left[\tfrac{1}{2}e^{2x}\right]_0^1 = \tfrac{1}{2}e^2 - \tfrac{1}{2} = 3{\cdot}195.$

Example 5 $\displaystyle\int x^2 e^{x^3}\,dx = \int \tfrac{1}{3}\,3x^2 e^{x^3}\,dx = \tfrac{1}{3}e^{x^3} + c.$

Derivative of a^x and $a^{f(x)}$

$$y = a^x \Leftrightarrow \ln y = x \ln a \Rightarrow \frac{1}{y}\frac{dy}{dx} = \ln a \Leftrightarrow \frac{dy}{dx} = y \ln a = a^x \ln a.$$

$$y = a^{f(x)} \Leftrightarrow \ln y = f(x) \ln a \Rightarrow \frac{1}{y}\frac{dy}{dx} = f'(x) \ln a$$

$$\Leftrightarrow \frac{dy}{dx} = f'(x)\, a^{f(x)} \ln a.$$

Alternative proof of product rule for differentiation

If $y = uv$ where y, u and v are differentiable functions of x,

$$\ln y = \ln(uv) = \ln u + \ln v \Rightarrow \frac{1}{y}\frac{dy}{dx} = \frac{1}{u}\frac{du}{dx} + \frac{1}{v}\frac{dv}{dx}.$$

$$\Leftrightarrow \frac{dy}{dx} = \frac{y}{u}\frac{du}{dx} + \frac{y}{v}\frac{dv}{dx} = v\frac{du}{dx} + u\frac{dv}{dx}.$$

These results can easily be extended to a product of three functions.

$$y = uvw \Leftrightarrow \ln y = \ln u + \ln v + \ln w$$

$$\Rightarrow \frac{dy}{dx} = vw\frac{dy}{dx} + uw\frac{dv}{dx} + uv\frac{dw}{dx}.$$

Derivative of ln $f(x)$

$$y = \ln f(x) \Leftrightarrow e^y = f(x) \Rightarrow e^y \frac{dy}{dx} = f'(x) \Rightarrow \frac{dy}{dx} = \frac{f'(x)}{f(x)}.$$

Example 6 $y = \ln x^3 \Rightarrow \dfrac{dy}{dx} = \dfrac{1}{x^3} 3x^2 = \dfrac{3}{x}$,

or $$y = \ln x^3 = 3 \ln x \Rightarrow \frac{dy}{dx} = 3 \frac{1}{x} = \frac{3}{x}.$$

Example 7 $y = \ln \sin x \Rightarrow \dfrac{dy}{dx} = \dfrac{\cos x}{\sin x} = \cot x.$

Example 8 $y = \ln \sec x \Rightarrow \dfrac{dy}{dx} = \dfrac{\sec x \tan x}{\sec x} = \tan x.$

Example 9 $y = \ln(\sec x + \tan x) \Rightarrow \dfrac{dy}{dx} = \dfrac{\sec x \tan x + \sec^2 x}{\sec x + \tan x}$

$$= \frac{\sec x(\tan x + \sec x)}{\sec x + \tan x}$$

$$= \sec x.$$

Applications in integration

Example 8 leads to the result

$$\int \cot x\, dx = \int \frac{\cos x\, dx}{\sin x} = \ln \sin x + \text{constant}.$$

This is an example of an integral whose numerator is the derivative of the denominator, and this integral can be performed directly.

Example 10 $\displaystyle\int \frac{2x}{x^2 - 1}\, dx = \ln(x^2 - 1) + c.$

Alternatively, by partial fractions, $\dfrac{2x}{x^2 - 1} = \dfrac{1}{x - 1} + \dfrac{1}{x + 1}.$

So $$\int \frac{2x}{x^2 - 1}\, dx = \int \left(\frac{1}{x - 1} + \frac{1}{x + 1}\right) dx$$

$$= \ln(x - 1) + \ln(x + 1) = \ln(x^2 - 1) + c.$$

In general $\displaystyle\int \frac{f'(x)}{f(x)}\, dx = \ln f(x) + c.$

Example 11 $\int \frac{x^2}{(x^3-2)}\,dx = \int \frac{1}{3}\frac{3x^2}{(x^3-2)}\,dx$

$$= \frac{1}{3}\ln(x^3-2)+k.$$

You may find this simpler using the method of substitution.

If $u = x^3 - 2$, then $\frac{du}{dx} = 3x^2 \Leftrightarrow x^2\,dx = \frac{du}{3}$.

$$\int \frac{x^2}{(x^3-2)}\,dx = \int \frac{1}{u}\frac{du}{3} = \frac{1}{3}\ln u = \frac{1}{3}\ln(x^3-2)+c.$$

If $u = f(x)$, $\frac{du}{dx} = f'(x)$, so $\int \frac{f'(x)}{f(x)}\,dx = \int \frac{du}{u}$

$$= \ln u + c = \ln f(x) + c.$$

Example 12

$$\int \frac{1}{x^2-a^2}\,dx = \int \frac{1}{(x+a)(x-a)}\,dx$$

$$= \int \left(\frac{1}{2a(x-a)} - \frac{1}{2a(x+a)}\right) dx \text{ by partial fractions}$$

$$= \frac{1}{2a}\ln|x-a| - \frac{1}{2a}\ln|x+a| + c$$

$$= \frac{1}{2a}\ln\left|\frac{x-a}{x+a}\right| + c.$$

Key terms

$y = \log_{10} x \Rightarrow \frac{dy}{dx} = \frac{1}{x}\log_{10} e$, $e = 2{\cdot}7183\ldots$

$y = \ln x \Rightarrow \frac{dy}{dx} = \frac{1}{x}$ and $\int \frac{1}{x}\,dx = \ln x + k$.

e^x and $\ln x$ are inverse functions; $e^{\ln x} = x$; $\ln^{e^x} = x$.

$y = e^{f(x)} \Rightarrow \frac{dy}{dx} = f'(x)\,e^{f(x)}$.

$y = a^{f(x)} \Rightarrow \frac{dy}{dx} = f'(x)\,a^{f(x)}\ln x$, where a is a constant.

Chapter 6
Coordinate Geometry

Introduction

Coordinate geometry is that branch of mathematics which studies the properties of lines, curves and planes using the techniques of algebra and calculus.

Coordinates

The position of a point P can be specified by two distances measured from two perpendicular intersecting lines. The lines are called the axes of coordinates and their point of intersection is the origin. The distances are usually denoted by x and y and as they take all possible values the point P moves to every position in the plane.

The values of x and y are called the **cartesian coordinates** of the point P and are written (x, y) (see figure 40a). These distances are known as the **abscissa** and **ordinate** respectively.

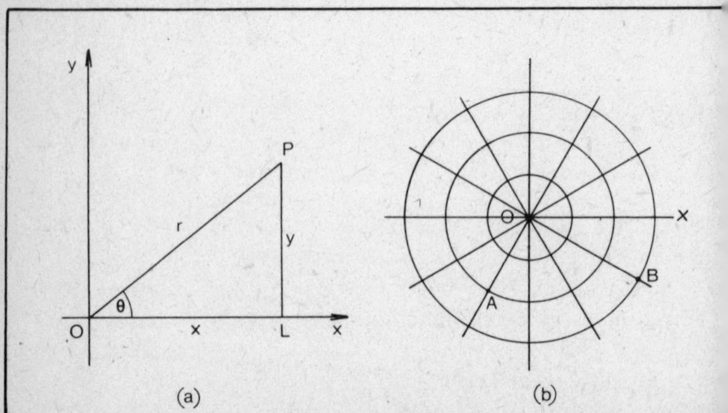

Figure 40

An alternative system is to specify the position of a point in a plane by using a distance from a fixed point together with an angle measured from a given fixed line (see figure 40a). Let O be the fixed point, called the **pole**, and OX the fixed line called the **initial line**. If OP $= r$ and the angle POX $= \theta$, then r and θ are called the **polar coordinates** of the point P and are written (r, θ). Note that θ is positive when measured anti-clockwise. The same point may be described in different ways as shown in figure 40(b). The point A may be written as (2, 240°) and the point B as (3, 330°), (3, −30°) or even (−3, 150°).

Relation between cartesian and polar coordinates

Consider a point P in a plane whose cartesian coordinates are given by (x, y) relative to axes OX and OY respectively. Let (r, θ) be the polar coordinates of the same point using OX as the initial line and O as the pole (see figure 40a).

Now in $\triangle$OPL $\quad \cos\theta = \dfrac{x}{r} \quad$ and $\quad \sin\theta = \dfrac{y}{r}$.

$\therefore\ x = r\cos\theta$ and $y = r\sin\theta$, which give x and y in terms of r and θ.

Squaring and adding these results we obtain

$$x^2 + y^2 = r^2\cos^2\theta + r^2\sin^2\theta = r^2.$$

Dividing the two original equations gives $\dfrac{y}{x} = \dfrac{r\sin\theta}{r\cos\theta} = \tan\theta.$

Thus r and θ are given in terms of x and y by

$$r = \sqrt{(x^2 + y^2)} \quad \text{and} \quad \theta = \tan^{-1}\left(\frac{y}{x}\right).$$

Example 1 Express the equation of the cardioid $r = \frac{1}{2}(1 + \cos\theta)$ in cartesian form.

Since $\quad r = \sqrt{(x^2 + y^2)} \quad$ and $\quad \cos\theta = \dfrac{x}{r} = \dfrac{x}{\sqrt{x^2 + y^2}}$,

we have $\quad \sqrt{x^2 + y^2} = \dfrac{1}{2}\left(1 + \dfrac{x}{\sqrt{x^2 + y^2}}\right).$

$$\therefore \quad 2(x^2 + y^2) = \sqrt{x^2 + y^2} + x.$$

Thus the cartesian equation is given by

$$\{2(x^2 + y^2) - x\}^2 = x^2 + y^2.$$

Example 2 Find the polar equation of the curve whose cartesian equation is given by $y^2 = 8(2 - x)$.

Since $x = r\cos\theta$ and $y = r\sin\theta$, we have, by substituting in the equation, $(r\sin\theta)^2 = 8(2 - r\cos\theta)$.

$$\therefore \quad r^2\sin^2\theta + 8r\cos\theta - 16 = 0$$
$$r^2(1 - \cos^2\theta) + 8r\cos\theta - 16 = 0$$
$$r^2 = 16 - 8r\cos\theta + r^2\cos^2\theta = (4 - r\cos\theta)^2.$$

Taking the positive square root we obtain $r = 4 - r\cos\theta$.

Hence $\qquad r(1 + \cos\theta) = 4.$

Thus the polar equation is given by $r = \dfrac{4}{(1 + \cos\theta)}$.

The length of a line segment

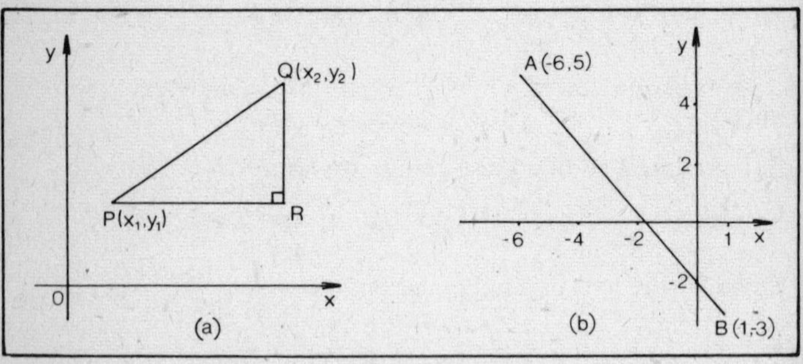

Figure 41

Consider the two points $P(x_1, y_1)$ and $Q(x_2, y_2)$ as shown in figure 41(a). By completing a right-angled triangle PQR with PR parallel to OX we have:

$$RP = x_2 - x_1 \quad \text{and} \quad QR = y_2 - y_1.$$

Using the theorem of Pythagoras, $PQ^2 = RP^2 + QR^2$.

$\therefore \quad PQ^2 = (x_2 - x_1)^2 + (y_2 - y_1)^2.$

Thus the length of PQ is given by

$$PQ = \sqrt{(x_2 - x_1)^2 + (y_2 - y_1)^2}.$$

The positive square root is taken for the length of a line segment.

Example 3 Find the distance between the two points A(−6, 5) and B(1, −3).

We have $x_2 - x_1 = 1 - (-6) = 7$ units

and $y_2 - y_1 = -3 - (5) = -8$ units;

∴ the length of AB is given by

$$AB = \sqrt{(x_2 - x_1)^2 + (y_2 - y_1)^2} = \sqrt{7^2 + (-8)^2}$$
$$= \sqrt{113} = 10{\cdot}6 \text{ to 1 decimal place.}$$

The straight line

If we consider a point P(x, y) in a plane where the values of x and y can vary, P could take any position in the plane. If, however, we define a relationship between x and y, we restrict the point P to a specific path. For example, if $y = x^2$, the point P can only take positions where the y-coordinate is the square of the x-coordinate. If $y = x^2 - 4x + 3$ the possible points are shown in figure 5(a), page 19. If the relationship is linear the graph will be a straight line.

Example 5 For the relationship (i) $y = 2x - 1$, (ii) $y = 2 - x$ find the values of y corresponding to $x = -3, -2, -1, 0, 1, 2, 3$. Plot these pairs of values (x, y) using cartesian axes.

Tables are constructed as follows:

(i)

x	−3	−2	−1	0	1	2	3
$2x$	−6	−4	−2	0	2	4	6
−1	−1	−1	−1	−1	−1	−1	−1
$2x - 1$	−7	−5	−3	−1	1	3	5

(ii)

x	−3	−2	−1	0	1	2	3
$-x$	3	2	1	0	−1	−2	−3
2	2	2	2	2	2	2	2
$2 - x$	5	4	3	2	1	0	−1

These pairs of points are plotted on the same set of axes as in figure 42.

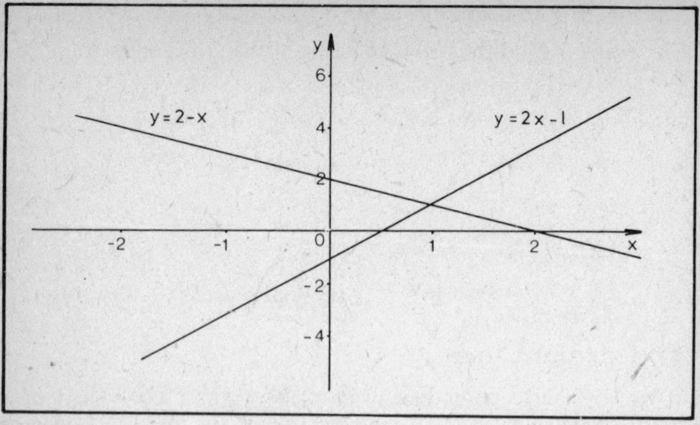

Figure 42

Gradient of a straight line

If the scales on both axes are equal, the gradient of a straight line can also be found by measuring the angle that it makes with the increasing direction of the x-axis and calculating its tangent (see figure 43a). Note that in figure 43(b) α will be obtuse and thus its tangent will be **negative**.

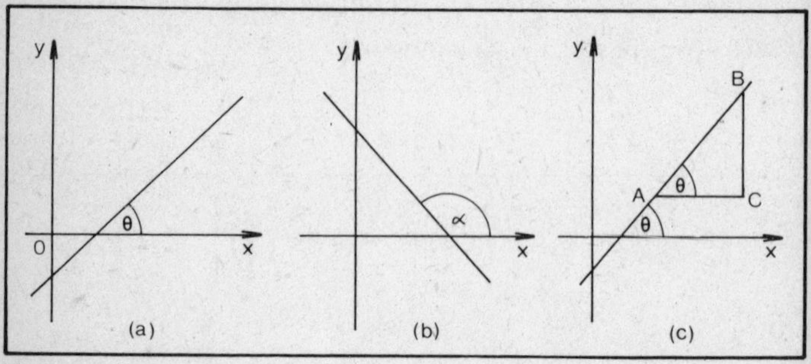

Figure 43

Check that the graphs of $y = 2x - 1$ and $y = 2 - x$ drawn in figure 42 have gradients of 2 and -1 respectively. The gradient can also be found by considering two points on the line and evaluating the ratio $\dfrac{BC}{AC}$ which is equivalent to $\tan\theta$ (see figure 43).

Example 5 Find the gradient of the line passing through the points P(−6, 4) and Q(4, −2).

$$\text{Gradient of PQ is } \frac{(y\text{-coordinate of Q}) - (y\text{-coordinate of P})}{(x\text{-coordinate of Q}) - (x\text{-coordinate of P})}$$

$$= \frac{-2-(4)}{4-(-6)} = -\frac{6}{10} = -\frac{3}{5}.$$

If a linear equation representing a straight line is written in the form $\boldsymbol{y = mx + c}$ the gradient is given by m and c defines the point at which the line cuts the y-axis. This is referred to as the **intercept** on the y-axis.

Special cases

(i) If $m = 0$, the line is parallel to the x-axis and its equation reduces to $y = c$.

(ii) If the line is parallel to the y-axis the equation $x = k$ defines the line.

Example 6 What is the equation of the straight line which cuts the y-axis at the point (0, −4) and whose gradient is $\frac{3}{2}$?

The general equation of a straight line is $y = mx + c$. In this case $m = \frac{3}{2}$ and $c = -4$ and the required equation is

$$y = \tfrac{3}{2}x - 4 \quad \text{or} \quad 2y = 3x - 8.$$

Alternative methods for finding the equation of a straight line are illustrated by the following cases.

Method (1) Find the equation of the straight line through the point (x_1, y_1) whose gradient is m.

Let A be the given point whose coordinates are (x_1, y_1), and consider a general point P whose coordinates are (x, y).

The gradient of AP is $\dfrac{y - y_1}{x - x_1} = m$.

Rearranging this gives the equation as $\boldsymbol{y - y_1 = m(x - x_1)}$.

This form of the equation is useful when the gradient and one point of the line are given.

Example 7 For a line whose gradient is −4, which passes through the point (3, −2) the equation is given by
$y - y_1 = m(x - x_1)$.

In this case $y - (-2) = -4(x - 3) \Rightarrow y + 2 = -4x + 12$.
$\therefore$ the equation is given by $y + 4x = 10$.

Method (2) Find the equation of the straight line joining the point $P(x_1, y_1)$ to the point $Q(x_2, y_2)$.

Consider a general point $R(x, y)$ on the line PQ. Since the points P, Q and R are collinear it follows that:
the gradient of PR = the gradient of QR.

$$\therefore \text{ the equation is } \frac{y - y_1}{x - x_1} = \frac{y - y_2}{x - x_2}.$$

Example 8 Find the equation of the line joining the points $P(-2, 1)$ and $Q(3, 4)$.

We use the result of method (2).

The equation of the line is given by $\dfrac{y - y_1}{x - x_1} = \dfrac{y - y_2}{x - x_2}$,

$$\text{i.e.} \quad \frac{y - 1}{x - (-2)} = \frac{y - 4}{x - 3}.$$

$$\begin{aligned} \text{Simplifying} \quad (y - 1)(x - 3) &= (x + 2)(y - 4) \\ xy - 3y - x + 3 &= xy + 2y - 4x - 8 \\ -5y + 3 &= -3x - 8 \\ \text{or} \quad 5y &= 3x + 11. \end{aligned}$$

The angle between two lines

Consider two intersecting straight lines with equations $y = m_1 x + c_1$ and $y = m_2 x + c_2$ as shown in figure 44. The angle between the two lines is the acute angle ACB. If the lines make

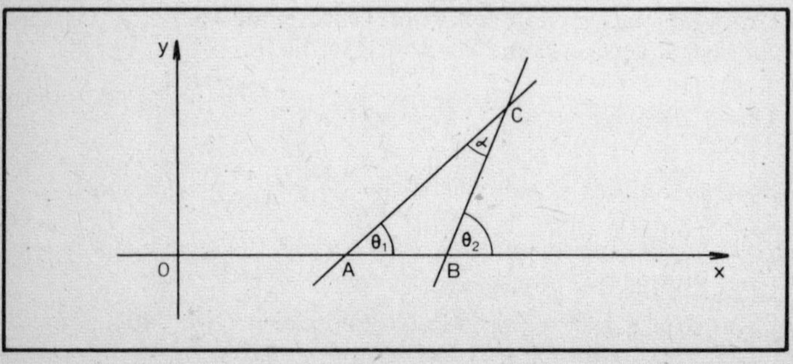

Figure 44

angles of θ_1 and θ_2 with the x-axis as shown we have angle $ACB = \theta_2 - \theta_1 = \alpha$ (say)

$$\therefore \tan \alpha = \tan (\theta_2 - \theta_1) = \frac{\tan \theta_2 - \tan \theta_1}{1 + \tan \theta_2 \tan \theta_1}.$$

Now $\tan \theta_1 = m_1$ (the gradient of AC) and $\tan \theta_2 = m_2$ (the gradient of BC).

$$\therefore \tan \alpha = \frac{m_2 - m_1}{1 + m_1 m_2} \Rightarrow \alpha = \tan^{-1}\left(\frac{m_2 - m_1}{1 + m_1 m_2}\right).$$

Therefore the angle between two lines whose gradients are m_1 and m_2 is given by

$$\tan^{-1}\left(\frac{m_2 - m_1}{1 + m_1 m_2}\right).$$

Special cases

(i) If $\alpha = \frac{\pi}{2}$ then $\tan \frac{\pi}{2} = \frac{m_2 - m_1}{1 + m_1 m_2}$.

But $\tan \frac{\pi}{2} = \infty$ and thus it follows that $1 + m_1 m_2 = 0$, i.e. $m_1 m_2 = -1$.

The product of the gradients of two lines perpendicular to each other is equal to -1.

(ii) If $\alpha = 0$, $m_1 = m_2$ and the lines are parallel.

Example 9 Find the angle between the two lines $y = 2x + 3$ and $y = -\frac{1}{4}x$.

In this case $m_1 = 2$ and $m_2 = -\frac{1}{4}$. Using the above result:

$$\alpha = \tan^{-1}\left(\frac{m_2 - m_1}{1 + m_1 m_2}\right)$$

$$= \tan^{-1}\left(\frac{-\frac{1}{4} - 2}{1 + 2(-\frac{1}{4})}\right) = \tan^{-1}(-4\tfrac{1}{2})$$

$$= 180° - 77{\cdot}47° = 102{\cdot}53°.$$

$\therefore$ The obtuse angle between the lines is 102·53°, and the acute angle between the lines is 77·47°.

Example 10 Find the coordinates of the point of intersection of the two perpendicular lines where one line has a gradient of 2 and cuts the x-axis at $x = -1$ and the other line cuts the x-axis at a point where $x = 19$.

The equation of the first line will be of the form

$$y - y_1 = m_1(x - x_1), \text{ i.e. } y - 0 = 2\{x - (-1)\}$$

as it passes through the point $(-1, 0)$, or $y = 2x + 2$.

The equation of the second line will be of the form

$$y - y_2 = m_2(x - x_2), \text{ i.e. } y - 0 = -\tfrac{1}{2}(x - 19),$$

since the lines are perpendicular, $m_2 = -\frac{1}{2}$ and it passes through $(19, 0)$.
Thus the equation of the second line is $2y = -x + 19$.
If the lines represented by these equations intersect at $P(a, b)$

then $b = 2a + 2$ and $2b = -a + 19$.

Solving gives $b = 8$ and $a = 3$. Thus the coordinates of the point of intersection are $(3, 8)$.

The distance of a general point $P(x_1, y_1)$ from a line whose equation is $ax + by + c = 0$.

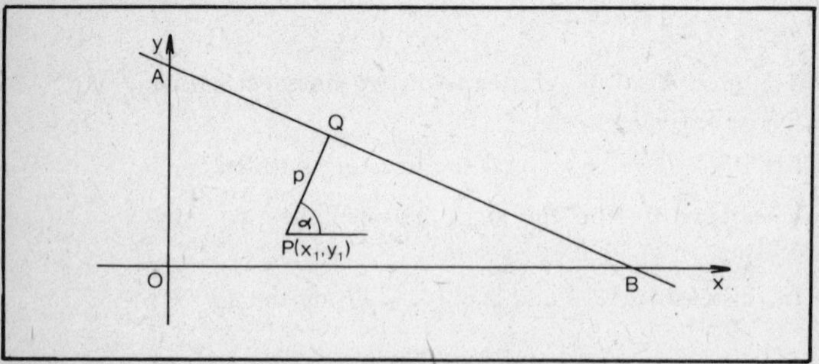

Figure 45

Since the gradient of the line AB is $-\dfrac{a}{b}$ (figure 45), the gradient of

PQ (the perpendicular) must be $\dfrac{b}{a}$, and $\tan\alpha = \dfrac{b}{a}$ giving

$$\cos\alpha = \pm\frac{a}{\sqrt{a^2 + b^2}} \text{ and } \sin\alpha = \pm\frac{b}{\sqrt{a^2 + b^2}}.$$

Then, if the length of the perpendicular PQ = p, the coordinates of Q will be $(x_1 + p\cos\alpha, y_1 + p\sin\alpha)$.
But Q lies on the given line $ax + by + c = 0$.

$\therefore\ a(x_1 + p\cos\alpha) + b(y_1 + p\sin\alpha) + c = 0.$
Thus $\quad p(a\cos\alpha + b\sin\alpha) = -(ax_1 + by_1 + c).$

$$\frac{p(a^2 + b^2)}{\pm\sqrt{a^2 + b^2}} = -(ax_1 + by_1 + c),$$

$$\therefore\quad p = \pm\frac{ax_1 + by_1 + c}{\sqrt{a^2 + b^2}},$$ the sign being chosen to make $p > 0$.

The sign, as evaluated, indicates whether the point (x_1, y_1) is on the same side of the line as the origin or not.

Example 11 Find the distance of the points A$(-3, 0)$ and B$(-5, 1)$ from the line $2x - 3y + 9 = 0$.

Substituting in the above result we have:

$$p_A = \frac{2(-3) - 3(0) + 9}{\sqrt{2^2 + 3^2}} = \frac{3}{\sqrt{13}}.$$

$$p_B = \frac{2(-5) - 3(1) + 9}{\sqrt{2^2 + 3^2}} = \frac{-4}{\sqrt{13}}.$$

Hence the distances are $\frac{3}{\sqrt{13}}$ and $\frac{4}{\sqrt{13}}$ respectively.

It can be seen from a sketch that the two points are actually on opposite sides of the line as indicated by the difference in sign in the calculation.

Example 12 Find the equation of the perpendicular bisector of the line segment joining the point P$(-3, -5)$ to the point Q$(-8, 2)$. If the bisector meets the y-axis at S, find the coordinates of S.

Let M be the mid-point of PQ.
The coordinates of P are $(-3, -5)$ and those of Q are $(-8, 2)$.

Thus the coordinates of M are given by

$$\left(\frac{-3 + (-8)}{2}, \frac{-5 + 2}{2}\right) = \left(\frac{-11}{2}, \frac{-3}{2}\right).$$

The gradient PQ $= \dfrac{2-(-5)}{-8-(-3)} = -\dfrac{7}{5}$.

Thus the gradient of the perpendicular bisector is $\frac{5}{7}$.
The equation of a line of gradient $\frac{5}{7}$ which passes through $(-\frac{11}{2}, -\frac{3}{2})$ is given by

$$y - (-\tfrac{3}{2}) = \tfrac{5}{7}\{x - (-\tfrac{11}{2})\}$$

$$y = \tfrac{5}{7}x + \tfrac{55}{14} - \tfrac{3}{2} = \tfrac{5}{7}x + \tfrac{17}{7}.$$

The equation of the perpendicular bisector is $7y = 5x + 17$.
The coordinates of S are $(0, \frac{17}{7})$.

The circle

The circle can be defined as a set of points equidistant from a fixed point, the centre of the circle.

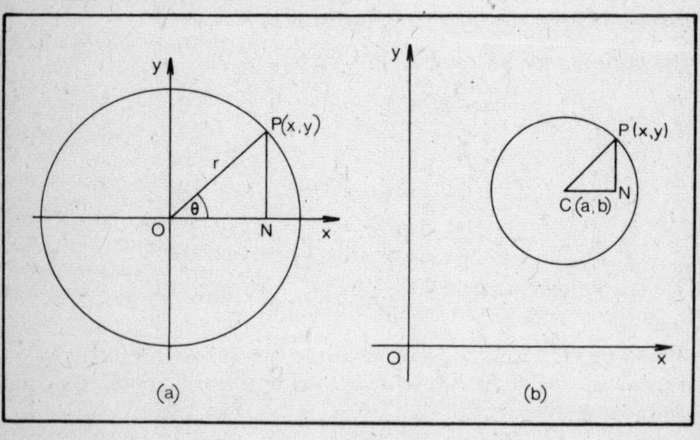

Figure 46

Figure 46(a) shows a circle, the centre of which has been chosen at the origin of coordinates. Let P(x, y) be any point on the circumference of the circle, radius r. Let PN be drawn perpendicular to the x-axis.
In triangle ONP: $x = r\cos\theta$ and $y = r\sin\theta$ where angle PON $= \theta$.
Squaring and adding these two results we obtain

$$x^2 + y^2 = r^2(\cos^2\theta + \sin^2\theta),$$

i.e.
$$\boldsymbol{x^2 + y^2 = r^2}.$$

This relationship between the coordinates of P(x, y) is satisfied by all points on the circumference of the circle and hence it represents the equation of the circle with centre the origin and radius r.

Example 13 Find the equation of a circle whose centre is at the origin of coordinates which passes through the point P(2, 3).

Since the point P(2, 3) lies on the circle, it follows that these coordinates will satisfy the equation $x^2 + y^2 = r^2$.

$$\therefore\ 2^2 + 3^2 = r^2 \quad \text{or} \quad r^2 = 13.$$

The equation of the circle is $x^2 + y^2 = 13$.

A circle of radius *r* whose centre is at C(*a*, *b*)

Figure 46(b) shows a circle with centre at C(a, b). Let P(x, y) be a general point on the circumference of the circle. Construct the right-angled triangle CPN as shown.

In ΔCPN $\quad CN^2 + NP^2 = CP^2$ (Pythagoras' theorem)

$$\therefore \qquad \mathbf{(x - a)^2 + (y - b)^2 = r^2.} \qquad (1)$$

This is one form for the equation of a circle, radius r and centre (a, b).

Example 14 What is the equation of a circle, centre $(-1, 4)$ whose radius is 2?

Using equation (1), the equation becomes

$$\{x - (-1)\}^2 + (y - 4)^2 = 2^2,$$

i.e.

$$(x + 1)^2 + (y - 4)^2 = 4.$$

This can be written as $\quad x^2 + y^2 + 2x - 8y + 13 = 0.$

We can expand equation (1) to give an alternative form,

$$(x - a)^2 + (y - b)^2 = r^2$$
$$x^2 - 2ax + a^2 + y^2 - 2by + b^2 = r^2$$
$$x^2 + y^2 - 2ax - 2by + a^2 + b^2 - r^2 = 0$$

This is usually written $\quad \mathbf{x^2 + y^2 + 2gx + 2fy + c = 0,} \qquad (2)$
where the values g, f and c are constants.
Note that in this form:
(i) the centre of the circle is at the point $(-g, -f)$,
(ii) the radius of the circle is r where $r^2 = g^2 + f^2 - c$.

Example 15 What are the coordinates of the centre and the radius of the circle given by $x^2 + y^2 + 6x - 9y + 4 = 0$?

Comparing with equation (2), it follows that the coordinates of the centre are given by $(-3, +4\frac{1}{2})$.
The radius is given by $c = g^2 + f^2 - r^2$.

$\therefore\ 4 = (-3)^2 + (4\frac{1}{2})^2 - r^2$ or $r^2 = 9 + \frac{81}{4} - 4 = \frac{101}{4}$.

$\therefore$ the radius is $\frac{1}{2}\sqrt{101}$.

Alternatively this problem can be solved by completing the square in x and y in the given equation.

$$x^2 + y^2 + 6x - 9y + 4 = 0$$
$$(x^2 + 6x + 3^2) + \{y^2 - 9y + (-4\frac{1}{2})^2\} + 4 = 3^2 + (-4\frac{1}{2})^2$$
$$(x + 3)^2 + (y - 4\frac{1}{2})^2 = \frac{101}{4}.$$

Hence the centre is $(-3, 4\frac{1}{2})$ and the radius $\frac{1}{2}\sqrt{101}$ as before.

Tangents

The idea of a tangent to a curve at a given point being the limiting case of a sequence of secants drawn from that point was introduced in the work on differentiation (see pages 42–3). Thus we can consider a tangent to a curve as the straight line which cuts the curve at two coincident points. The following example uses this fact to show that a given line touches a circle tangentially.

Example 16 What is the value of m if the line $y = mx$ is a tangent to the circle $x^2 + y^2 + 2x - 4y + 1 = 0$? What are the coordinates of the points at which the tangent touches the curve?

The points of intersection are found by solving simultaneously

$$x^2 + y^2 + 2x - 4y + 1 = 0 \quad \text{and} \quad y = mx.$$
$$\therefore \quad x^2 + m^2x^2 + 2x - 4mx + 1 = 0$$
$$\therefore \quad (1 + m^2)x^2 + (2 - 4m)x + 1 = 0. \tag{1}$$

Now since the general quadratic equation $ax^2 + bx + c = 0$ has equal roots if $b^2 = 4ac$, it follows that equation (1) has equal roots if

$$(2 - 4m)^2 = (4)(1 + m^2)$$
$$4 - 16m + 16m^2 = 4 + 4m^2$$

i.e. if $12m^2 - 16m = 0$ or $4m(3m - 4) = 0 \Rightarrow m = 0$ or $\frac{4}{3}$.

Thus if $m = 0$ or $m = \frac{4}{3}$, $y = mx$ forms a tangent to the circle. The equations are $y = 0$ and $y = \frac{4}{3}x$.

The coordinates of the point of contact can be found from the fact that the x-coordinates are given by the roots of equation (1) and since they are equal we can write:

$$x_1 = x_2 = -\frac{b}{2a} = -\frac{(2 - 4m)}{2(1 + m^2)}$$

(from the sum of the roots of the quadratic equation $ax^2 + bx + c = 0$).
When $m = 0$, the x-coordinate $= -1$,

$$\text{when } m = \tfrac{4}{3} \text{ the } x\text{-coordinate} = -\frac{\{2 - 4(\frac{4}{3})\}}{2\{1 + (\frac{4}{3})^2\}} = \frac{3}{5}.$$

Using $y = 0$ and $y = \frac{4}{3}x$, the points of contact are given by $(-1, 0)$ and $(\frac{3}{5}, \frac{4}{5})$.

Parametric form

It is possible to express the position of a general point on a circle and hence its equation in terms of a third arbitrary variable. Such a variable is called a **parameter**.

Example 17 A point P has coordinates $(a \cos \theta, a \sin \theta)$ where a is a constant. If θ varies, describe the path of P.

Let the coordinates of P be (x, y).
Then $x = a \cos \theta$ and $y = a \sin \theta$. Squaring and adding gives

$$x^2 + y^2 = a^2 \cos^2 \theta + a^2 \sin^2 \theta = a^2(\cos^2 \theta + \sin^2 \theta) = a^2.$$

Thus the relationship between the coordinates of P is $x^2 + y^2 = a^2$, the locus of P. Thus as θ varies, this relationship holds for all positions of P. Hence P moves on a circle, centre the origin, radius a.

When the coordinates of P are given in the form $(a \cos \theta, a \sin \theta)$ θ is called a parameter and $x = a \cos \theta$, $y = a \sin \theta$ are the **parametric equations** of a circle, centre the origin, radius a.

In this example θ is interpreted geometrically as the angle that the radius vector from the origin to P makes with the x-axis as

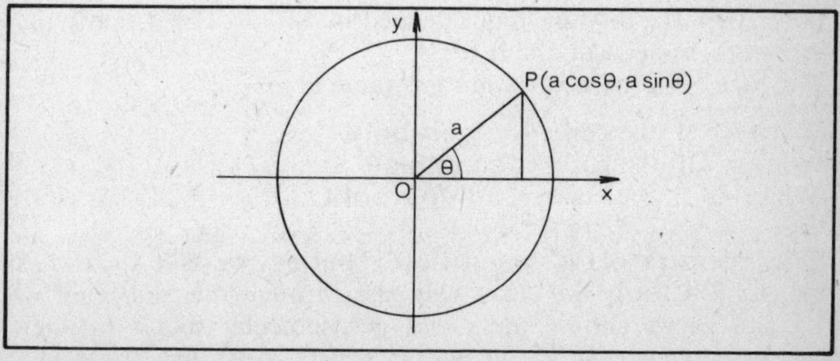

Figure 47

shown in figure 47. Note that by choosing specific values of θ, actual points will be defined.

For $\theta = 0$, P is the point $(a, 0)$.

For $\theta = \frac{\pi}{4}$, P is the point $\left(a \cos \frac{\pi}{4}, a \sin \frac{\pi}{4}\right) = \left(\frac{a}{\sqrt{2}}, \frac{a}{\sqrt{2}}\right)$.

The parabola

The locus of a point which moves so that its distance from a fixed point, S, is in a constant ratio, e, to its distance from a fixed line is called a **conic section**. If $e = 1$ this becomes the parabola.

The fixed point S is called the **focus**. The fixed line is called the **directrix**. The ratio e is called the **eccentricity**.

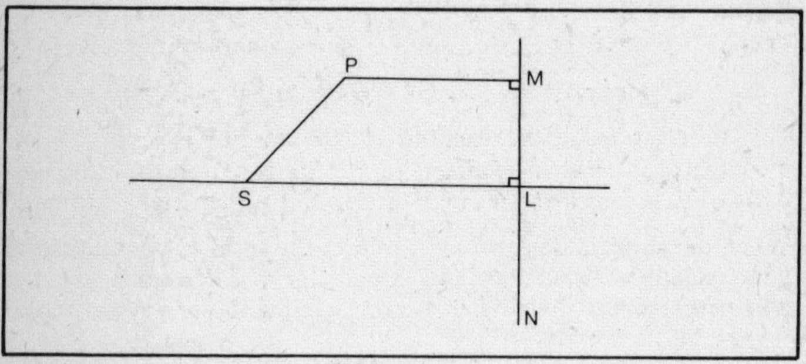

Figure 48

If in figure 48 MN is the given line, S is the focus and P a general point, then the relationship is defined by SP $= e$ PM. Clearly the locus is symmetrical about SL.
The locus of P depends upon the value of e.

When $e = 1$, the curve is a **parabola**.
When $e < 1$, the curve is an **ellipse**.
When $e > 1$, the curve is a **hyperbola**.

Since the parabola is symmetrical about SL, we take the x-axis along LS. Clearly the curve will pass through the mid-point of LS and so we choose the y-axis perpendicular to LS through this point thus ensuring the curve will pass through the origin of coordinates, O. This is called the **vertex** of the parabola.

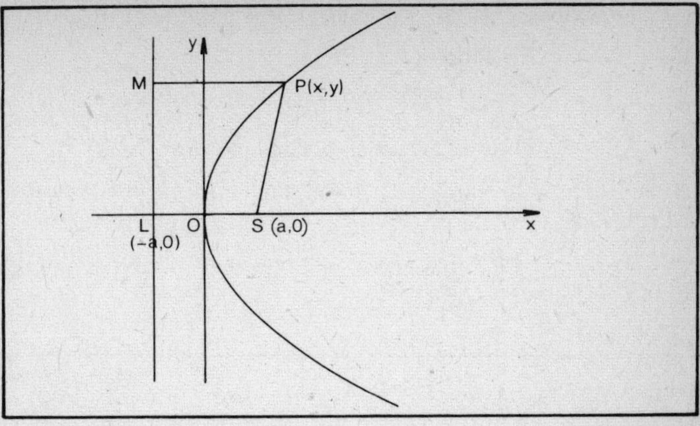

Figure 49

Let the distance of S from the origin be a, i.e. the focus is at the point $(a, 0)$. Let $P(x, y)$ be any point of the curve. From P draw PM parallel to SL to meet the line $x = -a$ (the directrix) at M. Join PS.

By definition SP = PM (since $e = 1$) giving $SP^2 = PM^2$.

$$\therefore \quad (y-0)^2 + (x-a)^2 = (x+a)^2$$
$$\therefore \quad y^2 + x^2 - 2ax + a^2 = x^2 + 2ax + a^2$$
$$\therefore \quad \mathbf{y^2 = 4ax}.$$

This is the **standard equation of a parabola**. When the equation is in this form it follows that:

(i) the focus S is the point $(a, 0)$,
(ii) the directrix is the line $x = -a$,
(iii) the vertex is at the origin of coordinates,
(iv) the curve is symmetrical about the x-axis,
(v) the y-axis is a tangent at the vertex.

A line through the focus parallel to the y-axis is called the **latus rectum**. If the equation is in the form $y^2 = 4ax$, then the length of the latus rectum is $4a$.

Example 18 Find the locus of the point $P(x, y)$ which moves such that its distance from the point $S(-3, 0)$ is equal to its distance from the fixed line $x = 3$.

Referring to figure 50(a):

$$PS^2 = (x+3)^2 + y^2 \quad \text{and} \quad PM^2 = (x-3)^2.$$

Since PS = PM it follows that $PS^2 = PM^2$.

$$\therefore \quad (x+3)^2 + y^2 = (x-3)^2$$
$$\therefore \quad x^2 + 6x + 9 + y^2 = x^2 - 6x + 9$$
$$\therefore \quad y^2 = -12x, \text{ which is the locus of P.}$$

If the equation is not already in the standard form it is possible to rearrange it.

Example 19 Find the focus and directrix of the parabola $y^2 = 8 - 2x$.

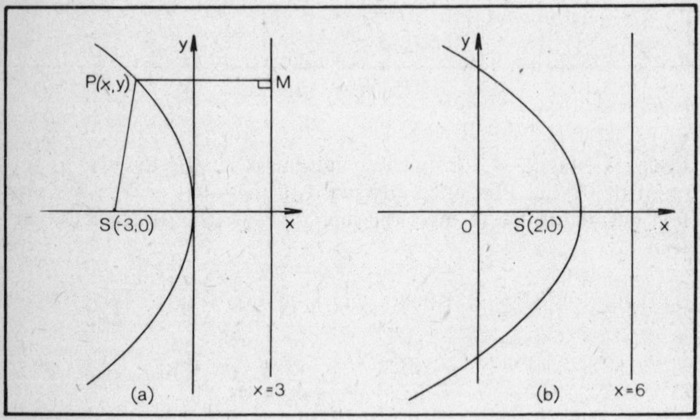

Figure 50

Rewrite the equation in the form $y^2 = 4(2 - \frac{1}{2}x)$.
By using the substitution $X = 2 - \frac{1}{2}x$, the equation of the parabola can be written as $y^2 = 4X$.
The focus is at the point where $X = 1$ and $y = 0$, i.e.

where $2 - \frac{1}{2}x = 1$ and $y = 0$,
or $x = 2$ and $y = 0$.

Thus the focus is the point (2, 0). The directrix is given by the equation $X = -1$, i.e. $2 - \frac{1}{2}x = -1$. Thus the directrix is the line $x = 6$. The curve is sketched in figure 50(b).

Example 20 Find the value of c which makes the straight line $y = mx + c$ a tangent to the parabola $y^2 = 4ax$.

Solving simultaneously we can form a quadratic equation in y.

$$y^2 = \frac{4a(y-c)}{m} \Rightarrow my^2 = 4ay - 4ac.$$

$$\therefore \qquad my^2 - 4ay + 4ac = 0.$$

The two roots of this equation will give the y-coordinates of the points of intersection of the line and the curve. A repeated root will be obtained when $(-4a)^2 = 4m(4ac)$.

$$\therefore \qquad a^2 = mac, \quad c = \frac{a}{m}.$$

$\therefore$ The line $y = mx + \left(\frac{a}{m}\right)$ will be a tangent to the parabola.

By using the substitution $t = 1/m$ this result can be written

$$y = (x/t) + at.$$

This suggests a possible parametric form for the equation. Any point on the parabola can be represented by the coordinates $(at^2, 2at)$. These values would satisfy the equation $y^2 = 4ax$. The **parametric equations** of the parabola are $\boldsymbol{x = at^2,\ y = 2at}$. Note that t has to take all values from $-\infty$ to ∞ if the complete curve is to be traced. Much of the detailed work on the parabola is done using parameters.

Example 21 Find the equation of the tangent and the normal to the parabola $y^2 = 4ax$ at the point $P(at^2, 2at)$.

Differentiating implicitly with respect to x,

$$2y\frac{dy}{dx} = 4a \Rightarrow \frac{dy}{dx} = \frac{2a}{2at} = \frac{1}{t} \quad \text{at the point } (at^2, 2at).$$

The gradient of the tangent is thus $1/t$ and since it passes through the point $(at^2, 2at)$, its equation is given by

$$y - 2at = \frac{1}{t}(x - at^2), \quad \text{or}$$

$$\boldsymbol{ty = x + at^2.}$$

Similarly the equation of the **normal** can be derived. If the gradient of the tangent at $(at^2, 2at)$ is $1/t$, the gradient of the normal is $-t$. The equation of the normal is

$$y - 2at = -t(x - at^2), \quad \text{or}$$

$$\boldsymbol{y + tx = 2at + at^3.}$$

Example 22 Find the equation of the chord joining the point $P(ap^2, 2ap)$ to the point $Q(aq^2, 2aq)$ on the parabola $y^2 = 4ax$.

$$\text{The gradient of the chord PQ} = \frac{2ap - 2aq}{ap^2 - aq^2} = \frac{2a(p-q)}{a(p^2-q^2)}$$

$$= \frac{2(p-q)}{(p+q)(p-q)} = \frac{2}{p+q}.$$

Since the chord passes through $(ap^2, 2ap)$, the equation is

$$y - 2ap = \frac{2}{(p+q)}(x - ap^2), \quad \text{or}$$

$$(p+q)y - 2ap^2 - 2apq = 2x - 2ap^2.$$

$$\therefore \qquad \boldsymbol{(p+q)y = 2x + 2apq}. \qquad (1)$$

From this equation of the chord we can deduce the equation of the tangent, for if $q \to p$ we have,

$$2py = 2x + 2ap^2 \Rightarrow py = x + ap^2 \quad \text{as before.}$$

A **focal chord** is a chord passing through the focus $(a, 0)$. Since this point lies on the line, these values satisfy equation (1) which gives the equation of a chord joining two points.

$$\therefore \qquad 0 = 2a + 2apq \Rightarrow pq = -1.$$

Thus **the condition that a chord should be a focal chord is $pq = -1$.**

Example 23 If tangents are drawn to the parabola $y^2 = 4ax$ at the ends of a focal chord show that they meet on the directrix.

Let P and Q be points with coordinates $(ap^2, 2ap)$ and $(aq^2, 2aq)$ respectively.
The equation of PQ is

$$(p+q)y = 2x + 2apq$$

(see equation (1) in example 22). Since PQ passes through the focus $S(a, 0)$ this equation gives

$$pq = -1. \qquad (1)$$

Now the equation of the tangent at $P(ap^2, 2ap)$ is

$$py = x + ap^2. \qquad (2)$$

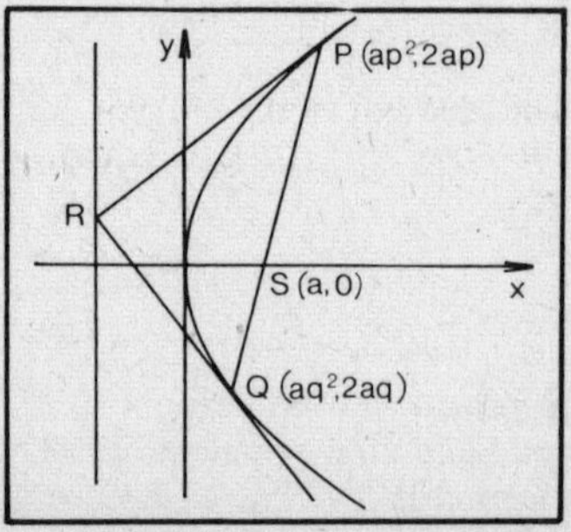

Figure 51

Also the equation of the tangent at $Q(aq^2, 2aq)$ is

$$qy = x + aq^2. \tag{3}$$

Solving equations (2) and (3) we find the point of intersection, R.

$$(p - q)y = ap^2 - aq^2 = a(p - q)(p + q) \Rightarrow y = a(p + q). \tag{4}$$

Substituting in equation (2):

$$pa(p + q) = x + ap^2 \Rightarrow x = apq.$$

Thus the tangents intersect at $R(apq, a(p + q))$.
But from equation (1), $pq = -1$ and hence $x = -a$, which is the equation of the directrix. Thus the tangents always intersect at a point on the directrix.

Example 24 If the normal at a point $P(ap^2, 2ap)$ on the parabola $y^2 = 4ax$ meets the curve again at $Q(aq^2, 2aq)$ prove that p and q are related by the equation $p^2 + pq + 2 = 0$.

Since the equation of the normal at $P(ap^2, 2ap)$ is given by

$$y + px = 2ap + ap^3 \quad \text{(see example 21)},$$

we have, solving simultaneously with the equation of the curve,

$$y + py^2/4a = 2ap + ap^3$$

$$\text{i.e.} \quad py^2 + 4ay - 4a(2ap + ap^3) = 0.$$

The roots of this equation give the y-coordinates of the points of intersection, say y_1 and y_2. Now $y_1 + y_2 = -4a/p$ (the sum of the roots of a quadratic equation).

But $y_1 = 2ap$ and hence $y_2 = -2ap - 4a/p$.

$$\therefore \quad y_2 = 2aq = (-4a - 2ap^2)/p \Rightarrow q = (-2 - p^2)/p,$$

i.e.

$$p^2 + pq + 2 = 0.$$

Example 25 Show that the normals drawn to the parabola $y^2 = 4ax$ at the points $P(ap^2, 2ap)$ and $Q(aq^2, 2aq)$ where PQ is a focal chord meet at a point $R(a(p^2 + q^2 + 1), a(p + q))$. Find the locus of R.

The equations of the normals at $P(ap^2, 2ap)$ and $Q(aq^2, 2aq)$ are $y + px = 2ap + ap^3$ and $y + qx = 2aq + aq^3$ respectively. Subtracting these equations we obtain:

$$(p - q)x = 2a(p - q) + a(p^3 - q^3),$$

$$\text{i.e.} \quad x = 2a + a(p^2 + pq + q^2), \quad \text{since } p - q \text{ is a factor.}$$

But since PQ is a focal chord $pq = -1$, thus $x = a(p^2 + q^2 + 1)$.

Substituting in the equation of the normal at P we have

$$y + ap(p^2 + q^2 + 1) = 2ap + ap^3$$
$$y = 2ap - apq^2 - ap = a(p + q), \quad \text{since } pq = -1.$$

Hence the point of intersection R is $(a(p^2 + q^2 + 1), a(p + q))$.
Let the coordinates of R be (x, y).

$\therefore \qquad x = a(p^2 + q^2 + 1) \qquad (1)$

and $\qquad y = a(p + q). \qquad (2)$

The relationship between x and y can be obtained by eliminating the parameters p and q.

Equation (1) gives $x = a\{(p + q)^2 - 2pq + 1\}$.
Since $pq = -1$, $x = a\{(p + q)^2 + 3\}$.

But $y = a(p + q) \quad \therefore \quad x = a\{(y^2/a^2) + 3\}$.

Thus the equation of the locus of R is $y^2 = ax - 3a^2$.

Key terms

Distance between (x_1, y_1) and (x_2, y_2) is

$$\sqrt{(x_2 - x_1)^2 + (y_2 - y_1)^2}.$$

The **equation of a straight line** can be written as:
(i) $y = mx + c$, where m is the gradient;
(ii) $y - y_1 = m(x - x_1)$; (iii) $\dfrac{y - y_1}{x - x_1} = \dfrac{y - y_2}{x - x_2}$.

The angle between two lines of gradients m and m' is $\tan^{-1}(m' - m)/(1 + mm')$. If the lines are perpendicular the product of the gradients is -1. The distance of $P(x_1, y_1)$ from the line $ax + by + c = 0$ is

$$\pm \frac{ax_1 + by_1 + c}{\sqrt{a^2 + b^2}}.$$

Circle $(x - a)^2 + (y - b)^2 = r^2$. The centre is (a, b) and radius r. The circle centred on the origin can be represented by the parametric equations $x = a \cos \theta$, $y = a \sin \theta$.

Parabola $y^2 = 4ax$.
Equation of the **tangent**: $ty = x + at^2$.
Equation of the **normal**: $y + tx = 2at + at^3$.
Equation of a **chord** joining points with parameters p and q is $(p + q)y = 2x + 2apq$.

Chapter 7
Matrices and Determinants

$$\begin{array}{lcc} & \textbf{Home} & \textbf{Away} \\ & \begin{matrix} W & D & L \end{matrix} & \begin{matrix} W & D & L \end{matrix} \\ \begin{matrix}\text{Liverpool} \\ \text{Arsenal}\end{matrix} & \begin{pmatrix} 3 & 2 & 1 \\ 5 & 0 & 1 \end{pmatrix} & \begin{pmatrix} 2 & 4 & 0 \\ 2 & 1 & 2 \end{pmatrix} \\ & \text{Matrix } H & \text{Matrix } A \end{array}$$

A **matrix** is a **rectangular array** of elements. In the example above, the elements are numbers and by the headings of the rows and columns it can be seen that the matrices represent the home and away playing records of two football teams.

Matrix **addition** can be achieved by adding together the corresponding elements in each matrix.

$$H + A = \begin{pmatrix} 3 & 2 & 1 \\ 5 & 0 & 1 \end{pmatrix} + \begin{pmatrix} 2 & 4 & 0 \\ 2 & 1 & 2 \end{pmatrix} = \begin{matrix} \\ \text{L} \\ \text{A} \end{matrix} \begin{matrix} W \quad D \quad L \\ \begin{pmatrix} 5 & 6 & 1 \\ 7 & 1 & 3 \end{pmatrix} \end{matrix} = \text{matrix } T.$$

T represents the total number of matches won, drawn or lost by each team.

Matrix **subtraction** is similar, but in this case the resulting matrix has little meaning except to compare the teams' home and away records.

$$H - A = \begin{pmatrix} 3 & 2 & 1 \\ 5 & 0 & 1 \end{pmatrix} - \begin{pmatrix} 2 & 4 & 0 \\ 2 & 1 & 2 \end{pmatrix} = \begin{pmatrix} 1 & -2 & 1 \\ 3 & -1 & -1 \end{pmatrix}.$$

The **order** of a matrix indicates its size or shape, and this is done by specifying the number of rows and columns. Matrix H has order 2×3 (2 by 3) since it has 2 rows and 3 columns. Matrices can only be **added** or **subtracted** if they have the same order.

Matrix **multiplication** is more involved.
The points awarded for football results are 2 for a win, 1 for a draw and 0 for a defeat. Liverpool's total points are $5 \times 2 + 6 \times 1 + 1 \times 0 = 16$. Similarly Arsenal's points are $7 \times 2 + 1 \times 1 + 3 \times 0 = 15$

$$\begin{array}{c} \begin{matrix} W & D & L \end{matrix} \\ \begin{matrix} L \\ A \end{matrix}\begin{pmatrix} 5 & 6 & 1 \\ 7 & 1 & 3 \end{pmatrix} \end{array} \begin{matrix} W \\ D \\ L \end{matrix}\begin{pmatrix} 2 \\ 1 \\ 0 \end{pmatrix} = \begin{array}{c} \text{Points} \\ \begin{pmatrix} 5 \times 2 + 6 \times 1 + 1 \times 0 \\ 7 \times 2 + 1 \times 1 + 3 \times 0 \end{pmatrix} \end{array} = \begin{array}{c} \text{Points} \\ \begin{matrix} L \\ A \end{matrix}\begin{pmatrix} 16 \\ 15 \end{pmatrix} \end{array}$$

$$= \text{Matrix } P.$$

The points matrix has been written as a column matrix (order 3 by 1) and the product matrix P is a column matrix (order 2 by 1) giving the total number of points for the two teams. At first sight the arrangement of the points matrix as a column matrix seems to complicate the matrix multiplication process, but there are advantages to this system and it is the universal convention. If we had a different points system P_2, of 3 points for a win, 1 for a draw and 0 for a defeat, the total points are given by

$$\begin{array}{c} \begin{matrix} W & D & L \end{matrix} \\ \begin{matrix} L \\ A \end{matrix}\begin{pmatrix} 5 & 6 & 1 \\ 7 & 1 & 3 \end{pmatrix} \end{array} \begin{array}{c} P_2 \\ \begin{matrix} W \\ D \\ L \end{matrix}\begin{pmatrix} 3 \\ 1 \\ 0 \end{pmatrix} \end{array} = \begin{array}{c} P_2 \\ \begin{matrix} L \\ A \end{matrix}\begin{pmatrix} 5 \times 3 + 6 \times 1 + 1 \times 0 \\ 7 \times 3 + 1 \times 1 + 3 \times 0 \end{pmatrix} \end{array} = \begin{array}{c} P_2 \\ \begin{matrix} L \\ A \end{matrix}\begin{pmatrix} 21 \\ 22 \end{pmatrix} \end{array}.$$

Both points systems can be incorporated in one matrix multiplication in the following way:

$$\begin{array}{c} \begin{matrix} W & D & L \end{matrix} \\ \begin{matrix} L \\ A \end{matrix}\begin{pmatrix} 5 & 6 & 1 \\ 7 & 1 & 3 \end{pmatrix} \\ T \end{array} \begin{array}{c} \begin{matrix} P_1 & P_2 \end{matrix} \\ \begin{matrix} W \\ D \\ L \end{matrix}\begin{pmatrix} 2 & 3 \\ 1 & 1 \\ 0 & 0 \end{pmatrix} \\ P \end{array}$$

$$= \begin{matrix} L \\ A \end{matrix}\begin{pmatrix} \overset{P_1}{5 \times 2 + 6 \times 1 + 1 \times 0} & \overset{P_2}{5 \times 3 + 6 \times 1 + 1 \times 0} \\ 7 \times 2 + 1 \times 1 + 3 \times 0 & 7 \times 3 + 1 \times 1 + 3 \times 0 \end{pmatrix}$$

$$= \begin{array}{c} \begin{matrix} P_1 & P_2 \end{matrix} \\ \begin{matrix} L \\ A \end{matrix}\begin{pmatrix} 16 & 21 \\ 15 & 22 \end{pmatrix} \\ S \end{array}.$$

In matrix multiplication each element of the product matrix comes from a row in the first matrix combined with a column in the second matrix. For example:

$$\begin{bmatrix} \text{row 1} \\ \text{in } T \end{bmatrix} \begin{matrix} \text{combined} \\ \text{with} \end{matrix} \begin{bmatrix} \text{column 2} \\ \text{in } P \end{bmatrix} \text{ gives } \begin{bmatrix} \text{the element of the} \\ \text{product } S \text{ which is in} \\ \text{the first row and} \\ \text{second column} \end{bmatrix}.$$

The number of columns in T must be equal to the number of rows in P.

Matrix	T	$\times$	P	$=$	S
Order	2 by 3	$\times$	3 by 2		2 by 2

Consider two more points systems $P_3(W=1, D=0, L=-1)$ and $P_4(W=2, D=0, L=-1)$

$$\begin{array}{c} \\ L \\ A \end{array}\overset{\begin{matrix} W & D & L \end{matrix}}{\begin{pmatrix} 5 & 6 & 1 \\ 7 & 1 & 3 \end{pmatrix}} \begin{array}{c} W \\ D \\ L \end{array}\overset{\begin{matrix} P_1 & P_2 & P_3 & P_4 \end{matrix}}{\begin{pmatrix} 2 & 3 & 1 & 2 \\ 1 & 1 & 0 & 0 \\ 0 & 0 & -1 & -1 \end{pmatrix}} = \begin{array}{c} L \\ A \end{array}\overset{\begin{matrix} P_1 & P_2 & P_3 & P_4 \end{matrix}}{\begin{pmatrix} 16 & 21 & 4 & 9 \\ 15 & 22 & 4 & 11 \end{pmatrix}}.$$

2 by 3 3 by 4 2 by 4

First row $\times$ third column gives (the element in the first row, third column).

If two matrices can be multiplied together they are **compatible**. A matrix of order $a \times b$ multiplied by a matrix of order $b \times c$ gives a product matrix of order $a \times c$.

In common with standard algebraic practice it is understood that if two matrices are written with no sign between them, they are to be multiplied. Many of the results derived below can apply to matrices of any order. The rows and columns may have headings, but it is usually clear what the elements refer to from the context of the application.

Unit matrices

$$\begin{pmatrix} 1 & 0 \\ 0 & 1 \end{pmatrix}\begin{pmatrix} 1 & 2 \\ 3 & 4 \end{pmatrix} = \begin{pmatrix} 1 & 2 \\ 3 & 4 \end{pmatrix}; \quad \begin{pmatrix} 1 & 2 \\ 3 & 4 \end{pmatrix}\begin{pmatrix} 1 & 0 \\ 0 & 1 \end{pmatrix} = \begin{pmatrix} 1 & 2 \\ 3 & 4 \end{pmatrix}.$$

The effect of premultiplying by $I_2 = \begin{pmatrix} 1 & 0 \\ 0 & 1 \end{pmatrix}$ is to leave the matrix $\begin{pmatrix} 1 & 2 \\ 3 & 4 \end{pmatrix}$ unchanged. I_2 is a **unit matrix**.

$$\begin{pmatrix} 1 & 0 & 0 \\ 0 & 1 & 0 \\ 0 & 0 & 1 \end{pmatrix}\begin{pmatrix} 1 & 2 \\ 3 & 4 \\ 5 & 6 \end{pmatrix} = \begin{pmatrix} 1 & 2 \\ 3 & 4 \\ 5 & 6 \end{pmatrix}; \quad I_3 = \begin{pmatrix} 1 & 0 & 0 \\ 0 & 1 & 0 \\ 0 & 0 & 1 \end{pmatrix} \text{ is a unit matrix.}$$

$\begin{pmatrix} 1 & 2 \\ 3 & 4 \\ 5 & 6 \end{pmatrix}\begin{pmatrix} 1 & 0 & 0 \\ 0 & 1 & 0 \\ 0 & 0 & 1 \end{pmatrix}$ is impossible, since the matrices are incompatible.

It is interesting to note the effect of multiplying by certain simple matrices.

$$\begin{pmatrix}2 & 0\\0 & 2\end{pmatrix}\begin{pmatrix}1 & 2\\3 & 4\end{pmatrix}=\begin{pmatrix}2 & 4\\6 & 8\end{pmatrix};\quad \begin{pmatrix}2 & 0\\0 & 2\end{pmatrix}$$ has the effect of doubling each element.

$$\begin{pmatrix}2 & 0\\0 & 1\end{pmatrix}\begin{pmatrix}1 & 2\\3 & 4\end{pmatrix}=\begin{pmatrix}2 & 4\\3 & 4\end{pmatrix};\quad \begin{pmatrix}2 & 0\\0 & 1\end{pmatrix}$$ doubles the top line.

$$\begin{pmatrix}1 & 0\\0 & -3\end{pmatrix}\begin{pmatrix}1 & 2\\3 & 4\end{pmatrix}=\begin{pmatrix}1 & 2\\-9 & -12\end{pmatrix};\quad \begin{pmatrix}1 & 0\\0 & -3\end{pmatrix}$$ multiplies the bottom line by -3.

$$\begin{pmatrix}1 & 1\\0 & 1\end{pmatrix}\begin{pmatrix}1 & 2\\3 & 4\end{pmatrix}=\begin{pmatrix}4 & 6\\3 & 4\end{pmatrix};\quad \begin{pmatrix}1 & 1\\0 & 1\end{pmatrix}$$ adds the bottom row to the top row.

$$\begin{pmatrix}0 & 1\\1 & 0\end{pmatrix}\begin{pmatrix}1 & 2\\3 & 4\end{pmatrix}=\begin{pmatrix}3 & 4\\1 & 2\end{pmatrix};\quad \begin{pmatrix}0 & 1\\1 & 0\end{pmatrix}$$ exchanges the top and bottom rows.

Multiplying a matrix by a number

$$\begin{pmatrix}1 & 2\\3 & 4\end{pmatrix}+\begin{pmatrix}1 & 2\\3 & 4\end{pmatrix}=\begin{pmatrix}2 & 4\\6 & 8\end{pmatrix}=2\times\begin{pmatrix}1 & 2\\3 & 4\end{pmatrix}$$

To multiply a matrix by a number you must multiply each element by that number.

If $I=\begin{pmatrix}1 & 0\\0 & 1\end{pmatrix}$, $A=\begin{pmatrix}1 & 2\\3 & 4\end{pmatrix}$ and $B=\begin{pmatrix}4 & 2\\1 & 3\end{pmatrix}$,

then $IA = AI = A$, and $IB = BI = B$.

The matrices I and A are **commutative** and so are I and B. However, this is not true for all matrices.

$$AB=\begin{pmatrix}1 & 2\\3 & 4\end{pmatrix}\begin{pmatrix}4 & 2\\1 & 3\end{pmatrix}=\begin{pmatrix}6 & 8\\16 & 18\end{pmatrix},$$

$$BA=\begin{pmatrix}4 & 3\\2 & 1\end{pmatrix}\begin{pmatrix}4 & 2\\1 & 3\end{pmatrix}=\begin{pmatrix}19 & 17\\9 & 7\end{pmatrix}.$$

$AB \neq BA$ so **matrix multiplication is not commutative.**

If $C=\begin{pmatrix}1 & 2 & 3\\2 & 0 & -1\end{pmatrix}$ and $D=\begin{pmatrix}1\\2\\3\end{pmatrix}$,

$$(AC)D = \left[\begin{pmatrix}1 & 2\\3 & 4\end{pmatrix}\begin{pmatrix}1 & 2 & 3\\2 & 0 & -1\end{pmatrix}\right]\begin{pmatrix}1\\2\\3\end{pmatrix}$$

$$= \begin{pmatrix}5 & 2 & 1\\11 & 6 & 5\end{pmatrix}\begin{pmatrix}1\\2\\3\end{pmatrix} = \begin{pmatrix}12\\38\end{pmatrix}.$$

$$A(CD) = \begin{pmatrix}1 & 2\\3 & 4\end{pmatrix}\begin{pmatrix}1 & 2 & 3\\2 & 0 & -1\end{pmatrix}\begin{pmatrix}1\\2\\3\end{pmatrix}$$

$$= \begin{pmatrix}1 & 2\\3 & 4\end{pmatrix}\begin{pmatrix}14\\-1\end{pmatrix} = \begin{pmatrix}12\\38\end{pmatrix}.$$

$(AC)D = A(CD)$ and this result is true in general, as long as the matrices are compatible. **Matrix multiplication is associative.**

The **transpose of a matrix** is the matrix formed by interchanging rows and columns. The transpose, C^T, of matrix C above is $\begin{pmatrix}1 & 2\\2 & 0\\3 & -1\end{pmatrix}$

Inverse matrices

$$\begin{pmatrix}3 & 4\\5 & 7\end{pmatrix}\begin{pmatrix}7 & -4\\-5 & 3\end{pmatrix} = \begin{pmatrix}1 & 0\\0 & 1\end{pmatrix} = \begin{pmatrix}7 & -4\\-5 & 3\end{pmatrix}\begin{pmatrix}3 & 4\\5 & 7\end{pmatrix}.$$

If $AB = BA = I$, then B is the **inverse** of matrix A and is denoted by A^{-1}. Does every matrix have an inverse?

Can we find the inverse of the matrix $G = \begin{pmatrix}a & b\\c & d\end{pmatrix}$?

Suppose $H = \begin{pmatrix}x & y\\s & t\end{pmatrix}$ and $HG = I$,

then $\begin{pmatrix}x & y\\s & t\end{pmatrix}\begin{pmatrix}a & b\\c & d\end{pmatrix} = \begin{pmatrix}xa + yc & xb + yd\\sa + tc & sb + td\end{pmatrix} = \begin{pmatrix}1 & 0\\0 & 1\end{pmatrix}.$

$xb + yd = 0 \Rightarrow \dfrac{x}{y} = \dfrac{-d}{b}$, so $x = \lambda d$ and $y = -\lambda b$.

$xa + yc = 1 \Rightarrow \lambda da - \lambda bc = 1$,

so $\lambda = \dfrac{1}{ad - bc}$ and $x = \dfrac{d}{ad - bc}$, $y = \dfrac{-b}{ad - bc}$.

Similarly $s = \dfrac{-c}{ad-bc}$, $t = \dfrac{a}{ad-bc}$.

$$\begin{pmatrix} x & y \\ s & t \end{pmatrix} = G^{-1} = \begin{pmatrix} \dfrac{d}{ad-bc} & \dfrac{-b}{ad-bc} \\ \dfrac{-c}{ad-bc} & \dfrac{a}{ad-bc} \end{pmatrix} = \frac{1}{ad-bc}\begin{pmatrix} d & -b \\ -c & a \end{pmatrix}$$

If $\begin{pmatrix} a & b \\ c & d \end{pmatrix} = G$, the quantity $\Delta = ad - bc = \begin{vmatrix} a & b \\ c & d \end{vmatrix}$ is the **determinant** of G.

An alternative method to find the inverse matrix follows.

$$\begin{pmatrix} x & y \\ s & t \end{pmatrix}\begin{pmatrix} a & b \\ c & d \end{pmatrix} = \begin{pmatrix} 1 & 0 \\ 0 & 1 \end{pmatrix}.$$

Try to find x and y so that the first row multiplied by the second column is 0. To achieve this we let $x = d$ and $y = -b$.

$$\begin{pmatrix} d & -b \\ * & * \end{pmatrix}\begin{pmatrix} a & b \\ c & d \end{pmatrix} = \begin{pmatrix} ad-bc & 0 \\ * & * \end{pmatrix}.$$

Let $s = -c$ and $t = a$, then

$$\begin{pmatrix} d & -b \\ -c & a \end{pmatrix}\begin{pmatrix} a & b \\ c & d \end{pmatrix} = \begin{pmatrix} ad-bc & 0 \\ 0 & -bc+ad \end{pmatrix}.$$

If we divide the final matrix by $\Delta = ad - bc$, we obtain $\begin{pmatrix} 1 & 0 \\ 0 & 1 \end{pmatrix}$.

If $G = \begin{pmatrix} a & b \\ c & d \end{pmatrix}$, $G^{-1} = \begin{pmatrix} \dfrac{d}{\Delta} & \dfrac{-b}{\Delta} \\ \dfrac{-c}{\Delta} & \dfrac{a}{\Delta} \end{pmatrix} = \dfrac{1}{(ad-bc)}\begin{pmatrix} d & -b \\ -c & a \end{pmatrix}$.

This form of the inverse of a 2 by 2 matrix is easy to memorize. The elements on the main diagonal (top left to bottom right) are interchanged, the elements on the reverse diagonal (top right to bottom left) are reversed in sign and each element is divided by the determinant, $\Delta = ad - bc$.

Example 1 to find the inverse of $\begin{pmatrix} 1 & 2 \\ 3 & 4 \end{pmatrix}$.

$\Delta = 1 \times 4 - 3 \times 2 = 4 - 6 = -2.$

$$\text{Inverse} = \frac{1}{-2}\begin{pmatrix} 4 & -2 \\ -3 & 1 \end{pmatrix} = \begin{pmatrix} -2 & 1 \\ 1\frac{1}{2} & -\frac{1}{2} \end{pmatrix}.$$

If the determinant of a matrix is zero, the matrix has no inverse.

Example 2 $A = \begin{pmatrix} 1 & 2 \\ 3 & 6 \end{pmatrix}$. $\Delta = \det A = 6 - 6 = 0.$

If $B = \begin{pmatrix} 6 & -2 \\ -3 & 1 \end{pmatrix}$, $BA = \begin{pmatrix} 6 & -2 \\ -3 & 1 \end{pmatrix}\begin{pmatrix} 1 & 2 \\ 3 & 6 \end{pmatrix} = \begin{pmatrix} 0 & 0 \\ 0 & 0 \end{pmatrix}.$

Any attempt to produce zeros on the reverse diagonal produces zeros also on the main diagonal.

The product of B and A gives $\begin{pmatrix} 0 & 0 \\ 0 & 0 \end{pmatrix}$ the **zero matrix**.

To prove matrix multiplication is not commutative for 2 by 2 matrices

$$\begin{pmatrix} a & b \\ c & d \end{pmatrix}\begin{pmatrix} e & f \\ g & h \end{pmatrix} = \begin{pmatrix} ae + bg & af + bh \\ ce + dg & cf + dh \end{pmatrix}$$

$$\begin{pmatrix} e & f \\ g & h \end{pmatrix}\begin{pmatrix} a & b \\ c & d \end{pmatrix} = \begin{pmatrix} ae + cf & be + df \\ ag + ch & bg + dh \end{pmatrix}.$$

The product matrices are in general not equal.

To prove matrix multiplication is **associative** for 2 by 2 matrices.

$$\left[\begin{pmatrix} a & b \\ c & d \end{pmatrix}\begin{pmatrix} e & f \\ g & h \end{pmatrix}\right]\begin{pmatrix} p & q \\ r & s \end{pmatrix}$$

$$= \begin{pmatrix} ae + bg & af + bh \\ ce + dg & cf + dh \end{pmatrix}\begin{pmatrix} p & q \\ r & s \end{pmatrix}$$

$$= \begin{pmatrix} aep + bgp + afr + bhr & aeq + bgq + afs + bhs \\ cep + dgp + cfr + dhr & ceq + dgq + cfs + dhs \end{pmatrix}$$

$$\begin{pmatrix} a & b \\ c & d \end{pmatrix}\left[\begin{pmatrix} e & f \\ g & h \end{pmatrix}\begin{pmatrix} p & q \\ r & s \end{pmatrix}\right]$$

$$= \begin{pmatrix} a & b \\ c & d \end{pmatrix}\begin{pmatrix} ep + fr & eq + fs \\ gp + hr & gq + hs \end{pmatrix}$$

$$= \begin{pmatrix} aep + afr + bgp + bhr & aeq + afs + bgq + bhs \\ cep + cfr + dgp + dhr & ceq + cfs + dgq + dhs \end{pmatrix}$$

These two matrices are identical term for term.

To prove **det(*AB*) = (det *A*) × (det *B*)** where A, B are two 2 by 2 matrices.

$$A = \begin{pmatrix} a & b \\ c & d \end{pmatrix} \quad B = \begin{pmatrix} e & f \\ g & h \end{pmatrix} \quad AB = \begin{pmatrix} ae + bg & af + bh \\ ce + dg & cf + dh \end{pmatrix}$$

$$\begin{aligned} \det(AB) &= (ae + bg)(cf + dh) - (af + bh)(ce + dg) \\ &= aecf + aedh + bgcf + bgdh - afce - afdg - bhce - bhdg \\ &= ad(eh - fg) - bc(he - gf) \\ &= (ad - bc)(he - fg) \\ &= (\det A) \times (\det B). \end{aligned}$$

This result is true for all square matrices. The determinant of a non-square matrix is not defined.

Simultaneous equations

Although the ideas of matrices and determinants are not necessary to solve simultaneous equations in two dimensions (2 equations and 2 unknowns) their study gives insight and understanding to simultaneous equations with 3 or more unknowns.

Consider first the equations: $ax + by = 0,$ (1)

$cx + dy = 0.$ (2)

These equations have the trivial solution $x = 0$ or $y = 0$. Geometrically the equations represent two straight lines through the origin. The only other solutions occur when each equation represent the same straight line. In this case equation (2) is a multiple of equation (1) or $ad - bc = 0$.

So $\begin{matrix} ax + by = 0 \\ cx + dy = 0 \end{matrix} \Leftrightarrow \begin{pmatrix} a & b \\ c & d \end{pmatrix}\begin{pmatrix} x \\ y \end{pmatrix} = 0,$

which has non-zero solutions when $ad - bc = 0$. We shall use this idea in three dimensions to define a 3×3 determinant.

Consider now the equations: $3x + 4y = 11,$ (3)

$x + 2y = 5.$ (4)

Elimination method

Equation (3) – {2 × equation (4)} gives $x = 1$.
Substitution in either equation then gives $y = 2$.

Matrix method $\begin{pmatrix} 3 & 4 \\ 1 & 2 \end{pmatrix}\begin{pmatrix} x \\ y \end{pmatrix} = \begin{pmatrix} 11 \\ 5 \end{pmatrix}$

The inverse of the matrix $\begin{pmatrix} 3 & 4 \\ 1 & 2 \end{pmatrix}$ is $\frac{1}{2}\begin{pmatrix} 2 & -4 \\ -1 & 3 \end{pmatrix}$.

Hence $\frac{1}{2}\begin{pmatrix} 2 & -4 \\ -1 & 3 \end{pmatrix}\begin{pmatrix} 3 & 4 \\ 1 & 2 \end{pmatrix}\begin{pmatrix} x \\ y \end{pmatrix} = \frac{1}{2}\begin{pmatrix} 2 & -4 \\ -1 & 3 \end{pmatrix}\begin{pmatrix} 11 \\ 5 \end{pmatrix}$

$$\begin{pmatrix} 1 & 0 \\ 0 & 1 \end{pmatrix}\begin{pmatrix} x \\ y \end{pmatrix} = \frac{1}{2}\begin{pmatrix} 2 \\ 4 \end{pmatrix}$$

$$\begin{pmatrix} x \\ y \end{pmatrix} = \begin{pmatrix} 1 \\ 2 \end{pmatrix}.$$

The matrix method may look longer but the solution is obtained only by multiplying $\begin{pmatrix} 11 \\ 5 \end{pmatrix}$ by the inverse of $\begin{pmatrix} 3 & 4 \\ 1 & 2 \end{pmatrix}$

General solution of $ax + by = p$,
$cx + dy = q$.

By matrices

$$\begin{pmatrix} x \\ y \end{pmatrix} = \frac{1}{(ad - bc)}\begin{pmatrix} d & -b \\ -c & a \end{pmatrix}\begin{pmatrix} p \\ q \end{pmatrix} = \frac{1}{(ad - bc)}\begin{pmatrix} pd - bq \\ -pc + aq \end{pmatrix}$$

$$x = \frac{pd - bq}{ad - bc} = \frac{\begin{vmatrix} p & b \\ q & d \end{vmatrix}}{\begin{vmatrix} a & b \\ c & d \end{vmatrix}}, \quad y = \frac{aq - pc}{ad - bc} = \frac{\begin{vmatrix} a & p \\ c & q \end{vmatrix}}{\begin{vmatrix} a & b \\ c & d \end{vmatrix}}.$$

This notation has an important extension in three dimensions, (three equations with three unknowns).

If $ad - bc = 0$ two important situations arise.

(i) $\left.\begin{array}{r} x + 2y = 3 \\ 3x + 6y = 9 \end{array}\right\}$ These are really the same equation and have many solutions; $x = 1$, $y = 1$ or $x = -1$, $y = 2$, etc.

Geometrically the two equations represent the same straight line in the xy-plane.

(ii) $\left.\begin{array}{r} x + 2y = 3 \\ 3x + 6y = 6 \end{array}\right\}$ There are no simultaneous solutions. The two equations represent two parallel straight lines in the xy-plane which do not intersect.

Determinant of a 3 by 3 matrix

The determinant, Δ, of the matrix $M = \begin{pmatrix} a_1 & b_1 & c_1 \\ a_2 & b_2 & c_2 \\ a_3 & b_3 & c_3 \end{pmatrix}$ is

$$\Delta = \begin{vmatrix} a_1 & b_1 & c_1 \\ a_2 & b_2 & c_2 \\ a_3 & b_3 & c_3 \end{vmatrix} = a_1 \begin{vmatrix} b_2 & c_2 \\ b_3 & c_3 \end{vmatrix} - b_1 \begin{vmatrix} a_2 & c_2 \\ a_3 & c_3 \end{vmatrix} + c_1 \begin{vmatrix} a_2 & b_2 \\ a_3 & b_3 \end{vmatrix}$$

$$= a_1(b_2 c_3 - b_3 c_2) - b_1(a_2 c_3 - a_3 c_2) + c_1(a_2 b_3 - a_3 b_2)$$

$$= a_1 b_2 c_3 - a_1 b_3 c_2 - a_2 b_1 c_3 + a_3 b_1 c_2 + a_2 b_3 c_1 - a_3 b_2 c_1. \qquad (1)$$

Example 3 $\begin{vmatrix} 1 & 2 & 6 \\ 3 & 5 & 7 \\ 4 & 8 & 9 \end{vmatrix} = +1 \begin{vmatrix} 5 & 7 \\ 8 & 9 \end{vmatrix} - 2 \begin{vmatrix} 3 & 7 \\ 4 & 9 \end{vmatrix} + 6 \begin{vmatrix} 3 & 5 \\ 4 & 8 \end{vmatrix}$

$$= (45 - 56) - 2(27 - 28) + 6(24 - 20)$$

$$= -11 + 2 + 24 = 15,$$

from equation (1)

$$\Delta = -a_2(b_1 c_3 - b_3 c_1) + b_2(a_1 c_3 - a_3 c_1) - c_2(a_1 b_3 - a_3 b_1)$$

$$= -a_2 \begin{vmatrix} b_1 & c_1 \\ b_3 & c_3 \end{vmatrix} + b_2 \begin{vmatrix} a_1 & c_1 \\ a_3 & c_3 \end{vmatrix} - c_2 \begin{vmatrix} a_1 & b_1 \\ a_3 & b_3 \end{vmatrix} \qquad (2)$$

$\Delta = \det M$, is also given by

$$\Delta = +a_3 \begin{vmatrix} b_1 & c_1 \\ b_2 & c_2 \end{vmatrix} - b_3 \begin{vmatrix} a_1 & c_1 \\ a_2 & c_2 \end{vmatrix} - c_3 \begin{vmatrix} a_1 & b_1 \\ a_2 & b_2 \end{vmatrix}. \qquad (3)$$

A determinant can be calculated using the first row to expand with as in (1), or the second row as in (2), or the third row as in (3), or even by using any column with + or − assigned to the column as in the row expansion. Example 1 using the second column is

$$\begin{vmatrix} 1 & 2 & 6 \\ 3 & 5 & 7 \\ 4 & 8 & 9 \end{vmatrix} = -2\begin{vmatrix} 3 & 7 \\ 4 & 9 \end{vmatrix} + 5\begin{vmatrix} 1 & 6 \\ 4 & 9 \end{vmatrix} - 8\begin{vmatrix} 1 & 6 \\ 3 & 7 \end{vmatrix}$$

$$= -2(27-28) + 5(9-24) - 8(7-18)$$

$$= +2 - 75 + 88 = 15, \quad \text{as before.}$$

The accepted notation for the elements of a matrix are

$$N = \begin{vmatrix} a_{11} & a_{12} & a_{13} \\ a_{21} & a_{22} & a_{23} \\ a_{31} & a_{32} & a_{33} \end{vmatrix}$$ so that a_{ij} represents the element in the ith row and jth column.

In the expression for det N, the elements of the row or column being used for expansion are taken in order with alternate signs and each is multiplied by the 2 by 2 matrix remaining when the row and column containing the element are removed. (Remember that the second row begins with a_{21} negative).

The transpose M' or M^T of the matrix M is the matrix formed by interchanging rows and columns.

$$M = \begin{pmatrix} 1 & 2 & 6 \\ 3 & 5 & 7 \\ 4 & 8 & 9 \end{pmatrix} \Rightarrow M' = \begin{pmatrix} 1 & 3 & 4 \\ 2 & 5 & 8 \\ 6 & 7 & 9 \end{pmatrix}.$$

$$N = \begin{pmatrix} a_{11} & a_{12} & a_{13} \\ a_{21} & a_{22} & a_{23} \\ a_{31} & a_{32} & a_{33} \end{pmatrix} \Rightarrow N' = \begin{pmatrix} a_{11} & a_{21} & a_{31} \\ a_{12} & a_{22} & a_{32} \\ a_{13} & a_{23} & a_{33} \end{pmatrix}.$$

Laws of determinants

Various laws and processes help in working with determinants which ease the arithmetic. The rules are true in general but we will prove them only for the third-order determinant

$$M = \begin{pmatrix} a_1 & b_1 & c_1 \\ a_2 & b_2 & c_2 \\ a_3 & b_3 & c_3 \end{pmatrix}, \quad \det M = \Delta.$$

Rule 1 Δ is changed in sign if two rows are interchanged.

$$\begin{vmatrix} a_2 & b_2 & c_2 \\ a_1 & b_1 & c_1 \\ a_3 & b_3 & c_3 \end{vmatrix} = a_2(b_1 c_3 - b_3 c_1) - b_2(a_1 c_3 - a_3 c_1) + c_2(a_1 b_3 - a_3 b_1)$$

$$= -a_1(b_2 c_3 - b_3 c_2) + b_1(a_2 c_3 - a_3 c_2) - c_1(a_2 b_3 - a_3 b_2) = -\Delta.$$

The same result is true if two columns are interchanged.

Rule 2 $\Delta = 0$ if two rows are identical.
From rule 1, interchanging the rows changes the sign of Δ, but this produces the same determinant so $\Delta = -\Delta \Rightarrow \Delta = 0$.

Rule 3 If each element of one row, or column is multiplied by k, then Δ is multiplied by k.

$$\begin{vmatrix} ka_1 & kb_2 & kc_3 \\ a_2 & b_2 & c_2 \\ a_3 & b_3 & c_3 \end{vmatrix} = k\Delta$$ since there will be a factor of k in each term.

An important consequence is that if every element of a row or column has a factor k, then Δ has a factor k.

Rule 4 Δ is unchanged if a multiple of a row is added to another row.

$$\begin{vmatrix} a_1 + ka_2 & b_1 + kb_2 & c_1 + kc_2 \\ a_2 & b_2 & c_2 \\ a_3 & b_3 & c_3 \end{vmatrix}$$

$$= (a_1 + ka_2)\begin{vmatrix} b_2 & c_2 \\ b_3 & c_3 \end{vmatrix} - (b_1 + kb_2)\begin{vmatrix} a_2 & c_2 \\ a_3 & c_3 \end{vmatrix} + (c_1 + kc_2)\begin{vmatrix} a_2 & b_2 \\ a_3 & b_3 \end{vmatrix}$$

$$= a_1\begin{vmatrix} b_2 & c_2 \\ b_3 & c_3 \end{vmatrix} - b_1\begin{vmatrix} a_2 & c_2 \\ a_3 & c_3 \end{vmatrix} + c_1\begin{vmatrix} a_2 & b_2 \\ a_3 & b_3 \end{vmatrix}$$

$$+ka_2\begin{vmatrix} b_2 & c_2 \\ b_3 & c_3 \end{vmatrix} - kb_2\begin{vmatrix} a_2 & c_2 \\ a_3 & c_3 \end{vmatrix} + kc_2\begin{vmatrix} a_2 & b_2 \\ a_3 & b_3 \end{vmatrix}$$

$$= \begin{vmatrix} a_1 & b_1 & c_1 \\ a_2 & b_2 & c_2 \\ a_3 & b_3 & c_3 \end{vmatrix} + k\begin{vmatrix} a_2 & b_2 & c_2 \\ a_2 & b_2 & c_2 \\ a_3 & b_3 & c_3 \end{vmatrix} = \Delta$$

since the second determinant has two identical rows and is zero.

Rule 4 gives a way of adding or subtracting determinants

$$\begin{vmatrix} a_1 + a & b_1 + b & c_1 + c \\ a_2 & b_2 & c_2 \\ a_3 & b_3 & c_3 \end{vmatrix} = \begin{vmatrix} a_1 & b_1 & c_1 \\ a_2 & b_2 & c_2 \\ a_3 & b_3 & c_3 \end{vmatrix} + \begin{vmatrix} a & b & c \\ a_2 & b_2 & c_2 \\ a_3 & b_3 & c_3 \end{vmatrix}.$$

Notice that two rows in the determinants are identical.

Example 4 Using these rules

$$\begin{vmatrix} 20 & 17 & 16 \\ 21 & 19 & 12 \\ 23 & 20 & 14 \end{vmatrix} = \begin{vmatrix} 3 & 17 & 16 \\ 2 & 19 & 12 \\ 3 & 20 & 14 \end{vmatrix} \quad \text{subtracting column 2 from column 1}$$

$$= \begin{vmatrix} 3 & 1 & 16 \\ 2 & 7 & 12 \\ 3 & 6 & 14 \end{vmatrix} \quad \text{subtracting column 3 from column 2}$$

$$= \begin{vmatrix} 1 & -6 & 4 \\ 2 & 7 & 12 \\ 1 & -1 & 2 \end{vmatrix} \quad \text{subtracting row 2 from rows 1 and 3}$$

$$= 1(14 + 12) + 6(4 - 12) + 4(-2 - 7)$$

$$= 26 - 48 - 36 = -58.$$

$$\begin{vmatrix} a & b & c \\ a^2 & b^2 & c^2 \\ a^3 & b^3 & c^3 \end{vmatrix} = a \begin{vmatrix} 1 & b & c \\ a & b^2 & c^2 \\ a^2 & b^3 & c^3 \end{vmatrix} = ab \begin{vmatrix} 1 & 1 & c \\ a & b & c^2 \\ a^2 & b^2 & c^3 \end{vmatrix}$$

$$= abc \begin{vmatrix} 1 & 1 & 1 \\ a & b & c \\ a^2 & b^2 & c^2 \end{vmatrix} = abc \begin{vmatrix} 0 & 0 & 1 \\ a-b & b-c & c \\ a^2-b^2 & b^2-c^2 & c^2 \end{vmatrix}$$

$$= abc(a-b)(b-c) \begin{vmatrix} 0 & 0 & 1 \\ 1 & 1 & c \\ a+b & b+c & c^2 \end{vmatrix}$$

$$= abc(a-b)(b-c)(c-a).$$

Cofactors

$$\Delta = \begin{vmatrix} a_1 & b_1 & c_1 \\ a_2 & b_2 & c_2 \\ a_3 & b_3 & c_3 \end{vmatrix} = a_1 \begin{vmatrix} b_2 & c_2 \\ b_3 & c_3 \end{vmatrix} - b_1 \begin{vmatrix} a_2 & c_2 \\ a_3 & c_3 \end{vmatrix} + c_1 \begin{vmatrix} a_2 & b_2 \\ a_3 & b_3 \end{vmatrix}$$

$$= a_1(b_2 c_3 - b_3 c_2) - b_1(a_2 c_3 - a_3 c_2) + c_1(a_2 b_3 - a_3 b_2)$$

$$= a_1 A_1 + b_1 B_1 + c_1 C_1,$$

where $A_1 = \begin{vmatrix} b_2 & c_2 \\ b_3 & c_3 \end{vmatrix}$, $B_1 = -\begin{vmatrix} a_2 & c_2 \\ a_3 & c_3 \end{vmatrix}$, $C_1 = \begin{vmatrix} a_2 & b_2 \\ a_3 & b_3 \end{vmatrix}$.

A_1, B_1, ... are the cofactors of a_1, b_1, ... and their numerical values are the determinants left by omitting the row and column containing the respective elements. The signs of the cofactors are alternately + and − starting with A_1 positive and proceeding row by row in turn.

So $A_2 = -\begin{vmatrix} b_1 & c_1 \\ b_3 & c_3 \end{vmatrix}$, $B_2 = \begin{vmatrix} a_1 & c_1 \\ a_3 & c_3 \end{vmatrix}$, $C_2 = -\begin{vmatrix} a_1 & b_1 \\ a_3 & b_3 \end{vmatrix}$

$A_3 = \begin{vmatrix} b_1 & c_1 \\ b_2 & c_2 \end{vmatrix}$, $B_3 = -\begin{vmatrix} a_1 & c_1 \\ a_2 & c_2 \end{vmatrix}$, $C_3 = \begin{vmatrix} a_1 & b_1 \\ a_2 & b_2 \end{vmatrix}$

$$\Delta = -a_2\begin{vmatrix} b_1 & c_1 \\ b_3 & c_3 \end{vmatrix} + b_2\begin{vmatrix} a_1 & c_1 \\ a_3 & c_3 \end{vmatrix} - c_2\begin{vmatrix} a_1 & b_1 \\ a_3 & b_3 \end{vmatrix}$$

by expanding from the second row.

$\Delta = a_2 A_2 + b_2 B_2 + c_2 C_2$,
$\Delta = a_3 A_3 + b_3 B_3 + c_3 C_3$ by expanding from the third row,
$\Delta = a_1 A_1 + a_2 A_2 + a_3 A_3$ by expanding from the first column.
$\Delta =$ the sum of the products of any row or column with the corresponding cofactors.

If the products are taken with the cofactors of a different row or column, the result is zero.

$$a_1 A_2 + b_1 B_2 + c_1 C_2 = -a_1\begin{vmatrix} b_1 & c_1 \\ b_3 & c_3 \end{vmatrix} + b_1\begin{vmatrix} a_1 & c_1 \\ a_3 & c_3 \end{vmatrix} - c_1\begin{vmatrix} a_1 & b_1 \\ a_3 & b_3 \end{vmatrix}$$

$$= \begin{vmatrix} a_1 & b_1 & c_1 \\ a_1 & b_1 & c_1 \\ a_3 & b_3 & c_3 \end{vmatrix} = 0,$$ since 2 rows are identical, from rule 2

The results for cofactors can be summarized as:

by rows	by columns
$a_1 A_1 + b_1 B_1 + c_1 C_1 = \Delta$	$a_1 A_1 + a_2 A_2 + a_3 A_3 = \Delta$,
$a_1 A_2 + b_1 B_2 + c_1 C_2 = 0$	$a_1 C_1 + a_2 C_2 + a_3 C_3 = 0$.

The definitions can be extended to larger determinants and the results are still true.

Equations in three unknowns

$$\begin{pmatrix} a_1 & b_1 & c_1 \\ a_2 & b_2 & c_2 \\ a_3 & b_3 & c_3 \end{pmatrix}\begin{pmatrix} x \\ y \\ z \end{pmatrix} = \begin{pmatrix} d_1 \\ d_2 \\ d_3 \end{pmatrix}$$

$a_1 x + b_1 y + c_1 z = d_1$, (1)
$a_2 x + b_2 y + c_2 z = d_2$, (2)
$a_3 x + b_3 y + c_3 z = d_3$. (3)

Multiply (1) by A_1, (2) by A_2 and (3) by A_3 and add to give

$$(a_1 A_1 + a_2 A_2 + a_3 A_3)x + (b_1 A_1 + b_2 A_2 + b_3 A_3)y + (c_1 A_1 + c_2 A_2 + c_3 A_3)z = d_1 A_1 + d_2 A_2 + d_3 A_3$$

But $a_1 A_1 + a_2 A_2 + a_3 A_3 = \Delta$

and $b_1 A_1 + b_2 A_2 + b_3 A_3 = c_1 A_1 + c_2 A_2 + c_3 A_3 = 0.$

So $\Delta x = d_1 A_1 + d_2 A_2 + d_3 A_3$

$$= d_1 \begin{vmatrix} b_2 & c_2 \\ b_3 & c_3 \end{vmatrix} - d_2 \begin{vmatrix} b_1 & c_1 \\ b_3 & c_3 \end{vmatrix} + d_3 \begin{vmatrix} b_1 & c_1 \\ b_2 & c_2 \end{vmatrix}$$

$$= \begin{vmatrix} d_1 & b_1 & c_1 \\ d_2 & b_2 & c_2 \\ d_3 & b_3 & c_3 \end{vmatrix}$$

Similarly $\Delta y = d_1 B_1 + d_2 B_2 + d_3 B_3 = \begin{vmatrix} a_1 & d_1 & c_1 \\ a_2 & d_2 & c_2 \\ a_3 & d_3 & c_3 \end{vmatrix}$

and $\Delta z = d_1 C_1 + d_2 C_2 + d_3 C_3 = \begin{vmatrix} a_1 & b_1 & d_1 \\ a_2 & b_2 & d_2 \\ a_3 & b_3 & d_3 \end{vmatrix}.$

These results can be summarized as

$$\frac{x}{\begin{vmatrix} d_1 & b_1 & c_1 \\ d_2 & b_2 & c_2 \\ d_3 & b_3 & c_3 \end{vmatrix}} = \frac{y}{\begin{vmatrix} a_1 & d_1 & c_1 \\ a_2 & d_2 & c_2 \\ a_3 & d_3 & c_3 \end{vmatrix}} = \frac{z}{\begin{vmatrix} a_1 & b_1 & d_1 \\ a_2 & b_2 & d_2 \\ a_3 & b_3 & d_3 \end{vmatrix}} = \frac{1}{\begin{vmatrix} a_1 & b_1 & c_1 \\ a_2 & b_2 & c_2 \\ a_3 & b_3 & c_3 \end{vmatrix}},$$

where the determinants are taken from Δ with the d_1, d_2, d_3 column replacing the first column under x, the second column under y and the third column under z. In this way the results are easy to remember.

Example 5 Solve

$$\begin{aligned} x + 2y + 3z &= 14, \\ 2x + 5y - z &= 9, \\ 3x + 8y - 4z &= 7. \end{aligned}$$

$$\frac{x}{\begin{vmatrix} 14 & 2 & 3 \\ 9 & 5 & -1 \\ 7 & 8 & -4 \end{vmatrix}} = \frac{y}{\begin{vmatrix} 1 & 14 & 3 \\ 2 & 9 & -1 \\ 3 & 7 & -4 \end{vmatrix}} = \frac{z}{\begin{vmatrix} 1 & 2 & 14 \\ 2 & 5 & 9 \\ 3 & 8 & 7 \end{vmatrix}}$$

$$= \frac{1}{\begin{vmatrix} 1 & 2 & 3 \\ 2 & 5 & -1 \\ 3 & 8 & -4 \end{vmatrix}}.$$

$$\begin{vmatrix} 14 & 2 & 3 \\ 9 & 5 & -1 \\ 7 & 8 & -4 \end{vmatrix} = \begin{vmatrix} 41 & 17 & 0 \\ 9 & 5 & -1 \\ -29 & -12 & 0 \end{vmatrix} = \begin{vmatrix} 12 & 5 & 0 \\ 9 & 5 & -1 \\ -29 & -12 & 0 \end{vmatrix}$$

$$= \begin{vmatrix} 12 & 5 & 0 \\ 9 & 5 & -1 \\ 7 & 3 & 0 \end{vmatrix} = 1(36 - 35) = 1.$$

$$\begin{vmatrix} 1 & 14 & 3 \\ 2 & 9 & -1 \\ 3 & 7 & -4 \end{vmatrix} = \begin{vmatrix} 1 & 14 & 3 \\ 0 & -19 & -7 \\ 0 & -35 & -13 \end{vmatrix} = \begin{vmatrix} 1 & 14 & 3 \\ 0 & 19 & -7 \\ 0 & 3 & 1 \end{vmatrix}$$

$$= 1(-19 + 21) = 2.$$

$$\begin{vmatrix} 1 & 2 & 14 \\ 2 & 5 & 9 \\ 3 & 8 & 7 \end{vmatrix} = \begin{vmatrix} 1 & 2 & 14 \\ 0 & 1 & -19 \\ 0 & 2 & -35 \end{vmatrix} = 1(-35 + 38) = 3.$$

$$\begin{vmatrix} 1 & 2 & 3 \\ 2 & 5 & -1 \\ 3 & 8 & -4 \end{vmatrix} = \begin{vmatrix} 1 & 2 & 3 \\ 0 & 1 & -7 \\ 0 & 2 & -13 \end{vmatrix} = 1(-13 + 14) = 1.$$

$$\frac{x}{1} = \frac{y}{2} = \frac{z}{3} = \frac{1}{1} \Rightarrow x = 1,\ y = 2,\ z = 3.$$

This method involves working out four determinants. We shall compare this method with others.

The inverse of a 3 by 3 matrix

$$\begin{pmatrix} a_1 & b_1 & c_1 \\ a_2 & b_2 & c_2 \\ a_3 & b_3 & c_3 \end{pmatrix}\begin{pmatrix} x \\ y \\ z \end{pmatrix} = \begin{pmatrix} d_1 \\ d_2 \\ d_3 \end{pmatrix} \Rightarrow \begin{aligned} \Delta x &= d_1 A_1 + d_2 A_2 + d_3 A_3, \\ \Delta y &= d_1 B_1 + d_2 B_2 + d_3 B_3, \\ \Delta z &= d_1 C_1 + d_2 C_2 + d_3 C_3. \end{aligned}$$

$$\Delta\begin{pmatrix} x \\ y \\ z \end{pmatrix} = \begin{pmatrix} A_1 & A_2 & A_3 \\ B_1 & B_2 & B_3 \\ C_1 & C_2 & C_3 \end{pmatrix}\begin{pmatrix} d_1 \\ d_2 \\ d_3 \end{pmatrix}$$

or

$$\begin{pmatrix} x \\ y \\ z \end{pmatrix} = \frac{1}{\Delta}\begin{pmatrix} A_1 & A_2 & A_3 \\ B_1 & B_2 & B_3 \\ C_1 & C_2 & C_3 \end{pmatrix}\begin{pmatrix} d_1 \\ d_2 \\ d_3 \end{pmatrix}.$$

Writing $M = \begin{pmatrix} a_1 & b_1 & c_1 \\ a_2 & b_2 & c_2 \\ a_3 & b_3 & c_3 \end{pmatrix}$, $\det M = \Delta$, $\begin{pmatrix} x \\ y \\ z \end{pmatrix} = \mathbf{r}$ and $\begin{pmatrix} d_1 \\ d_2 \\ d_3 \end{pmatrix} = \mathbf{d}$

$$M\mathbf{r} = \mathbf{d} \Leftrightarrow \mathbf{r} = M^{-1}\mathbf{d},$$

where M^{-1} is the inverse of the matrix M, i.e.

$$M^{-1}M = MM^{-1} = I_3.$$

$$M^{-1} = \frac{1}{\Delta}\begin{pmatrix} A_1 & A_2 & A_3 \\ B_1 & B_2 & B_3 \\ C_1 & C_2 & C_3 \end{pmatrix} = \frac{1}{\Delta} M_c^1$$

where M_c^1 is the transpose of M_c, the matrix of cofactors of M.

As a check

$$M^{-1}M = \frac{1}{\Delta}\begin{pmatrix} A_1 & A_2 & A_3 \\ B_1 & B_2 & B_3 \\ C_1 & C_2 & C_3 \end{pmatrix}\begin{pmatrix} a_1 & b_1 & c_1 \\ a_2 & b_2 & c_2 \\ a_3 & b_3 & c_3 \end{pmatrix} = \frac{1}{\Delta}\begin{pmatrix} \Delta & 0 & 0 \\ 0 & \Delta & 0 \\ 0 & 0 & \Delta \end{pmatrix}$$

$$= \begin{pmatrix} 1 & 0 & 0 \\ 0 & 1 & 0 \\ 0 & 0 & 1 \end{pmatrix} = I_3 .$$

Example 5 repeated

$$\begin{aligned} x + 2y + 3z &= 14 \\ 2x + 5y - 1z &= 9 \\ 3x + 8y - 4z &= 7 \end{aligned} \Leftrightarrow \begin{pmatrix} 1 & 2 & 3 \\ 2 & 5 & -1 \\ 3 & 8 & -4 \end{pmatrix}\begin{pmatrix} x \\ y \\ z \end{pmatrix} = \begin{pmatrix} 14 \\ 9 \\ 7 \end{pmatrix} \Leftrightarrow Mr = \begin{pmatrix} 14 \\ 9 \\ 7 \end{pmatrix}.$$

$$\Delta = \det M = \begin{vmatrix} 1 & 2 & 3 \\ 2 & 5 & -1 \\ 3 & 8 & -4 \end{vmatrix} = \begin{vmatrix} 1 & 2 & 3 \\ 0 & 1 & -7 \\ 0 & 2 & -13 \end{vmatrix}$$

$$= 1(-13 + 14) = 1.$$

$$M^{-1} = \frac{1}{\Delta}\begin{pmatrix} A_1 & A_2 & A_3 \\ B_1 & B_2 & B_3 \\ C_1 & C_2 & C_3 \end{pmatrix} = \begin{pmatrix} -12 & 32 & -17 \\ 5 & -13 & 7 \\ 1 & -2 & 1 \end{pmatrix}$$

$$\begin{pmatrix} x \\ y \\ z \end{pmatrix} = \begin{pmatrix} -12 & 32 & -17 \\ 5 & -13 & 7 \\ 1 & -2 & 1 \end{pmatrix}\begin{pmatrix} 14 \\ 9 \\ 7 \end{pmatrix} = \begin{pmatrix} 1 \\ 2 \\ 3 \end{pmatrix}.$$

Transformations in three dimensions

Three-dimensional transformations are described with reference to a set of three mutually perpendicular axes Ox, Oy and Oz. Points are described by three coordinates (x, y, z). The transformation can be described by a 3 × 3 matrix.

If $P(x, y, z)$ is transformed to $P_1(x_1, y_1, z_1)$,

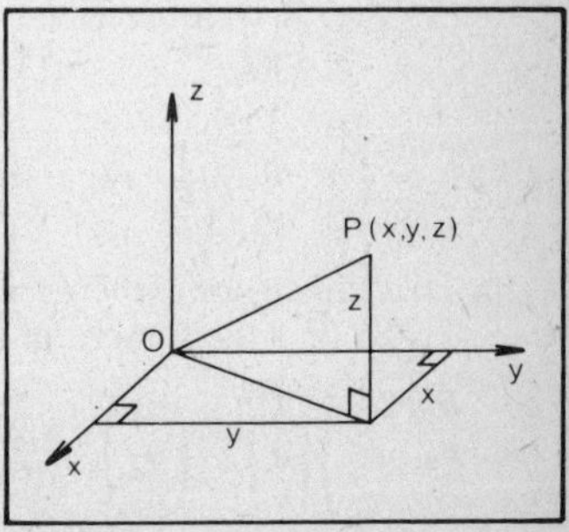

Figure 52

$$\begin{pmatrix} x_1 \\ y_1 \\ z_1 \end{pmatrix} = \begin{pmatrix} a_1 & b_1 & c_1 \\ a_2 & b_2 & c_2 \\ a_3 & b_3 & c_3 \end{pmatrix} \begin{pmatrix} x \\ y \\ z \end{pmatrix},$$

where $M = \begin{pmatrix} a_1 & b_1 & c_1 \\ a_2 & b_2 & c_2 \\ a_3 & b_3 & c_3 \end{pmatrix}$ is the transformation matrix.

The mirror for a reflection in three dimensions is a plane, (instead of a line in two dimensions) and a rotation takes place about a line (instead of a point as in two dimensions).

Example 6 Reflection in the plane $z = 0$.

The point (x, y, z) is reflected to $(x, y, -z)$.

$$\begin{pmatrix} x \\ y \\ z \end{pmatrix} \text{ becomes } \begin{pmatrix} 1 & 0 & 0 \\ 0 & 1 & 0 \\ 0 & 0 & -1 \end{pmatrix} \begin{pmatrix} x \\ y \\ z \end{pmatrix} = \begin{pmatrix} x \\ y \\ -z \end{pmatrix}$$

and $M_1 = \begin{pmatrix} 1 & 0 & 0 \\ 0 & 1 & 0 \\ 0 & 0 & -1 \end{pmatrix}$

is the matrix for a reflection in the plane $z = 0$.

Example 7 Reflection in the plane $x = y$.

(x, y, z) is reflected to (y, x, z).

$$\begin{pmatrix} x \\ y \\ z \end{pmatrix} \text{ becomes } \begin{pmatrix} 0 & 1 & 0 \\ 1 & 0 & 0 \\ 0 & 0 & 1 \end{pmatrix} \begin{pmatrix} x \\ y \\ z \end{pmatrix} = \begin{pmatrix} y \\ x \\ z \end{pmatrix}$$

and $M_2 = \begin{pmatrix} 0 & 1 & 0 \\ 1 & 0 & 0 \\ 0 & 0 & 1 \end{pmatrix}$ represents reflection in $y = x$.

As in two dimensions, the easiest way to find a transformation matrix is to find the images of the unit vectors along the axes.

$\begin{pmatrix} a_1 & b_1 & c_1 \\ a_2 & b_2 & c_2 \\ a_3 & b_3 & c_3 \end{pmatrix} \begin{pmatrix} 1 \\ 0 \\ 0 \end{pmatrix} = \begin{pmatrix} a_1 \\ a_2 \\ a_3 \end{pmatrix}$ and $(1, 0, 0)$ is transformed to (a_1, a_2, a_3).

Similarly $(0, 1, 0)$ goes to (b_1, b_2, b_3) and $(0, 0, 1)$ goes to (c_1, c_2, c_3).

Example 8 Rotation of 180° about Oz.

(1, 0, 0) rotates to (−1, 0, 0)
(0, 1, 0) rotates to (0, −1, 0)
(0, 0, 1) is invariant

so $M_3 = \begin{pmatrix} -1 & 0 & 0 \\ 0 & -1 & 0 \\ 0 & 0 & 1 \end{pmatrix}$ is the matrix for a rotation of 180° about Oz.

A **translation** necessarily moves the origin. Consequently since the image of (0, 0, 0) is always (0, 0, 0) with a 3 × 3 matrix transformation, a translation cannot be represented by a 3 × 3 matrix alone. It is sufficient to say

$$\begin{pmatrix} x \\ y \\ z \end{pmatrix} \text{ is translated to } \begin{pmatrix} x+1 \\ y+2 \\ z+3 \end{pmatrix} = \begin{pmatrix} x \\ y \\ z \end{pmatrix} + \begin{pmatrix} 1 \\ 2 \\ 3 \end{pmatrix}$$

for the translation represented by the vector $\begin{pmatrix} 1 \\ 2 \\ 3 \end{pmatrix}$.

An **enlargement** of scale factor 3, centre the origin, will take (1, 0, 0) to (3, 0, 0), (0, 1, 0) to (0, 3, 0) and (0, 0, 1) to (0, 0, 3).

$$\begin{pmatrix} x \\ y \\ z \end{pmatrix} \text{ becomes } \begin{pmatrix} 3 & 0 & 0 \\ 0 & 3 & 0 \\ 0 & 0 & 3 \end{pmatrix}\begin{pmatrix} x \\ y \\ z \end{pmatrix} = \begin{pmatrix} 3x \\ 3y \\ 3z \end{pmatrix}$$

so $M_4 = \begin{pmatrix} 3 & 0 & 0 \\ 0 & 3 & 0 \\ 0 & 0 & 3 \end{pmatrix}$ is the matrix for an enlargement of scale factor 3, centre (0, 0, 0).

An enlargement of scale factor −1 has matrix

$$\begin{pmatrix} -1 & 0 & 0 \\ 0 & -1 & 0 \\ 0 & 0 & -1 \end{pmatrix}$$

Its effect on the unit cube is shown in figure 53. Det $M_6 = -1$, so the transformation is not a rotation and it is not a reflection as the lines joining corresponding points are not parallel.

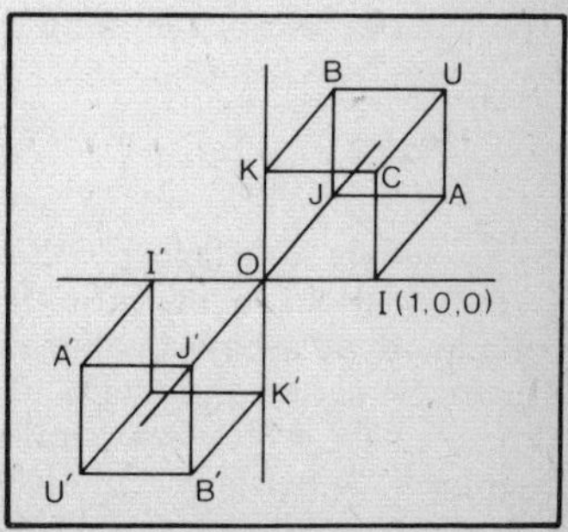

Figure 53

A **shear** in the direction of the x-axis with **invariant plane** $z = 0$ so that points in the plane $z = 1$ (i.e. distance 1 unit from invariant plane) move 3 units will take (0, 0, 1) to (3, 0, 1) and leave (1, 0, 0) and (0, 1, 0) invariant.

$$\begin{pmatrix} x \\ y \\ z \end{pmatrix} \text{ becomes } \begin{pmatrix} 1 & 0 & 3 \\ 0 & 1 & 0 \\ 0 & 0 & 1 \end{pmatrix}\begin{pmatrix} x \\ y \\ z \end{pmatrix} = \begin{pmatrix} x + 3z \\ y \\ z \end{pmatrix}$$

so $\begin{pmatrix} 1 & 0 & 3 \\ 0 & 1 & 0 \\ 0 & 0 & 1 \end{pmatrix}$ is the matrix for this shear.

The 3×3 matrices considered above are quite simple. More complicated transformations can be represented by a 3×3 matrix but it is difficult to interpret these in terms of simple transformations. It is most helpful to consider their effect on the unit cube whose vertices are O(0, 0, 0), I(1, 0, 0), J(0, 1, 0), K(0, 0, 1), A(1, 1, 0), B(0, 1, 1), C(1, 0, 1), U(1, 1, 1) (figure 53). You can check that the determinant of $M_1 = -1$ and det $M_2 = -1$, while det $M_3 = 1$. These results are similar to those for reflections and rotations in two dimensions and in fact the determinant of a 3×3 transformation matrix represents the ratio of the volume of the image to the volume of the object.

Det $M_4 = 27$ which means that the image volume is 27 times the object volume in an enlargement scale factor 3.

Det $M_5 = 1$ which confirms the important shearing property that volume is invariant.

Volume scale factor

The determinant of the matrix

$$M = \begin{pmatrix} a_1 & b_1 & c_1 \\ a_2 & b_2 & c_2 \\ a_3 & b_3 & c_3 \end{pmatrix}$$

is the volume scale factor of the transformation, i.e. the ratio of image volume to object volume. Figure 54 shows the image of the unit cube which is a parallelepiped (squashed cuboid) generated by the vectors $\mathbf{OI}_1$, $\mathbf{OJ}_1$, $\mathbf{OK}_1$.

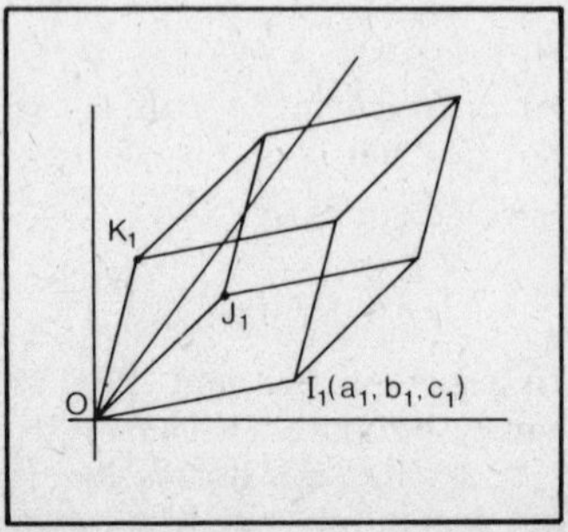

Figure 54

Opposite faces are parallel and the parallelepiped has three other edges parallel to $\mathbf{OI}_1$, three parallel to $\mathbf{OJ}_1$ and three parallel to $\mathbf{OK}_1$. It can be proved that the volume of this parallelepiped is equal to

$$\det M = \begin{pmatrix} a_1 & b_1 & c_1 \\ a_2 & b_2 & c_2 \\ a_3 & b_3 & c_3 \end{pmatrix}.$$

Key terms

A matrix of order p by q multiplied with a matrix of order q by r gives a matrix of order p by r.

Matrix multiplication is associative but not commutative.

For a matrix M, the identity I satisfies $IM = MI = M$.

The inverse, M^{-1}, of the matrix M satisfies $MM^{-1} = M^{-1}M = I$.

The inverse of $M = \begin{pmatrix} a & b \\ c & d \end{pmatrix}$ is $\frac{1}{\Delta}\begin{pmatrix} d & -b \\ -c & a \end{pmatrix}$,
where $\Delta = \det M = ad - bc$.

The inverse of a 3×3 matrix N is the transpose of the matrix of cofactors of N, divided by $\det N$.

Chapter 8
Series

Sequences

A **sequence** of numbers can be obtained by writing a set of values in a given order such that there is a definite rule by which each term may be calculated.

For example:

(i) 1, 3, 5, 7, ... (iii) 1, −1, 1, −1, 1, ...
(ii) 1, 4, 9, 16, ... (iv) 2, 5, 8, 11, 14, 17, ...

In general we write the terms of a sequence as $u_1, u_2, u_3, \ldots, u_n, \ldots$. There is a more convenient way of defining a sequence which saves having to write out the terms as we have above. We can use an **inductive definition** which gives an initial term and a rule for determining each term from a previous one. So (i) above could be defined as

$$u_1 = 1, \quad u_{k+1} = u_k + 2.$$

This gives all the terms starting with $u_1 = 1$, $u_2 = u_1 + 2 = 3$, $u_3 = u_2 + 2 = 5$ and so on.
Alternatively we can define a sequence by giving an **algebraic formula** for a general term.
So the sequence (i) above could be written as $u_n = 2n - 1$ for $n = 1, 2, 3, \ldots$.

Two important sequences can be formed. In the first, each term is formed by adding a constant to the previous term and in the second, each term is formed by multiplying the previous term by a constant.
The first case gives an **arithmetic progression**, A.P., and the constant may be positive, negative or fractional.
The second case gives a **geometric progression**, G.P., and again the constant may be a positive or negative integer or a fraction.
Examples of arithmetic progressions are:

(i) $u_n = 2n - 5$ for $n = 0, 1, 2, \ldots$.
(ii) $-2, -4, -6, -8, \ldots$.
(iii) $u_1 = -3$, $u_{k+1} = u_k + 4$.

Examples of geometric progressions are:

(i) $u_n = 3^n$ for $n = 0, 1, 2, 3, \ldots$
(ii) $1, -\frac{1}{2}, +\frac{1}{4}, -\frac{1}{8}, +\frac{1}{16}, \ldots$
(iii) $u_1 = 5, u_{k+1} = 2u_k$.

We may consider the terms of a sequence as a sum, in which case we write it as $u_1 + u_2 + u_3 + \cdots + u_n$.
This is called a **series** and may be finite or infinite.

For example $3 + 7 + 11 + 15 + 19$ is a finite series, but $1 + 2 + 3 + 4 + 5 + \cdots$ is infinite. The greek letter $\sum$ (sigma) is used to denote a summation and we can write the general series above as $\sum_{r=1}^{n} u_r$.

The arithmetic series would be written as $\sum_{r=1}^{7} (2r - 1)$ which is equivalent to $1 + 3 + 5 + 7 + 9 + 11 + 13$.
This is a finite series. If we have an infinite series we write $\sum_{r=2}^{\infty} r^2 = 2^2 + 3^2 + 4^2 + 5^2 + \cdots$

Example 1 Find the terms of the series $\sum_{r=1}^{4} \frac{1}{r(r+1)}$.

Clearly $\sum_{r=1}^{4} \frac{1}{r(r+1)} = \frac{1}{1 \times 2} + \frac{1}{2 \times 3} + \frac{1}{3 \times 4} + \frac{1}{4 \times 5}$.

Example 2 Write the series $1 - 2 + 4 - 8 + 16 - 32$ in the $\sum$ notation.
We must find an algebraic expression for the general term. In this case each term can be obtained from $(-1)^r 2^r$ by substituting $r = 0, 1, 2, \ldots, 5$.

Hence $1 - 2 + 4 - 8 + 16 - 32 = \sum_{r=0}^{5} (-1)^r 2^r$.

The arithmetic series

An A.P. is formed by adding a positive or negative constant to each term to obtain the next term of the series. If a is the first term and d is the value to be added, we can write the series as

$$a + (a + d) + (a + 2d) + (a + 3d) + \cdots$$

If we take n terms, the nth term will be $a + (n - 1)d$. Thus if the sum to n terms is S_n we have:

$$S_n = a + (a + d) + (a + 2d) + \cdots + \{a + (n - 1)d\},$$

or by rewriting in the opposite order:

$$S_n = \{a + (n - 1)d\} + \{a + (n - 2)d\} + \cdots + a.$$

Adding $2S_n = \{2a + (n - 1)d\} \times n$

Thus the **sum of an A.P.** is given by $\boldsymbol{S_n = \frac{1}{2}n\{2a + (n - 1)d\}}$, where a is the first term, **d is the common difference** and n is the number of terms.
The **value of the nth term** is given by $\boldsymbol{a + (n - 1)d}$

Example 3 Find the value of the 15th term and the sum of the first 10 terms of the series $8 + 5 + 2 + \cdots$

The value of the 15th term $= a + 14d = 8 + 14(-3) = -34$.

The sum of the first 10 terms $S_{10} = \frac{1}{2} \times 10\{(2 \times 8) + 9(-3)\}$
$= 5 \times (-11)$
$= -55$.

Example 4 Find the number of terms in the arithmetic progression $2{\cdot}8 + 3{\cdot}3 + 3{\cdot}8 + \cdots + 17{\cdot}8$.

Now $a = 2{\cdot}8$ and the value of the nth term $= 17{\cdot}8$.
$\therefore \quad 17{\cdot}8 = 2{\cdot}8 + (n - 1)0{\cdot}5 \Rightarrow \frac{1}{2}(n - 1) = 15$.
Thus $n - 1 = 30 \Rightarrow n = 31$ and there are 31 terms in the series.

Note that if we denote the last term by l then the sum to n terms can be written $S_n = \frac{1}{2}n\{2a + (n - 1)d\}$

$$= \tfrac{1}{2}n[a + \{a + (n - 1)d\}] = \tfrac{1}{2}n(a + l).$$

This can be a useful alternative when the first and last terms of the series are given.

Geometric series

A G.P. is formed by multiplying each term by a constant factor to produce the next term. If a is the first term and r is the given factor, the series can be written as

$$a + ar + ar^2 + ar^3 + \cdots$$

If we take n terms, the nth term is ar^{n-1}. Thus the sum to n terms is given by

$$S_n = a + ar + ar^2 + ar^3 + \cdots + ar^{n-1}. \qquad (1)$$

Multiplying by r

$$rS_n = ar + ar^2 + ar^3 + \cdots + ar^{n-1} + ar^n. \qquad (2)$$

Subtracting (2) from (1) we have

$$(1-r)S_n = a - ar^n \Rightarrow S_n = \frac{a(1-r^n)}{1-r}.$$

If $r > 1$, this is best expressed as $S_n = \dfrac{a(r^n - 1)}{r - 1}$,

and r is called the **common ratio** of the series.

Hence the **value of the nth term** $= ar^{n-1}$.

The **sum to n terms** $= \dfrac{a(1-r^n)}{1-r} = \dfrac{a(r^n-1)}{r-1}$.

Example 5 Find the sum of the first 10 terms of the geometric progression $3 + 6 + 12 + 24 + \cdots$

The sum of the G.P. is $S_{10} = \dfrac{3(2^{10}-1)}{2-1} = 3(2^{10}-1)$

$$= 3 \times 1023 = 3069.$$

If more terms of the series in example 5 were taken, the sum would be correspondingly larger. In fact, the value of S_n increases without limit and we say that $S_n \to \infty$ as $n \to \infty$. However, this is not always true and we have the idea of a limiting value.

A limit

Consider the function $f(r) = 1 + (1/r)$. As r takes values of 1, 10, 100, 1000, ... and so on, $f(r)$ takes the set of values 2, 1·1, 1·01, 1·001, ... and so on. Although $f(r)$ decreases as r tends to infinity, it is approaching the value of 1 more closely. In fact, we can make $f(r)$ as near to 1 as we like by choosing r to be sufficiently large.

We say that $f(r) \to 1$ as $r \to \infty$, or $\lim\limits_{r \to \infty} f(r) = 1$.

We can define this mathematically by saying that any function S_n of n will tend to a limit S (which is independent of n) if for all $n \geq N$, $|S_n - S|$ is less than some arbitrary small positive value, ε.

That is there exists an N such that $|S_n - S| < \varepsilon$ for all $n \geq N$.

If in an infinite series, $S_n = u_1 + u_2 + u_3 + \cdots + u_n$ is the sum of the first n terms of $\sum\limits_{r=1}^{\infty} u_r$, and if as n tends to infinity, S_n tends to a finite limit S, then the infinite series $\sum\limits_{r=1}^{\infty} u_r$

is said to be **convergent** and S is called its **sum to infinity**. So for a G.P. we have already shown that the sum to n terms is given by $S_n = \dfrac{a(1-r^n)}{1-r}$. If $-1 < r < 1$, $r^n \to 0$ as $n \to \infty$ and hence S_n tends to a finite limit $S = \dfrac{a}{1-r}$. Hence an infinite G.P. converges if $|r| < 1$. Note that if $r > 1$, $r^n \to \infty$ as $n \to \infty$ and hence S_n does not tend to a finite limit and we say that the series **diverges**.

Example 6 Find the sum of the first n terms of the series $1 - \frac{1}{2} + \frac{1}{4} - \frac{1}{8} + \cdots$ and deduce its sum to infinity.

Since $a = 1$ and $r = -\frac{1}{2}$

$$S_n = \frac{a(1-r^n)}{1-r} \Rightarrow S_n = \frac{1\{1-(-\frac{1}{2})^n\}}{1-(-\frac{1}{2})} = \tfrac{2}{3}\{1-(-\tfrac{1}{2})^n\}.$$

Now as $n \to \infty$ $(-\frac{1}{2})^n \to 0$, and the sum to infinity $= \frac{2}{3}$.

Example 7 In a geometrical progression the sum of the third and fourth terms is 18 and the sixth term is eight times the third. Find the first term, the common ratio and the seventh term.

Now a G.P. can be written $a + ar + ar^2 + ar^3 + \cdots$

Hence $\quad ar^2 + ar^3 = 18, \quad (1)$

and $\quad ar^5 = 8ar^2. \quad (2)$

From (2) $r^3 = 8 \Rightarrow r = 2$.

$\therefore$ in equation (1) $4a + 8a = 18 \Rightarrow a = 1\frac{1}{2}$
and the seventh term $= ar^6 = \frac{3}{2}(2)^6 = 96$.

Other finite series

Many other finite series exist apart from the arithmetic and geometric progressions. The methods of finding the sums of these series differ, but there are a number of recognized techniques.

Example 8 Find the sum of the arithmetico-geometrical series

$$2 + 5x + 8x^2 + 11x^3 + \cdots + 29x^9.$$

There are 10 terms in the series so let

$$S_{10} = 2 + 5x + 8x^2 + 11x^3 + \cdots + 29x^9. \quad (1)$$

Multiplying by x gives:

$$xS_{10} = 2x + 5x^2 + 8x^3 + \cdots + 26x^9 + 29x^{10}. \quad (2)$$

Subtracting (2) from (1) we have

$$(1 - x)S_{10} = 2 + 3x + 3x^2 + 3x^3 + \cdots + 3x^9 - 29x^{10}$$
$$= 2 + \frac{3x(1 - x^9)}{1 - x} - 29x^{10}$$
$$\therefore \qquad S_{10} = \frac{2 - 29x^{10}}{1 - x} + \frac{3x(1 - x^9)}{(1 - x)^2}.$$

The **difference method** can be applied to evaluate the sum of a series $\sum_{r=1}^{n} u_r = u_1 + u_2 + u_3 + \cdots + u_r + \cdots + u_n$.

If u_r can be expressed as the difference of two functions, say, $f(r) - f(r - 1)$ then clearly.

$$\begin{aligned} u_n &= f(n) - f(n - 1), \\ u_{n-1} &= f(n - 1) - f(n - 2), \\ u_{n-2} &= f(n - 2) - f(n - 3), \\ &\vdots \\ u_2 &= f(2) - f(1), \\ u_1 &= f(1) - f(0). \end{aligned}$$

By adding these terms it follows that

$$u_1 + u_2 + u_3 + \cdots + u_{n-1} + u_n = f(n) - f(0).$$

Example 9 Find the sum of the series $\sum_{r=1}^{n} r(r + 1)(r + 2)$.

Now $u_r = r(r + 1)(r + 2)$. If we let $f(r) = r(r + 1)(r + 2)(r + 3)$, then

$$\begin{aligned} f(r) - f(r - 1) &= r(r + 1)(r + 2)(r + 3) \\ &\quad - r(r - 1)(r + 1)(r + 2) \\ &= r(r + 1)(r + 2)(r + 3 - r + 1) \\ &= 4r(r + 1)(r + 2) = 4u_r \end{aligned}$$

$$\therefore \qquad u_r = \tfrac{1}{4}\{f(r) - f(r - 1)\}.$$

Hence

$$\begin{aligned} \sum_{r=1}^{n} u_r &= \sum_{r=1}^{n} \tfrac{1}{4}\{f(r) - f(r - 1)\} \\ &= \tfrac{1}{4}\{f(n) - f(0)\} \\ &= \tfrac{1}{4}n(n + 1)(n + 2)(n + 3). \end{aligned}$$

The method of differences is useful in series where each term is a reciprocal of a product of a constant number of factors in A.P., the first factors of each term being in the same A.P.

Example 10 Find the sum to n terms of the series

$$\frac{1}{1\times 2}+\frac{1}{2\times 3}+\frac{1}{3\times 4}+\cdots$$

Clearly the series can be written $\sum_{r=1}^{n}\frac{1}{r(r+1)}$,

i.e. $u_r=\frac{1}{r(r+1)}=\frac{1}{r}-\frac{1}{r+1}$.

Hence $$\sum_{r=1}^{n}u_r=\left(\frac{1}{1}-\frac{1}{2}\right)+\left(\frac{1}{2}-\frac{1}{3}\right)+\left(\frac{1}{3}-\frac{1}{4}\right)+\cdots$$

$$+\left(\frac{1}{n-1}-\frac{1}{n}\right)+\left(\frac{1}{n}-\frac{1}{n+1}\right).$$

Adding $$\sum_{r=1}^{n}\frac{1}{r(r+1)}=1-\frac{1}{n+1}.$$

Or we could have defined $f(r)$ as $\frac{1}{r+1}$ in which case $-u_r=f(r)-f(r-1)$.

In this case $$\sum_{r=1}^{n}u_r=\sum_{r=1}^{n}-\{f(r)-f(r-1)\}$$

$$=-\{f(n)-f(0)\}=1-\frac{1}{n+1}\text{ as before.}$$

Series of natural numbers

Three very useful results, which should be memorized are: (i) the sum of the natural numbers, (ii) the sum of the squares of the natural numbers and (iii) the sum of the cubes of the natural numbers.

(i) Let $S_n=1+2+3+4+\cdots+n$

Since this is an A.P., $S_n=\frac{1}{2}n(n+1)$ (see page 148).

(ii) Let $S_n=1^2+2^2+3^2+4^2+\cdots+(n-1)^2+n^2$.

By putting $f(r)=r(r+1)(2r+1)$ and using the method of differences it can be proved that:

$$S_n=\tfrac{1}{6}n(n+1)(2n+1).$$

(iii) If $S_n=1^3+2^3+3^3+\cdots+n^3$.

In this case choose $f(r) = \{r(r+1)\}^2$ and the method of differences gives:

$$S_n = \tfrac{1}{4}n^2(n+1)^2 = \{\tfrac{1}{2}n(n+1)\}^2 = \left(\sum_1^n r\right)^2.$$

These results should be learnt since they can be used to sum other series.

Example 11 Find the sum of the series $\sum_{r=1}^{n} (r+1)(r+3)$.

This series is

$$(2 \times 4) + (3 \times 5) + (4 \times 6) + (5 \times 7) + \cdots + (n+1) \times (n+3)$$

and the general term is $r^2 + 4r + 3$.

$$\begin{aligned} \therefore \quad \sum_{r=1}^{n} (r^2 + 4r + 3) &= \sum_{r=1}^{n} r^2 + \sum_{r=1}^{n} 4r + \sum_{r=1}^{n} 3 \\ &= \tfrac{1}{6}n(n+1)(2n+1) + 4 \times \tfrac{1}{2}n(n+1) + 3n \\ &= \tfrac{1}{6}n\{(n+1)(2n+1) + 12(n+1) + 18\} \\ &= \tfrac{1}{6}n\{2n^2 + 3n + 1 + 12n + 12 + 18\} \\ &= \tfrac{1}{6}n(2n^2 + 15n + 31). \end{aligned}$$

Proof by induction

It sometimes happens that the sum of a series can be found by experiment or by a process that does not form a proof. In this situation we can prove the truth of the result by a special process called **induction**. There are two basic steps.

(i) To show that if it is true for some value of n then it is also true for the next integral value of n, and
(ii) to show that it is true for the first value.

To illustrate the method we consider the series of the squares of the natural numbers, i.e.

$$1^2 + 2^2 + 3^2 + 4^2 + \cdots + n^2 = \tfrac{1}{6}n(n+1)(2n+1). \qquad (1)$$

Suppose this result is true for some value of n, say $n = k$, then

$$1^2 + 2^2 + 3^2 + \cdots + k^2 = \tfrac{1}{6}k(k+1)(2k+1).$$

Adding the next term of the series to both sides of the equation we obtain:

$$\begin{aligned} 1^2 + 2^2 + 3^2 + \cdots + k^2 + (k+1)^2 &= \tfrac{1}{6}k(k+1)(2k+1) + (k+1)^2 \\ &= \tfrac{1}{6}(k+1)\{k(2k+1) + 6(k+1)\} \\ &= \tfrac{1}{6}(k+1)(2k^2 + 7k + 6) \\ &= \tfrac{1}{6}(k+1)(k+2)(2k+3). \end{aligned}$$

Now this is clearly the result for $n = k + 1$ and thus if the result is true for $n = k$, it is also true for $n = k + 1$.
Now when $n = 1$, considering (1)

L.H.S. $= 1^2 = 1$ and R.H.S. $= \frac{1}{6} \times 1 \times 2 \times 3 = 1$.

The result is true for $n = 1$ and hence for $n = 2$. From this it follows that it is also true for $n = 3$ and so on for all positive integral values of n.

This can be written out in a slightly different manner as follows.

(i) Proof that $k \in T \Rightarrow k + 1 \in T$, where T is the set of values for which the statement is true.

$$\begin{aligned} \text{Now } k \in T \Rightarrow S_k &= \tfrac{1}{6}k(k+1)(2k+1) \\ \Rightarrow S_{k+1} &= S_k + (k+1)^2 = \tfrac{1}{6}k(k+1)(2k+1) + (k+1)^2 \\ \Rightarrow S_{k+1} &= \tfrac{1}{6}(k+1)(k+2)(2k+3). \end{aligned}$$

This is the result with $n = k + 1$. Hence we have shown that $k \in T \Rightarrow (k + 1) \in T$.

(ii) Proof that $1 \in T$.
Now $S_1 = 1^2 = 1$. When $n = 1$, $\frac{1}{6}n(n + 1)(2n + 1) = 1 \Rightarrow 1 \in T$.

So both parts of the induction principle are satisfied and it follows that the result is true for all integral values of n.

Remember that to use induction we must have the answer first.

Permutations and combinations

We now consider certain problems in selection and arrangement. A **permutation** is an arrangement of items in a particular order.

Example 12 How many numbers can be made by using all of the digits 2, 3, 6 and 7?

The 1st digit chosen can be any one of 4 digits.
The 2nd digit must then be chosen from 3 digits, one having already been chosen.
The 3rd digit is then restricted to a choice of 2 digits and the 4th digit is then determined.

The total is obtained by multiplying these results since for each of the first 4 choices there are 3 choices for the next and so on. The number of different arrangements is $4 \times 3 \times 2 \times 1 = 24$, i.e.

2367	3267	6327	7326
2376	3276	6372	7362
2637	3627	6273	7263
2673	3672	6237	7236
2763	3726	6732	7632
2736	3762	6723	7623.

It follows that if we had to place 8 different colour counters in order that the total number of arrangements is given by $8 \times 7 \times 6 \times 5 \times 4 \times 3 \times 2 \times 1$.

We can use a dot to signify a product, for example, $8.7.6.5.4.3.2.1$, but an alternative notation can be adopted, namely $8!$ This is read factorial eight. Remember that it includes each integer from the highest down to 1. We write

$$n! = n(n-1)(n-2)(n-3)\cdots 4 \times 3 \times 2 \times 1.$$

Example 13 Simplify $\dfrac{15!}{11!4!} + \dfrac{15!}{12!3!}$

Now $\dfrac{15!}{11!4!} + \dfrac{15!}{12!3!} = \dfrac{12 \times 15!}{12!4!} + \dfrac{4 \times 15!}{12!4!} = \dfrac{16 \times 15!}{12!4!} = \dfrac{16!}{12!4!}$

Using this notation it means that there are $n!$ permutations of n unlike objects.
If we only select **r objects from n unlike objects** then:
the 1st is chosen in n ways, the 2nd is chosen in $(n-1)$ ways, the 3rd is chosen in $(n-2)$ ways and so on until we arrive at the last object which can be chosen in $(n-r+1)$ ways.

Thus the total number of arrangements

$$= n(n-1)(n-2)\cdots(n-r+1)$$

$$= \frac{n!}{(n-r)!}$$

We use the notation $\boldsymbol{{}^{n}P_{r} = \dfrac{n!}{(n-r)!} \quad (r \leq n)}$.

The following examples illustrate the methods which can be used.

Example 14 How many three-letter arrangements can be formed from the first eight letters of the alphabet?

Since we are choosing 3 objects from 8 the number of different arrangements is $^8P_3 = \frac{8!}{5!} = 8 \times 7 \times 6 = 336$.

When some items are repeated, some modification is necessary.

Example 15 In how many ways can the letters of the word EXPERIMENT be arranged?

Since there are three Es call them E_1, E_2, E_3. Treating the Es as different there are 10! ways of arranging the letters. But in every distinct arrangement the three Es can be rearranged amongst themselves without altering the other letters. So, TNIEXEPREM, for example, has been included 3! times in the total 10! arrangements. Hence the number of distinct arrangements of the letters of the word EXPERIMENT $= \frac{10!}{3!}$.

Sometimes it pays to find the exact opposite of what is asked in the question and subtract this from the number of unrestricted arrangements.

Example 16 How many arrangements of eight stories in a book can be made if the longest and the shortest must not come together.

If there is no restriction, the number of arrangements $= 8!$
If we consider the longest and the shortest to be together so that they cannot separate then we effectively have seven stories to arrange. This can be done in 7! ways but the longest and shortest can be placed in either order. Thus the total number of arrangements when the longest and the shortest are together is $2 \times 7!$
Thus the number of arrangements when they are not together is clearly $8! - 2 \times 7! = 7!(8 - 2) = 6 \times 7!$

A **combination** is a selection of objects where the order is unimportant.

We have already seen that we can arrange r objects from n in $\frac{n!}{(n-r)!}$ different ways. Clearly for each of these arrangements we can arrange the r objects amongst themselves in $r!$ ways, and hence if the order is unimportant we can form $\frac{n!}{(n-r)!r!}$ different groups of the r objects.

We use the notation $^nC_r = \dfrac{n!}{(n-r)!r!}$

This is sometimes written as $_nC_r$ or $\binom{n}{r}$.

Example 17 In how many ways can a committee of 4 men and 3 women be formed from 10 men and 8 women.

Since the order is not important:

4 men can be chosen from 10 men in $^{10}C_4$ ways,

3 women can be chosen from 8 women in $^8C_3$ ways.

Since with each selection of men we could put each selection of the women, the total ways of forming the committee is obtained by multiplying these results, i.e.

$$^{10}C_4 \times {}^8C_3 = \frac{10!}{6!4!} \times \frac{8!}{5!3!}$$

$$= \frac{10 \times 9 \times 8 \times 7}{4 \times 3 \times 2 \times 1} \times \frac{8 \times 7 \times 6}{3 \times 2 \times 1}$$

$$= 210 \times 56 = 11\,760.$$

The binomial theorem for a positive integral index

Consider the expansions $(1 + x)^2 = 1 + 2x + x^2$.

$$(1 + x)^3 = (1 + x)(1 + x)^2 = 1 + 3x + 3x^2 + x^3.$$

$$(1 + x)^4 = (1 + x)(1 + x)^3 = 1 + 4x + 6x^2 + 4x^3 + x^4. \quad (1)$$

Now in the expansion of $(1 + x)^4$ we have

$$(1 + x)^4 = (1 + x)(1 + x)(1 + x)(1 + x).$$

Each term in the expansion (1) is obtained by selecting one element from each bracket so that the complete expansion consists of the sum of all such combinations.
For example, the term $4x^3$ is obtained by selecting three xs from four which can be done in $^4C_3$ ways. Thus the coefficient is $\dfrac{4!}{3!\,1!} = 4.$

Extending this process it suggests that

$$(1 + x)^n = 1 + {}^nC_1x + {}^nC_2x^2 + \cdots + {}^nC_rx^r + \cdots + {}^nC_nx^n.$$

This can be proved by induction.
In general, the expansion of $(a + x)^n$ can be obtained by writing $a^n\left(1 + \frac{x}{a}\right)^n$.

$$\therefore \quad (a + x)^n = a^n\left\{1 + {}^nC_1\left(\frac{x}{a}\right) + {}^nC_2\left(\frac{x}{a}\right)^2 + {}^nC_3\left(\frac{x}{a}\right)^3 + \cdots \right.$$

$$\left. + {}^nC_r\left(\frac{x}{a}\right)^r + \cdots + \left(\frac{x}{a}\right)^n\right\}$$

$$= a^n + {}^nC_1 a^{n-1}x + {}^nC_2 a^{n-2}x^2 + \cdots$$

$$+ {}^nC_r a^{n-r}x^r + \cdots + x^n.$$

Note that:
(i) there are $n + 1$ terms in this finite expansion,
(ii) the sum of the indices of a and x in each term is equal to n,
(iii) the term in x^r is the $(r + 1)$ term in the expansion in ascending powers of x.

The coefficients in the binomial expansion using different values of n form a pattern and it is useful to remember this for fairly small values. They can be written in the form of a triangle and it is known as **Pascal's triangle.**

						1							$(1 + x)^0$
					1		1						$(1 + x)^1$
				1		2		1					$(1 + x)^2$
			1		3		3		1				$(1 + x)^3$
		1		4		6		4		1			$(1 + x)^4$
	1		5		10		10		5		1		$(1 + x)^5$
1		6		15		20		15		6		1	$(1 + x)^6$
.	.	.	.	.	.	.	.	.	.	.	.	.	

Apart from the first and last terms which are always unity, each term of a line is formed by adding the two terms on either side of it in the row above, for example:

$$5 + 10 = 15 \text{ and } 10 + 10 = 20.$$

Example 18 Expand $(1 + 2x)^7$ in ascending powers of x.

Making use of Pascal's triangle the coefficients required will be given by the next line in the above table, i.e.

1 7 21 35 35 21 7 1

$$\therefore \quad (1+2x)^7 = 1 + 7(2x) + 21(2x)^2 + 35(2x)^3 + 35(2x)^4 + 21(2x)^5 + 7(2x)^6 + (2x)^7$$
$$= 1 + 14x + 84x^2 + 280x^3 + 560x^4 + 672x^5 + 448x^6 + 128x^7.$$

Example 19 Expand, by the binomial theorem, $(2 + \frac{1}{2}x)^{10}$, giving the first five terms of the expansion in ascending powers of x.

$$(2 + \tfrac{1}{2}x)^{10} = 2^{10}\left(1 + \frac{x}{4}\right)^{10}$$
$$= 2^{10}\left\{1 + {}^{10}C_1\left(\frac{x}{4}\right) + {}^{10}C_2\left(\frac{x}{4}\right)^2 + {}^{10}C_3\left(\frac{x}{4}\right)^3 + {}^{10}C_4\left(\frac{x}{4}\right)^4 + \cdots\right\}$$
$$= 2^{10}\left\{1 + 10\left(\frac{x}{4}\right) + \frac{10!}{8!2!}\left(\frac{x}{4}\right)^2 + \frac{10!}{7!3!}\left(\frac{x}{4}\right)^3 + \frac{10!}{6!4!}\left(\frac{x}{4}\right)^4 + \cdots\right\}$$
$$= 2^{10}\left\{1 + \frac{5x}{2} + \frac{45x^2}{16} + \frac{120x^3}{64} + \frac{210x^4}{256} + \cdots\right\}$$
$$= 1024 + 2560x + 2880x^2 + 1920x^3 + 840x^4 + \cdots$$

Example 20 Find the term in x^6 in the expansion of $\left(x - \dfrac{2}{x}\right)^8$.

In this problem it is not necessary to write out the whole series. The term in x^r is given by ${}^nC_r a^{n-r} x^r$ in the expansion of $(a + x)^n$.

In this case $n = 8$, $a = x$ and x is replaced by $\left(-\dfrac{2}{x}\right)$.

$\therefore$ The term in x^6 is formed when $n - r - r = 6$, i.e.

$$8 - 2r = 6 \Rightarrow r = 1.$$

$\therefore$ the term in x^6 is given by ${}^8C_1 x^7\left(-\dfrac{2}{x}\right)^1 = -16x^6$.

Example 21 Find the first four terms of the expansion of $(1-x)^{12}$ in ascending powers of x and hence find the value of $(0{\cdot}997)^{12}$ correct to three decimal places.

Now $\quad (1-x)^{12} = 1 + {}^{12}C_1(-x) + {}^{12}C_2(-x)^2 + {}^{12}C_3(-x)^3$

$$= 1 - 12x + \frac{12 \times 11}{2 \times 1}x^2 - \frac{12 \times 11 \times 10}{3 \times 2 \times 1}x^3$$

$$= 1 - 12x + 66x^2 - 220x^3$$

Now if $x = 0{\cdot}003$:

$$\begin{aligned}(1-x)^{12} &= (1-0{\cdot}003)^{12}\\ &= 0{\cdot}997^{12}\\ &= 1 - 12(0{\cdot}003) + 66(0{\cdot}003)^2 - 220(0{\cdot}003)^3 + \cdots\\ &= 1 - 0{\cdot}036 + 0{\cdot}000594 - 0{\cdot}00000594\\ &= 0{\cdot}964594\\ &= 0{\cdot}965 \text{ to three decimal places.}\end{aligned}$$

The binomial theorem for any index

We have seen that if the index is a positive integer the series terminates after $n + 1$ terms. This is not the case if n is a positive or negative fraction or negative integer. In addition, the ${}^nC_r$ notation has no meaning and so the series is given by

$$(1+x)^n = 1 + nx + \frac{n(n-1)}{2!}x^2 + \frac{n(n-1)(n-2)}{3!}x^3 + \cdots$$
$$+ \frac{n(n-1)(n-2)\cdots(n-r+1)}{r!}x^r + \cdots$$

This form of the series does not produce a sum to infinity for all values of x, only if $-1 < x < 1$.

Example 22 Expand $(1-2x)^{-1/2}$ in ascending powers of x as far as the term in x^3.

By using the binomial theorem

$$(1-2x)^{-1/2} = 1 + (-\tfrac{1}{2})(-2x) + \frac{(-\frac{1}{2})(-\frac{3}{2})}{2!}(-2x)^2$$
$$+ \frac{(-\frac{1}{2})(-\frac{3}{2})(-\frac{5}{2})}{3!}(-2x)^3 + \cdots$$
$$= 1 + x + \tfrac{3}{2}x^2 + \tfrac{5}{2}x^3 + \cdots$$

Example 23 Find the sum to infinity of the series stating the range of validity of x.

$$1 - x + \frac{1 \times 3}{1 \times 2}x^2 - \frac{1 \times 3 \times 5}{1 \times 2 \times 3}x^3 + \frac{1 \times 3 \times 5 \times 7}{1 \times 2 \times 3 \times 4}x^4 - \cdots$$

By rearranging the coefficients in each term, the series can be written

$$1 + (-\tfrac{1}{2})(2x) + \frac{(-\frac{1}{2})(-\frac{3}{2})}{2!}(2x)^2 + \frac{(-\frac{1}{2})(-\frac{3}{2})(-\frac{5}{2})}{3!}(2x)^3 + \frac{(-\frac{1}{2})(-\frac{3}{2})(-\frac{5}{2})(-\frac{7}{2})}{4!}(2x)^4,$$

which is clearly the expansion of $(1 + 2x)^{-1/2}$.
Hence the sum to infinity is given by $(1 + 2x)^{-1/2}$. The series will possess this sum if $-1 < 2x < 1$, i.e. $|x| < \frac{1}{2}$.

Taylor's theorem

A general result known as Taylor's theorem can be used to expand a function $f(x)$.

If $f(x)$ and its first $(n - 1)$ derivatives are continuous in $a \leq x \leq b$ and its nth derivative exists in $a < x < b$, then

$$f(a + h) = f(a) + h f'(a) + \frac{h^2}{2!} f''(a) + \cdots + \frac{h^{n-1}}{(n-1)!} f^{(n-1)}(a) + R_n,$$

where $R_n = \dfrac{h^n}{n!} f^{(n)}(a + \theta h), \quad 0 < \theta < 1.$

The last term is usually called the remainder R_n and this form is known as the **Lagrange remainder**. We have expanded $f(a + h)$ as a polynomial in h of degree n. To find R_n precisely, a value for θ has to be found, and this is not possible in general. We can, however, find an approximate value of the expansion by neglecting R_n when it tends to zero giving a **Taylor series**, i.e.

$$f(a + h) = f(a) + hf'(a) + \frac{h^2}{2!} f''(a) + \cdots$$

As a particular case, we take $h = x$ and $a = 0$ which gives the **Maclaurin series**, i.e.

$$f(x) = f(0) + xf'(0) + \frac{x^2}{2!}f''(0) + \cdots + \frac{x^n}{n!}f^{(n)}(0) + \cdots$$

Series for sin x

$$f(x) = \sin x \Rightarrow f^{(n)}(x) = \sin(x + \tfrac{1}{2}n\pi)$$
$$f^{(n)}(0) = \sin \tfrac{1}{2}n\pi.$$

Hence $f^{(n)}(0)$ takes successively values of 0, 1, 0, -1, ... and the series is given by

$$\sin x = x - \frac{x^3}{3!} + \frac{x^5}{5!} - \frac{x^7}{7!} + \cdots + \frac{(-1)^r x^{2r+1}}{(2r+1)!} + \cdots$$

Note that for this series, since $f^{(n)}x = \sin(x + \frac{1}{2}n\pi)$, the Lagrange form of the remainder $R_n = \sin(\theta x + \frac{1}{2}n\pi)\dfrac{x^n}{n!}$ $(a = 0)$.

Now since $|\sin(\theta x + \frac{1}{2}n\pi)| \leq 1$ and for all large n, $\left|\dfrac{x^n}{n!}\right| \to 0$, $R_n \to 0$.

Series for cos x

A similar process gives $f^{(n)}(0) = \cos(\frac{1}{2}n\pi)$ and the series is given by

$$\cos x = 1 - \frac{x^2}{2!} + \frac{x^4}{4!} - \frac{x^6}{6!} + \cdots + (-1)^r \frac{x^{2r}}{(2r)!} + \cdots$$

The series for $\sin x$ and $\cos x$ converge for all values of x and hence the expansions are valid for all values of x.

The exponential series

If $f(x) = e^x$, then $f^{(n)}(x) = e^x$. Thus $f^{(n)}(0) = 1$.
Hence the series can be written using Maclaurin's theorem

$$e^x = 1 + x + \frac{x^2}{2!} + \frac{x^3}{3!} + \frac{x^4}{4!} + \cdots + \frac{x^r}{r!} + \cdots$$

This series is convergent for all values of x and is sometimes written $\exp(x)$. By putting $x = 1$ we can find the value of e, i.e.

$$\begin{aligned} e &= 1 + 1 + \frac{1}{2!} + \frac{1}{3!} + \frac{1}{4!} + \frac{1}{5!} + \frac{1}{6!} + \frac{1}{7!} + \cdots \\ &= 2 + 0{\cdot}5 + 0{\cdot}16667 + 0{\cdot}04167 + 0{\cdot}00833 + 0{\cdot}00139 \\ &\quad + 0{\cdot}00020 + 0{\cdot}00002 + 0{\cdot}00000 \\ &\simeq 2{\cdot}71828. \end{aligned}$$

Example 24 Find the first three terms of the expansion $(1+2x)e^{-2x}$

$$(1+2x)e^{-2x}$$

$$= (1+2x)\left\{1 + (-2x) + \frac{(-2x)^2}{2!} + \frac{(-2x)^3}{3!} + \frac{(-2x)^4}{4!} + \cdots\right\}$$

$$= (1+2x)\left(1 - 2x + 2x^2 - \frac{4x^3}{3} + \frac{2x^4}{3} - \cdots\right).$$

Multiplying out and retaining only terms up to and including x^3

$$(1+2x)e^{-2x} = 1 - 2x + 2x^2 - \tfrac{4}{3}x^3 + 2x - 4x^2 + 4x^3 + \cdots$$
$$= 1 - 2x^2 + \tfrac{8}{3}x^3 + \cdots$$

The logarithmic series

If $f(x) = \ln(1+x)$, $\quad f'(x) = \dfrac{1}{1+x} = (1+x)^{-1}$.

Hence $\quad f''(x) = -(1+x)^{-2}$, $\; f'''(x) = (-1)(-2)(1+x)^{-3}$

and $\quad f^{(n)}(x) = \dfrac{(-1)^{n-1}(n-1)!}{(1+x)^n}$.

When $x = 0$, $f(0) = 0$, $f'(0) = 1$, $f''(0) = -1$, $f'''(0) = 2$ and, in general, $f^{(n)}(0) = (-1)^{n-1}(n-1)!$.

Using Maclaurin's theorem:

$$\ln(1+x) = x - \frac{x^2}{2} + \frac{x^3}{3} - \frac{x^4}{4} + \cdots + \frac{(-1)^{r-1}x^r}{r} + \cdots$$

The range of convergence is $-1 < x \leq 1$.
Note that we have given a series for $\ln(1+x)$ and not $\ln x$. This is because $\ln x$ and its derivatives do not exist at $x = 0$, and hence $\ln x$ cannot be represented as a power series.

By replacing x with $-x$, we can form the alternative series

$$\ln(1-x) = -x - \frac{x^2}{2} - \frac{x^3}{3} - \frac{x^4}{4} - \cdots - \frac{x^r}{r} - \cdots$$

provided $-1 \leq x < 1$.

By combining these two results we form yet another two logarithmic expansions.

$$\ln(1+x)+\ln(1-x)=\ln(1-x^2).$$

$$\ln(1-x^2)=2\left(-\frac{x^2}{2}-\frac{x^4}{4}-\frac{x^6}{6}-\cdots-\frac{x^{2r}}{2r}-\cdots\right).$$

Also

$$\ln\left(\frac{1+x}{1-x}\right)=2\left(x+\frac{x^3}{3}+\frac{x^5}{5}+\cdots+\frac{x^{2r-1}}{2r-1}+\cdots\right).$$

These expansions are valid if $-1<x\leq 1$ and $-1\leq x<1$ and so for both to be valid $-1<x<1$.

Example 25 Expand $\ln(2-5x)$ as far as term in x^3 and give the general term.

$$\ln(2-5x)=\ln 2(1-\tfrac{5}{2}x)=\ln 2+\ln(1-\tfrac{5}{2}x)$$

$$\ln(2-5x)=\ln 2+\left\{-\frac{5}{2}x-\frac{1}{2}\left(-\frac{5}{2}x\right)^2+\frac{1}{3}\left(-\frac{5}{2}x\right)^3+\cdots\right.$$

$$\left.+\frac{(-1)^{r-1}}{r}\left(-\frac{5}{2}\right)^r+\cdots\right\}$$

$$=\ln 2-\frac{5}{2}x-\frac{25}{8}x^2-\frac{125}{24}x^3+\cdots-\frac{5^r x^r}{2^r\times r}-\cdots$$

This expansion is valid if $-1\leq\frac{5}{2}x<1$, i.e. $-\frac{2}{5}\leq x<\frac{2}{5}$.

Key terms

Arithmetical progression $a+(a+d)+(a+2d)+\cdots+\{a+(n-1)d\}$.

Sum to n terms $\frac{n}{2}\{2a+(n-1)d\}$.

Geometrical progression $a+ar+ar^2+ar^3+\cdots+ar^n$.

Sum to n terms $\frac{a(1-r^n)}{1-r}$.

Sum to infinity $\frac{a}{1-r}$, $|r|<1$.

Series of **natural numbers**

(i) $\sum_{1}^{n} r = \frac{1}{2}n(n+1)$. (ii) $\sum_{1}^{n} r^2 = \frac{1}{6}n(n+1)(2n+1)$.

(iii) $\sum_{1}^{n} r^3 = \frac{1}{4}n^2(n+1)^2$.

A **permutation** is an arrangement of items in a particular order; r items can be chosen from n different items in ${}^n\mathrm{P}_r = \dfrac{n!}{(n-r)!}$ ways.

A **combination** is a selection where the order is not important. We can form groups of r objects from n in ${}^n\mathrm{C}_r = \dfrac{n!}{(n-r)!\,r!}$ ways.

The **binomial theorem**

$$(1+x)^n = 1 + nx + \frac{n(n-1)}{2!}x^2 + \frac{n(n-1)(n-2)}{3!}x^3 + \cdots$$

The **Taylor series**

$$f(a+h) = f(a) + h\,f'(a) + \frac{h^2}{2!}\,f''(a) + \cdots + \frac{h^n}{n!}\,f^{(n)}(a) + \cdots$$

A special case is the **Maclaurin series**

$$f(x) = f(0) + x\,f'(0) + \frac{x^2}{2!}\,f''(0) + \cdots + \frac{x^n}{n!}\,f^{(n)}(0) + \cdots$$

$$\sin x = x - \frac{x^3}{3!} + \frac{x^5}{5!} - \frac{x^7}{7!} + \cdots$$

$$\cos x = 1 - \frac{x^2}{2!} + \frac{x^4}{4!} - \frac{x^6}{6!} + \cdots$$

$$e^x = 1 + x + \frac{x^2}{2!} + \frac{x^3}{3!} + \frac{x^4}{4!} + \cdots$$

$$\ln(1+x) = x - \frac{x^2}{2} + \frac{x^3}{3} - \frac{x^4}{4} + \frac{x^5}{5} - \cdots, \quad -1 < x \le 1.$$

Chapter 9
Complex Numbers

In the algebraic work on polynomials, we have looked at methods by which they can be factorized. Clearly, by inspection, the function $P(x) = x^2 - x - 6$ can be written as $(x - 3)(x + 2)$. For a cubic or polynomial of higher degree it may be necessary to use the factor theorem (see page 12). If $F(x) = x^3 - 3x^2 - 6x + 8$, we see that $F(1) = 0$ and hence $x - 1$ is a factor. The second factor can be obtained by division and we find

$$F(x) = (x - 1)(x^2 - 2x - 8).$$

The quadratic factor can also be factorized, so that $F(x)$ can be completely factorized into three linear factors, i.e.

$$F(x) = x^3 - 3x^2 - 6x + 8 = (x - 1)(x + 2)(x - 4).$$

It is not always possible to go as far as this and, for example,

$$G(x) = x^3 + 4x^2 + 4x + 3 = (x + 3)(x^2 + x + 1).$$

The quadratic factor $x^2 + x + 1$ cannot be factorized further over the real numbers.

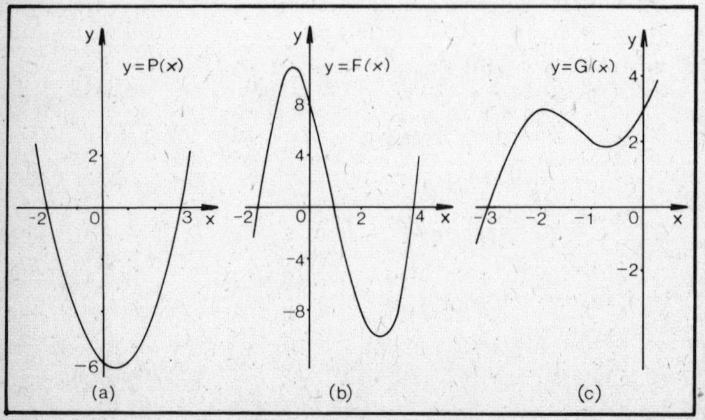

Figure 55

If we consider the equations $P(x) = 0$, $F(x) = 0$ and $G(x) = 0$ we have (only over the real field):

$$\begin{aligned} P(x) &= x^2 - x - 6 = (x-3)(x+2) \\ &= 0 \Rightarrow x = 3 \quad \text{or} \quad -2. \\ F(x) &= x^3 - 3x^2 - 6x + 8 = (x-1)(x+2)(x-4) \\ &= 0 \Rightarrow x = 1, -2 \quad \text{or} \quad 4. \\ G(x) &= x^3 + 4x^2 + 4x + 3 = (x+3)(x^2+x+1) \\ &= 0 \Rightarrow x = -3 \end{aligned}$$

These results can be illustrated graphically for the solutions give the cutting points on the x-axis. For $P(x)$ and $F(x)$ there are two and three distinct points respectively, whereas for $G(x)$ there is only one real point (see figure 55).

Thus we see that a quadratic may have 0 or 2 real roots and a cubic may have 1 or 3 real roots (it cannot have two real roots since this would necessitate there being a third). It would be much more convenient if we could simply say that every quadratic has two roots, every cubic has three roots and, generally, every equation of degree n has n roots. To do this we have to consider equations such as $x^2 + 4 = 0$ and $x^2 + x + 1 = 0$ which have no real solutions.

We define a new number, denoted by j where $j^2 = -1$, i.e. $j = \sqrt{-1}$. If this new number is combined with the real numbers we form a more general number in the form $a + bj$ where a and b are real numbers. Such a number is called a **complex number**. Remember that a and b may be negative, fractional or irrational, but must be real.
Thus $3 + 2j$, $4 - 7j$, $-\frac{1}{2} + \frac{1}{3}j$ are examples of complex numbers.
To return to the equation $x^2 + 4 = 0$. This can be written as

$$x^2 = -4 \text{ or } x^2 = (-1)(4) \Rightarrow x = \sqrt{-1} \times \sqrt{4} = \pm 2j.$$

Similarly we can now find solutions for the equation $x^2 + x + 1 = 0$ but we need the formula solution. The solution of $ax^2 + bx + c = 0$ is $x = \dfrac{-b \pm \sqrt{b^2 - 4ac}}{2a}$, and hence, in this case $a = 1$, $b = 1$, $c = 1$.

$$\therefore \qquad x = \frac{-1 \pm \sqrt{1 - 4(1)(1)}}{2} \Rightarrow x = \tfrac{1}{2}(-1 \pm \sqrt{-3}).$$

The solutions of the equation $x^2 + x + 1 = 0$ are $x = \frac{1}{2}(-1 \pm \sqrt{3}j)$.

Example 1 Solve completely the equation

$$x^3 + 2x^2 - 3x - 10 = 0.$$

Let $f(x) = x^3 + 2x^2 - 3x - 10$.
$f(1) = 1 + 2 - 3 - 10 \neq 0 \quad \therefore\ x - 1$ is not a factor.
$f(2) = 8 + 8 - 6 - 10 = 0 \quad \therefore\ x - 2$ is a factor.
The quadratic factor can be found by division.

$$\begin{array}{r}
x^2 + 4x + 5 \\
x - 2\overline{)x^3 + 2x^2 - 3x - 10} \\
\underline{x^3 - 2x^2} \qquad\qquad \\
4x^2 - 3x \qquad \\
\underline{4x^2 - 8x} \qquad \\
5x - 10 \\
5x - 10
\end{array}$$

Hence $x^3 + 2x^2 - 3x - 10 = (x - 2)(x^2 + 4x + 5) = 0$.

$\therefore \quad x - 2 = 0 \quad \text{or} \quad x^2 + 4x + 5 = 0.$

$$\therefore \quad x = 2 \quad \text{or by the formula} \quad x = \frac{-4 \pm \sqrt{16 - 20}}{2}$$

$$= \tfrac{1}{2}(-4 \pm 2j) = -2 \pm j.$$

Hence the roots of $x^3 + 2x^2 - 3x - 10 = 0$ are $x = 2,\ -2 \pm j$.

The algebra of complex numbers

As with all algebras, we need to define our rules for combining elements (cf. the algebra of vectors, page 194, and the algebra of matrices, page 125). We adopt the same operations for complex numbers as for real numbers, i.e. addition, subtraction, multiplication and division. We apply the same rules of algebra to give:

$$\begin{aligned}
(a + bj) + (c + dj) &= (a + c) + (b + d)j \\
(a + bj) - (c + dj) &= (a - c) + (b - d)j \\
(a + bj)(c + dj) &= ac + adj + bcj + bdj^2 \\
&= (ac - bd) + (ad + bc)j \quad \text{since } j^2 = -1.
\end{aligned}$$

Before considering division we need to note a useful result.

Conjugate complex numbers

If $z = a + bj$ is a given complex number, then the number $\bar{z} = a - bj$ is called its **conjugate**. Note that the conjugate is obtained by changing the sign of the imaginary part of the complex number.

The product of two complex conjugates is always real.

$$(p + qj)(p - qj) = p^2 - q^2j^2 = p^2 + q^2.$$

We can use this to define the operation of division. Consider $\frac{(a + bj)}{(c + dj)}$ and multiply the numerator and denominator by $(c - dj)$ (the conjugate of the denominator) to give

$$\frac{a + bj}{c + dj} = \frac{(a + bj)(c - dj)}{(c + dj)(c - dj)} = \frac{(ac + bd) + j(bc - ad)}{c^2 + d^2}$$

$$= \left(\frac{ac + bd}{c^2 + d^2}\right) + \left(\frac{bc - ad}{c^2 + d^2}\right) j.$$

This is clearly another complex number in the form $x + yj$. Hence each of the standard rules of algebra can be applied to combine complex numbers, each giving a new complex number in the form $a + bj$.

Example 2 Evaluate the following:

(i) $(3 + 2j) + (4 - 3j) = 7 - j,$
(ii) $(2 - 3j) - (5 - 6j) = -3 + 3j,$
(iii) $(4 + 3j)(2 - 3j) = 8 - 12j + 6j - 9j^2 = 17 - 6j,$
(iv) $\frac{(2 + 3j)}{(1 - j)} = \frac{(2 + 3j)(1 + j)}{(1 - j)(1 + j)} = \frac{2 + 5j + 3j^2}{2} = \frac{-1}{2} + \frac{5j}{2}.$

When a complex number is written in the form $x + yj$, x and yj are known as the real and imaginary parts respectively. Two complex numbers $a + bj$ and $c + dj$ are equal if and only if, $a = c$ and $b = d$, i.e. if the real and imaginary parts are equal.

Remember that $j^3 = j^2 \times j = -j,$
$j^4 = j^2 \times j^2 = (-1)(-1) = 1,$
$j^5 = j^4 \times j = 1 \times j = j$ and so on.

Example 3 Evaluate $(2 + 3j)^4$.

We can use the binomial expansion (see page 157) to give

$$\begin{aligned}(2 + 3j)^4 &= 2^4 + 4 \times 2^3(3j) + 6 \times 2^2(3j)^2 + 4 \times 2(3j)^3 + (3j)^4\\ &= 16 + 96j + 216j^2 + 216j^3 + 81j^4\\ &= 16 + 96j - 216 - 216j + 81 = -119 - 120j.\end{aligned}$$

The Argand diagram

Since any complex number $a + bj$ is determined by the values of a and b, we can uniquely associate an ordered pair (a, b) with the complex number $a + bj$. The order is important since $a + bj \neq b + aj$. Thus, we can represent a complex number $a + bj$ by a point in a cartesian plane whose coordinates are (a, b). This representation is called an **Argand diagram** (after J. R. Argand) and the plane is called the **Argand plane** or **complex plane**.

Notice the similarity between the correspondence here of a complex number and a point in a plane and the one-to-one correspondence between the real numbers and a point on a straight line.

Figure 56 shows an Argand diagram with the points A, B and C representing the complex numbers $3 + 2j$, $-4 + 3j$ and $2 - 2j$ respectively.

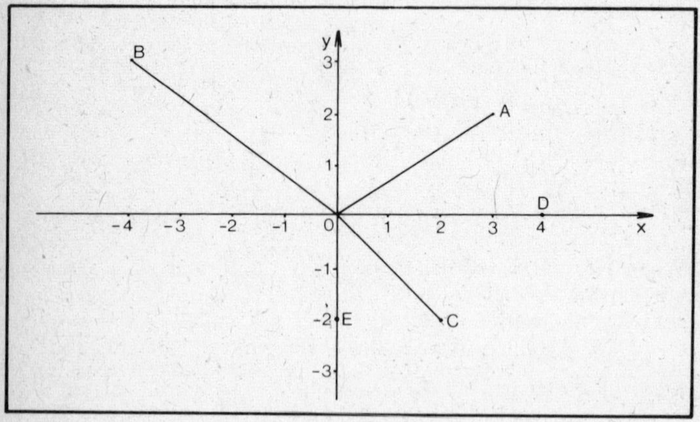

Figure 56

On the diagram, point D represents the complex number $4 + 0j$ (i.e. the real number 4) and point E represents the complex number $0 - 2j$ (i.e. $-2j$). Thus real numbers are represented by points on the x-axis and complex numbers in which the real part is zero are represented by points on the y-axis.

Notice that each point in the plane has an associated radius vector. The radius vectors in figure 56 above are **OA**, **OB**, **OC**, **OD** and **OE**.

Consider figure 57 in which a complex number $x + yj$ is represented by point P. Its radius vector is **OP**.

The magnitude of **OP** (usually denoted by r) is called the modulus of the complex number $x + yj$.
Thus $r = \sqrt{x^2 + y^2}$. A special notation is used for the modulus, namely $|x + yj|$.
Hence

$$|x + yj| = r = \sqrt{x^2 + y^2}.$$

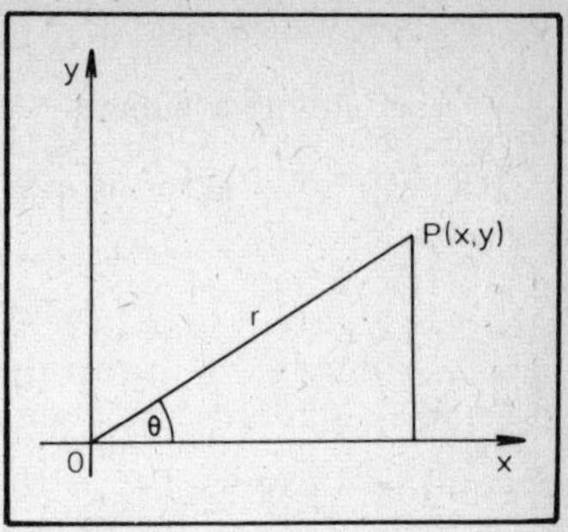

Figure 57

The angle that **OP** makes with the real axis is called the **argument** of the complex number $x + yj$. This is abbreviated to $\arg(x + yj)$. Since this is a many valued function we take, by convention, the principal value of the argument so that

$$-\pi < \arg(x + yj) \leq \pi.$$

In figure 58 three complex numbers have been represented by points in the Argand diagram. Their associated radius vectors are drawn from the origin.
In each case the modulus and argument has been calculated. We can adopt the technique used in vectors by using a single letter to describe a complex number, e.g. $z = 3 + 4j$.

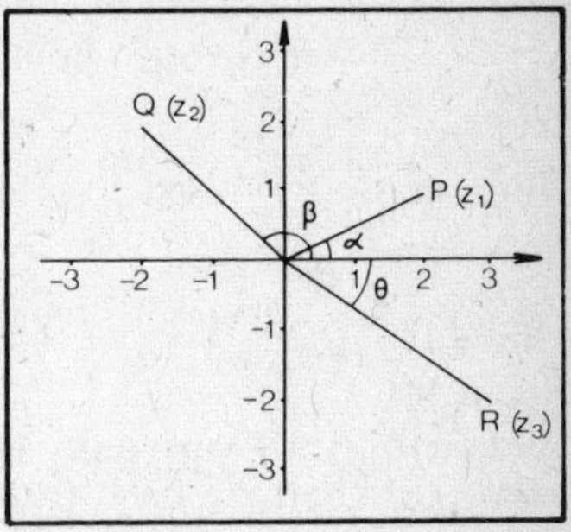

Figure 58

(i) $z_1 = 2 + j \quad \therefore |z_1| = \sqrt{2^2 + 1^2} = \sqrt{5}$

$$\arg z_1 = \alpha = \tan^{-1}(\tfrac{1}{2}) = 26^\circ\ 34'.$$

(ii) $z_2 = -2 + 2j \quad \therefore |z_2| = \sqrt{(-2)^2 + 2^2} = 2\sqrt{2}$

$$\arg z_2 = \beta = \pi - \tan^{-1}(1) = \frac{3\pi}{4} = 135^\circ.$$

(iii) $z_3 = 3 - 2j \quad \therefore |z_3| = \sqrt{3^2 + (-2)^2} = \sqrt{13}$

$$\arg z_3 = \theta = \tan^{-1}(-\tfrac{2}{3}) = -33^\circ\ 41'.$$

Since any complex number $z = x + yj$ has a modulus, r, given by $\sqrt{x^2 + y^2}$ and an argument θ given by $\tan^{-1}\left(\frac{y}{x}\right)$ it follows that z can be expressed in terms of r and θ, i.e.

$$z = x + yj = r(\cos\theta + \sin\theta j),$$

where $x = r\cos\theta$ and $y = r\sin\theta$ (see figure 57 on page 171). This is known as the **modulus-argument** or **polar form** of a complex number.

Example 4 Express (i) $z = 1 + j$ and (ii) $z = 1 - j$ in polar form.

(i) Since $z = 1 + j$, $|z| = \sqrt{2}$ and $\arg z = \theta = \tan^{-1}(1) = \frac{\pi}{4}$.

$\therefore$ $1 + j$ can be written in the form $\sqrt{2}\left\{\cos\left(\frac{\pi}{4}\right) + j\sin\left(\frac{\pi}{4}\right)\right\}$.

(ii) For $z = 1 - j$, $|z| = \sqrt{2}$ and $\arg z = \alpha = \tan^{-1}(-1) = -\frac{\pi}{4}$.

$\therefore$ $1 - j$ can be written in the form $\sqrt{2}\left\{\cos\left(-\frac{\pi}{4}\right) + j\sin\left(-\frac{\pi}{4}\right)\right\}$.

But $\cos\left(-\frac{\pi}{4}\right) = \cos\left(\frac{\pi}{4}\right)$ and $\sin\left(-\frac{\pi}{4}\right) = -\sin\left(\frac{\pi}{4}\right)$

$\therefore$ $1 - j = \sqrt{2}\left\{\cos\left(\frac{\pi}{4}\right) - \sin\left(\frac{\pi}{4}\right)\right\}$.

Geometrical representation for the addition and subtraction of two complex numbers

Consider figure 59 which shows two complex numbers z_1 and z_2 with their associated radius vectors **OP** and **OQ**, respectively.

Let R be the fourth vertex of the completed parallelogram OPRQ. If $z_1 = x_1 + y_1 j$ and $z_2 = x_2 + y_2 j$, then P is the point (x_1, y_1) and Q the point (x_2, y_2). Since the diagonals bisect each other, L has coordinates $(\frac{1}{2}(x_1 + x_2), \frac{1}{2}(y_1 + y_2))$.

$\therefore$ The coordinates of R are $((x_1 + x_2), (y_1 + y_2))$ and hence the point R represents the complex number $(x_1 + x_2) + (y_1 + y_2)j$, i.e. $z_1 + z_2$.
Note that the length of **OR** $= |z_1 + z_2|$.

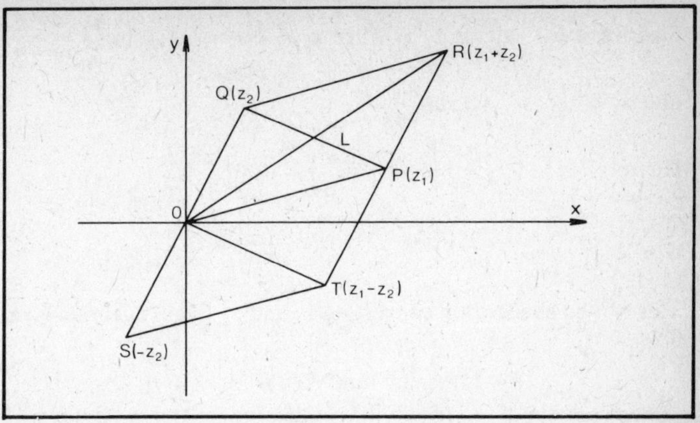

Figure 59

In order to subtract two complex numbers, say $z_1 - z_2$, we use the fact that $z_1 - z_2 = z_1 + (-z_2)$.
Now $(-z_2)$ is represented by the point S in figure 59 and hence if this is added to z_1 we complete the parallelogram OSTP and point T represents $z_1 - z_2$.

Note that the length of **OT** = $|z_1 - z_2|$.
Since in figure 59 **OT** = **QP** (opposite sides of parallelogram OTPQ) we have the length of **QP** = $|z_1 - z_2|$, that is, in the original parallelogram OPRQ the length of longer diagonal, **OR**, represents $|z_1 + z_2|$ and the length of shorter diagonal, **QP**, represents $|z_1 - z_2|$.

Multiplication and division in polar form

Consider two complex numbers written in modulus-argument form.

$$z_1 = r_1(\cos\theta + j\sin\theta) \quad \text{and} \quad z_2 = r_2(\cos\phi + j\sin\phi).$$

$$\begin{aligned} \text{Now} \quad z_1 z_2 &= r_1 r_2(\cos\theta + j\sin\theta)(\cos\phi + j\sin\phi) \\ &= r_1 r_2\{\cos\theta\cos\phi - \sin\theta\sin\phi \\ &\quad + j(\cos\theta\sin\phi + \sin\theta\cos\phi)\}. \\ \text{Thus} \quad z_1 z_2 &= r_1 r_2\{\cos(\theta + \phi) + j\sin(\theta + \phi)\}. \end{aligned}$$

This product $z_1 z_2$ is in the standard polar form with a modulus $r_1 r_2$ and an argument $(\theta + \phi)$.
Hence to multiply two complex numbers in polar form, we multiply the moduli and add the arguments.
Note that it can be extended to any number of terms.

Example 5 Evaluate $z_1 z_2$ when $z_1 = 2\left(\cos\frac{\pi}{3} + j\sin\frac{\pi}{3}\right)$ and $z_2 = 4\left(\cos\frac{\pi}{6} + j\sin\frac{\pi}{6}\right)$

Hence $$z_1 z_2 = 8\left\{\cos\left(\frac{\pi}{3}+\frac{\pi}{6}\right) + j\sin\left(\frac{\pi}{3}+\frac{\pi}{6}\right)\right\}$$
$$= 8\left(\cos\frac{\pi}{2} + j\sin\frac{\pi}{2}\right) = 8j.$$

This can be checked by converting z_1 and z_2 to an algebraic form first.

$$z_1 = 1 + \sqrt{3}j \quad \text{and} \quad z_2 = 2(\sqrt{3} + j)$$
$$\therefore \quad z_1 z_2 = 2(1 + \sqrt{3}j)(\sqrt{3} + j) = 2\{(\sqrt{3} - \sqrt{3}) + j(3 + 1)\} = 8j.$$

Example 6 Evaluate $(1 + j)^7$.

This can be written in polar form as $(\sqrt{2})^7\left(\cos\frac{\pi}{4} + j\sin\frac{\pi}{4}\right)^7$, since its modulus is $\sqrt{2}$ and its argument is $\frac{\pi}{4}$.

$$\therefore \quad (1 + j)^7 = (\sqrt{2})^7\left(\cos\frac{7\pi}{4} + j\sin\frac{7\pi}{4}\right) = 8\sqrt{2}\left(\frac{1}{\sqrt{2}} - \frac{1}{\sqrt{2}}j\right).$$
$$\therefore \quad (1 + j)^7 = 8 - 8j.$$

For division of two complex numbers in polar form we make use of the complex conjugate.
If $z_1 = r_1(\cos\theta + j\sin\theta)$ and $z_2 = r_2(\cos\phi + j\sin\phi)$ we have:

$$\frac{z_1}{z_2} = \frac{r_1}{r_2}\frac{(\cos\theta + j\sin\theta)}{(\cos\phi + j\sin\phi)}.$$

Multiplying the numerator and denominator by the complex conjugate of the denominator, we obtain

$$\frac{z_1}{z_2} = \frac{r_1}{r_2}\frac{(\cos\theta + j\sin\theta)(\cos\phi - j\sin\phi)}{(\cos\phi + j\sin\phi)(\cos\phi - j\sin\phi)}$$
$$= \frac{r_1}{r_2}\frac{\cos\theta\cos\phi + \sin\theta\sin\phi + j(\sin\theta\cos\phi - \cos\theta\sin\phi)}{\cos^2\phi + \sin^2\phi}.$$

Thus $$\frac{z_1}{z_2} = \frac{r_1}{r_2}\{\cos(\theta - \phi) + j\sin(\theta - \phi)\}.$$

Hence to divide two complex numbers in polar form we divide the moduli and subtract the arguments. Remember that the order is important in division.

Example 7 Find the quotient $\frac{z_1}{z_2}$, where $z_1 = 1 - 2j$ and $z_2 = 2 + j$ by converting to polar form.

Since $z_1 = 1 - 2j$, $|z_1| = \sqrt{5}$ and $\arg z_1 = -63° 26'$.
Also $z_2 = 2 + j$, $|z_2| = \sqrt{5}$ and $\arg z_2 = 26° 34'$.

Hence $z_1 = \sqrt{5}\{\cos(-63° 26') + j \sin(-63° 26')\}$
and $z_2 = \sqrt{5}\{\cos(26° 34') + j \sin(26° 34')\}$.

$$\therefore \quad \frac{z_1}{z_2} = 1\{\cos(-90°) + j \sin(-90°)\} = -j.$$

This can be checked by algebraic division.

$$\frac{z_1}{z_2} = \frac{1 - 2j}{(2 + j)} = \frac{(1 - 2j)(2 - j)}{(2 + j)(2 - j)} = -\frac{5j}{5} = -j.$$

Geometrical representation for multiplication and division of two complex numbers

In order to find a point R representing the product of two complex numbers z_1 and z_2 where $z_1 = r_1(\cos\theta + j \sin\theta)$ and $z_2 = r_2(\cos\phi + j \sin\phi)$ we need a point whose polar coordinates are $(r_1 r_2, \theta + \phi)$.

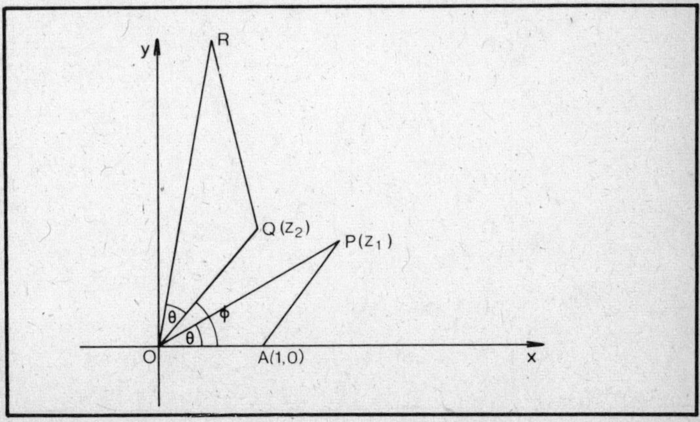

Figure 60

Figure 60 shows z_1 and z_2 with their associated radius vectors **OP** and **OQ** and the construction necessary to find R.

Take a point A(1, 0) on the real axis and complete the triangle OAP. Now construct triangle OQR similar to triangle OAP. The vertex, R, of this triangle represents the complex number $z_1 z_2$, since its argument is clearly $(\theta + \phi)$ and its modulus is given by the length of the radius vector **OR**.

But the triangles are similar $\Rightarrow \dfrac{\text{OR}}{\text{OQ}} = \dfrac{\text{OP}}{\text{OA}} \Rightarrow \text{OR} = \dfrac{|z_1| \times |z_2|}{1}$

Hence, since $\text{OP} = |z_1| = r_1$ and $\text{OQ} = |z_2| = r_2$, $\text{OR} = r_1 r_2$.

For division, we need to find a point whose polar coordinates are $\left(\dfrac{r_1}{r_2}, \theta - \phi\right)$. Two complex numbers z_1 and z_2 are shown in figure 61 together with their associated radius vectors **OP** and **OQ**.

The construction can be completed by taking a point A(1, 0) as before and forming triangle OAS similar to triangle OQP.
The vertex, S, of this triangle represents the complex number $\dfrac{z_1}{z_2}$, since its argument is clearly $(\theta - \phi)$ and its modulus is given by the length of the radius vector **OS**.

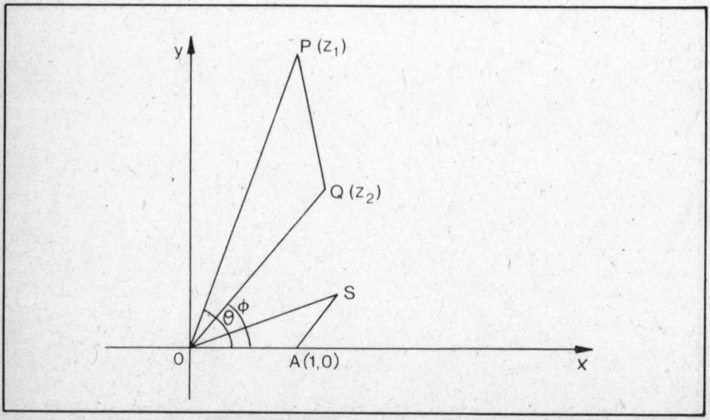

Figure 61

But the triangles are similar $\Rightarrow \dfrac{OS}{OA} = \dfrac{OP}{OQ} \Rightarrow OS = \dfrac{|z_1|}{|z_2|} \times 1.$

Hence, since $OP = |z_1| = r_1$ and $OQ = |z_2| = r_2$, $OS = \dfrac{r_1}{r_2}$.

Note that figure 60 is, in fact, illustrating the product of

$$z_1 = 2(\cos 30° + j \sin 30°) \quad \text{and} \quad z_2 = \tfrac{3}{2}(\cos 50° + j \sin 50°).$$

Similarly, figure 61 shows the quotient of $z_1 = 3(\cos 70° + j\sin 70°)$ and $z_2 = 2(\cos 50° + j \sin 50°)$.

De Moivre's theorem

This states that for any integral or rational value of n

$$(\cos \theta + j \sin \theta)^n = \cos n\theta + j \sin n\theta.$$

The proof is in three stages.

(i) Use induction for the case when n is a positive integer. Assume that the result is true for a value $n = k$, i.e.

$$\begin{aligned}
(\cos \theta + j \sin \theta)^k &= \cos k\theta + j \sin k\theta \\
\therefore \quad (\cos \theta + j \sin \theta)^{k+1} &= (\cos k\theta + j \sin k\theta)(\cos \theta + j \sin \theta) \\
&= (\cos k\theta \cos \theta - \sin k\theta \sin \theta) \\
&\quad + j(\sin k\theta \cos \theta + \cos k\theta \sin \theta) \\
&= \cos(k + 1)\theta + j \sin(k + 1)\theta.
\end{aligned}$$

Thus if the result is true for $n = k$, it also holds for $n = k + 1$. But the result is clearly true for $n = 1$ and therefore for $n = 2$ and so for any positive integral value of n.

(ii) If n is a negative integer, let $n = -m$ where m is a positive integer.

$$\begin{aligned}
\text{Thus } (\cos \theta + j \sin \theta)^n = (\cos \theta + j \sin \theta)^{-m} &= \frac{1}{(\cos \theta + j \sin \theta)^m} \\
&= \frac{1}{(\cos m\theta + j \sin m\theta)},
\end{aligned}$$

since m is a positive integer. Multiply numerator and denominator by $\cos m\theta - j \sin m\theta$.

$$\begin{aligned}
\therefore \quad (\cos \theta + j \sin \theta)^n &= \frac{\cos m\theta - j \sin m\theta}{1} \\
&= \cos(-m\theta) + j \sin(-m\theta) \\
&= \cos n\theta + j \sin n\theta.
\end{aligned}$$

Thus the result holds if n is a negative integer.

(iii) If n is a rational value, let $n = \frac{p}{q}$, where p and q are integral and coprime (i.e. have no common factor). Assume that q is positive.

Now as $q \to 0$, $\left\{\cos\left(\frac{\theta}{q}\right) + j\sin\left(\frac{\theta}{q}\right)\right\}^q = \cos\theta + j\sin\theta$,

i.e. $\cos\left(\frac{\theta}{q}\right) + j\sin\left(\frac{\theta}{q}\right)$ is one qth root of $\cos\theta + j\sin\theta$, or

$\cos\left(\frac{\theta}{q}\right) + j\sin\left(\frac{\theta}{q}\right)$ is one value of $(\cos\theta + j\sin\theta)^{1/q}$.

Taking the pth power of each side:

$\cos\left(\frac{p\theta}{q}\right) + j\sin\left(\frac{p\theta}{q}\right)$ is one value of $(\cos\theta + j\sin\theta)^{p/q}$.

Hence we see that De Moivre's theorem holds for any positive or negative rational value of n.

The cube roots of ±1

The solution of $z^3 - 1 = 0$ over the complex numbers gives three roots, i.e.

$$z = 1,\ z = \cos\left(\frac{2\pi}{3}\right) + j\sin\left(\frac{2\pi}{3}\right),\ z = \cos\left(\frac{4\pi}{3}\right) + j\sin\left(\frac{4\pi}{3}\right),$$

$$\text{or} \quad z = 1, \quad \frac{1}{2} + \left(\frac{\sqrt{3}}{2}\right)j \quad \text{and} \quad -\frac{1}{2} - \left(\frac{\sqrt{3}}{2}\right)j.$$

The cube roots of unity are $z = 1,\ -\tfrac{1}{2}(1 + \sqrt{3}j),\ -\tfrac{1}{2}(1 - \sqrt{3}j)$, and are usually denoted by $1, \omega_1, \omega_2$.

We can also derive these results in this special case by solving the algebraic equation $z^3 - 1 = (z - 1)(z^2 + z + 1) = 0$.

$$\therefore \quad z = 1 \quad \text{or} \quad z^2 + z + 1 = 0 \Rightarrow z = \frac{-1 \pm \sqrt{1 - 4}}{2}$$

$$= -\tfrac{1}{2}(1 \pm \sqrt{3}j).$$

i.e. $1, \omega_1, \omega_2$ as before.

Note that $\omega_1^2 = \{-\tfrac{1}{2}(1 + \sqrt{3}j)\}^2 = \tfrac{1}{4}(1 - 3 + 2\sqrt{3}j)$

$$= -\tfrac{1}{2}(1 - \sqrt{3}j) = \omega_2$$

Likewise $\omega_2^2 = \omega_1$ and so the cube roots of unity are usually given as 1, ω and ω^2.

Since ω is a root of $z^3 = 1$, we observe, by considering the roots of a cubic equation, that

(i) $\omega^3 = 1$, since the product of the roots is 1.
(ii) $1 + \omega + \omega^2 = 0$, since the sum of the roots $= 0$ (the coefficient of the term z^2).

Similarly the cube roots of -1 can be found by solving the equation $z^3 + 1 = (z+1)(z^2 - z + 1) = 0$.

$$\Rightarrow z = -1 \quad \text{or} \quad z^2 - z + 1 = 0 \Rightarrow z = \tfrac{1}{2}(1 \pm \sqrt{3}j).$$

Hence the roots are $-1, \tfrac{1}{2}(1 \pm \sqrt{3}j)$.

Trigonometrical identities

We can use De Moivre's to prove some trigonometrical identities by making use of the fact that if two complex numbers are equal, the real and imaginary parts of each can be equated, i.e.

$$a + bj = c + dj \Rightarrow a = c \quad \text{and} \quad b = d.$$

Example 8 Prove that (i) $\cos 3\theta = 4\cos^3\theta - 3\cos\theta$ and (ii) $\sin 3\theta = 3\sin\theta - 4\sin^3\theta$.

Now by De Moivre's Theorem

$$\begin{aligned}\cos 3\theta + j\sin 3\theta &= (\cos\theta + j\sin\theta)^3 \\ &= \cos^3\theta + 3j\cos^2\theta\sin\theta \\ &\quad + 3j^2\cos\theta\sin^2\theta + j^3\sin^3\theta \\ &= \cos^3\theta - 3\cos\theta\sin^2\theta \\ &\quad + j(3\cos^2\theta\sin\theta - \sin^3\theta).\end{aligned}$$

Thus equating real and imaginary parts:

$$\cos 3\theta = \cos^3\theta - 3\cos\theta\sin^2\theta, \qquad (1)$$

and

$$\sin 3\theta = 3\cos^2\theta\sin\theta - \sin^3\theta. \qquad (2)$$

Using the identity $\cos^2\theta + \sin^2\theta = 1$ we have

$$\cos 3\theta = 4\cos^3\theta - 3\cos\theta \quad \text{and} \quad \sin 3\theta = 3\sin\theta - 4\sin^3\theta.$$

Notice that an expression for $\tan 3\theta$ can be obtained by dividing the result (2) by (1), i.e.

$$\tan 3\theta = \frac{3\cos^2\theta\sin\theta - \sin^3\theta}{\cos^3\theta - 3\cos\theta\sin^2\theta} = \frac{3\tan\theta - \tan^3\theta}{1 - 3\tan^2\theta}$$

after dividing by $\cos^3\theta$.

Powers of $\sin\theta$ and $\cos\theta$ in terms of multiple angles

The above process is best when expressing the sine or cosine of a multiple angle in terms of powers of sine or cosine, but an alternative method is suitable when expressing powers in terms of multiple angles.

By De Moivre's theorem,

$$z = \cos\theta + j\sin\theta \Rightarrow z^n = \cos n\theta + j\sin n\theta,$$

and $\quad z^{-1} = \dfrac{1}{z} = \cos\theta - j\sin\theta \Rightarrow z^{-n} = \dfrac{1}{z^n} = \cos n\theta - j\sin n\theta$

Hence

$$z + \frac{1}{z} = 2\cos\theta \tag{3}$$

$$z - \frac{1}{z} = 2j\sin\theta \tag{4}$$

$$z^n + \frac{1}{z^n} = 2\cos n\theta \tag{5}$$

$$z^n - \frac{1}{z^n} = 2j\sin n\theta \tag{6}$$

Example 9 Prove that

$$\cos^6\theta = \tfrac{1}{32}(\cos 6\theta + 6\cos 4\theta + 15\cos 2\theta + 10).$$

Using relation (3): $(2\cos\theta)^6 = \left(z + \dfrac{1}{z}\right)^6,$

$$\therefore \quad 64\cos^6\theta = z^6 + 6z^4 + 15z^2 + 20 + \frac{15}{z^2} + \frac{6}{z^4} + \frac{1}{z^6}$$

$$= \left(z^6 + \frac{1}{z^6}\right) + 6\left(z^4 + \frac{1}{z^4}\right) + 15\left(z^2 + \frac{1}{z^2}\right) + 20.$$

Using (5)

$$64\cos^6\theta = 2\cos 6\theta + 6(2\cos 4\theta) + 15(2\cos 2\theta) + 20.$$

$$\therefore \quad \cos^6\theta = \tfrac{1}{32}(\cos 6\theta + 6\cos 4\theta + 15\cos 2\theta + 10).$$

Key terms

A complex number is of the form $a + bj$, where a and b are real. If $z = a + bj$, then $\bar{z} = a - bj$ is the complex conjugate of z.

Argand diagram Any complex number $x + yj$ can be represented by a point in a cartesian plane whose coordinates are (a, b). A complex number can be written in modulus-argument form $z = r(\cos\theta + j\sin\theta)$.

The modulus of $z = x + yj$ is $|z| = \sqrt{x^2 + y^2}$.

The argument of z is given by $\arg z = \tan^{-1}(y/x)$.

De Moivre's theorem

$(\cos\theta + j\sin\theta)^n = \cos n\theta + j\sin n\theta \quad$ for n rational.

Chapter 10
Differential Equations

Differential equations are equations in the usual sense but with the additional characteristic that at least one term contains a differential coefficient.
The following equations are all differential equations:

(i) $\dfrac{dy}{dx} = 2x^2 + 3$, (ii) $(1 + x)^2 \dfrac{dy}{dx} + y^2 = 1$,

(iii) $\dfrac{d^2x}{dt^2} - 3\dfrac{dx}{dt} + 2x = 0$.

The **order** of the differential equation is that of the highest derivative involved. Hence equations (i) and (ii) are first order and equation (iii) is second order.
The **degree** of a differential equation is the power of the highest derivative involved. Hence the equations above are all of the first degree. The equation $\left(\dfrac{dx}{dt}\right)^2 = \omega^2(a^2 - x^2)$ is of second degree.

Now consider a very simple differential equation $\dfrac{dy}{dx} = 4$. Clearly this is obtained from the equation $y = 4x + c$ which is the equation of all straight lines of gradient 4. We have obtained the result $y = 4x + c$ by direct integration which is possible in this trivial case. The result $y = 4x + c$ is called the **general solution** of the differential equation $\dfrac{dy}{dx} = 4$, i.e. the solution represents a family of lines all having a specific gradient. If some further data is given, usually called **boundary or initial conditions**, it is possible to determine the value of c.

For example, if $x = 1$ when $y = 7$, then substituting in the general solution we have $7 = 4(1) + c \Rightarrow c = 3$. Thus in this case the **particular solution** is $y = 4x + 3$. Compare the process shown on page 67 concerning basic integration. We can see that

(i) a **differential equation** defines some property common to a family of curves,

(ii) the **general solution** is the equation representing any member of the family and this must contain arbitrary constants equal in number to the order,

(iii) the **particular solution** gives the equation of one specific member of the family and does not contain any arbitrary constants.

We now consider the various types of differential equation and the methods for solving them.

First-order equations

Type 1: Equations of the form $\dfrac{dy}{dx} = f(x)$. These can be solved by straightforward integration.

Example 1 Solve the differential equation $\dfrac{dy}{dx} = x^3 + \cos x + 2.$

Thus the general solution is $y = \frac{1}{4}x^4 + \sin x + 2x + c.$

Type 2: Equations of the form $\dfrac{dy}{dx} = f(y)$. This can be written $\dfrac{1}{f(y)}\dfrac{dy}{dx} = 1 \Rightarrow \dfrac{dx}{dy} = \dfrac{1}{f(y)}$. Integrating with respect to y

$$\int \frac{1}{f(y)}\,dy = \int 1\,dx, \qquad \text{i.e.} \quad \int \frac{1}{f(y)}\,dy = x + c.$$

Example 2 Solve the equation $\dfrac{dy}{dx} = 2 \sec y.$

Rearranging $\dfrac{1}{\sec y}\dfrac{dy}{dx} = 2 \Rightarrow \cos y \dfrac{dy}{dx} = 2.$

Integrating with respect to x: $\displaystyle\int \cos y\,dy = \int 2\,dx.$

The general solution is $\sin y = 2x + c.$

Type 3: Variables separable. This type of equation contains variables in x and y but it is possible to separate them into terms, one a function of x only and the other a function of y only. This would be in the form

$$f(y)\frac{dy}{dx} + g(x) = 0.$$

Integrating with respect to x yields

$$\int f(y)\,dy + \int g(x)\,dx = 0.$$

Example 3 Find the general solution of the equation $(x+1)\dfrac{dy}{dx} - y = 0$, and sketch the family of solution curves. If $y = 3$ when $x = 0$, give the particular solution.

Separating the variables: $\dfrac{1}{y}\dfrac{dy}{dx} = \dfrac{1}{x+1}$.

Integrating with respect to x gives: $\displaystyle\int \frac{1}{y}\,dy = \int \frac{1}{x+1}\,dx.$

$\Rightarrow \ln y = \ln c(x+1)$ writing the constant $= \ln c$.

The general solution is $y = c(x+1)$.

When $x = 0$, $y = 3$, hence $3 = c(0+1) \Rightarrow c = 3$.
The particular solution is $y = 3x + 3$.

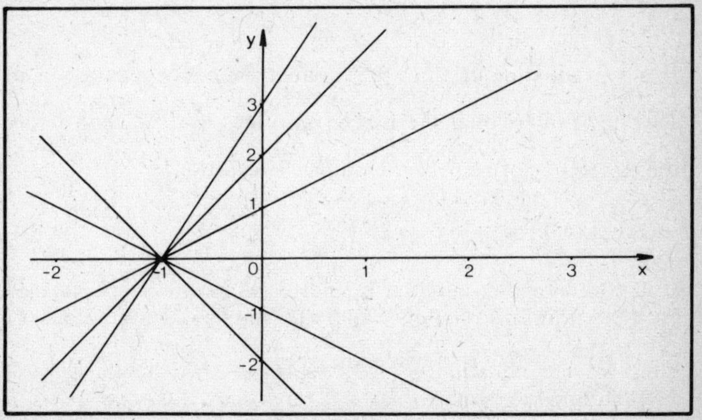

Figure 62

Figure 62 shows the family of solutions giving the cases when $c = -2, -1, 0, 1, 2, 3$. Notice the particular solution $y = 3x + 3$ passes through $(0, 3)$.

Example 4 Solve the equation $\frac{dy}{dx} = \frac{y(y+1)}{2x}$.

Separating the variables: $\frac{1}{y(y+1)}\frac{dy}{dx} = \frac{1}{2x}$.

Integrating with respect to x: $\int \frac{1}{y(y+1)}\,dy = \int \frac{1}{2x}\,dx$.

$$\Rightarrow \int \left(\frac{1}{y} - \frac{1}{y+1}\right) dy = \int \frac{1}{2x}\,dx,$$

i.e. $\ln y - \ln(y+1) = \frac{1}{2}\ln x + c$.

$$\ln\left(\frac{y}{y+1}\right) = \ln\sqrt{x} + c.$$

$$\ln\left(\frac{y}{y+1}\right) = \ln k\sqrt{x} \text{ writing } c \text{ as } \ln k.$$

Hence the general solution is $\frac{y}{y+1} = k\sqrt{x}$.

Type 4: Exact equations. Consider the differential equation $3xy^2\frac{dy}{dx} + y^3 = e^x$.

This is an equation in which the variables cannot be separated, but it will be noticed that the L.H.S. is, in fact, $\frac{d}{dx}(xy^3)$ and hence the equation can be solved by direct integration.

i.e. $\frac{d}{dx}(xy^3) = e^x \Rightarrow xy^3 = e^x + c$.

Thus a differential equation is said to be **exact** if it is formed from the primitive relating x and y by simple differentiation.

Thus for the equation $x^4 + 3y^4 + 4xy = c$ we obtain by differentiation with respect to x:

$$4x^3 + 12y^3\frac{dy}{dx} + 4x\frac{dy}{dx} + 4y = 0,$$

$$\Rightarrow (3y^3 + x)\frac{dy}{dx} + x^3 + y = 0.$$

This can be written in the form $(x^3 + y) + (3y^3 + x)\dfrac{dy}{dx} = 0$, and is an exact differential equation.

Example 5 Solve the equation $(x^3 + y) + (3y^3 + x)\dfrac{dy}{dx} = 0$.

We can rewrite the equation as

$$x^3 + \left(y + x\frac{dy}{dx}\right) + 3y^3\frac{dy}{dx} = 0,$$

$$\Rightarrow x^3 + \frac{d}{dx}(xy) + \frac{d}{dx}\left(\frac{3y^4}{4}\right) = 0.$$

Integrating with respect to x:

$$\frac{x^4}{4} + xy + \frac{3y^4}{4} = k.$$

The general solution is $x^4 + 4xy + 3y^4 = c$.

Type 5: Equations requiring an integrating factor. Some differential equations are not exact as they stand but can be made exact by multiplying each side by a suitable factor, called the **integrating factor**.

Consider the equation $\dfrac{dy}{dx} + Py = Q$, where P and Q are constants or functions of x only. Such equations are called **linear differential equations.**

If the equation is not exact, assume that we can make it exact by multiplying by a function of x, say R.

Thus
$$R\frac{dy}{dx} + RPy = RQ. \tag{1}$$

Now
$$\frac{d}{dx}(Ry) = R\frac{dy}{dx} + y\frac{dR}{dx}.$$

Thus equation (1) can be written

$$\frac{d}{dx}(Ry) - y\frac{dR}{dx} + RPy = RQ,$$

i.e.
$$\frac{d}{dx}(Ry) + y\left(PR - \frac{dR}{dx}\right) = RQ. \tag{2}$$

Now choose R so that $\frac{dR}{dx} = PR$, i.e. the coefficient of $y = 0$.

$$\therefore \qquad \frac{1}{R}\frac{dR}{dx} = P \Rightarrow \ln R = \int P\,dx.$$

Therefore, R, the integrating factor is given by $R = e^{\int P\,dx}$ and equation (2) becomes

$$\frac{d}{dx}(e^{\int P\,dx}y) = Qe^{\int P\,dx},$$

which can be integrated. Note that the integrating factor itself does not contain any arbitrary constants.

Example 6 Solve the equation $\frac{dy}{dx} + 4y = e^{3x}$. (1)

The integrating factor is $e^{\int P\,dx} = e^{\int 4\,dx} = e^{4x}$.
Multiplying equation (1) by this factor we have:

$$e^{4x}\frac{dy}{dx} + 4e^{4x}y = e^{4x}e^{3x} = e^{7x}.$$

$$\therefore \qquad \frac{d}{dx}(e^{4x}y) = e^{7x}.$$

Integrating with respect to x:

$$ye^{4x} = \tfrac{1}{7}e^{7x} + c,$$

i.e.

$$y = \tfrac{1}{7}e^{3x} + ce^{-4x}.$$

Example 7 Solve the equation $\frac{dy}{dx} + y\cot x = \sec x$.

The integrating factor is $e^{\int P\,dx} = e^{\int \cot x\,dx} = e^{\ln \sin x} = \sin x$.

Hence, multiplying by $\sin x$ the equation becomes

$$\sin x\frac{dy}{dx} + y\cos x = \tan x,$$

$$\frac{d}{dx}(y\sin x) = \tan x.$$

Integrating with respect to x

$$y \sin x = \int \tan x \, dx$$

$$y \sin x = -\ln \cos x + k = \ln \sec x + k.$$

The general solution is $y = k \operatorname{cosec} x + \operatorname{cosec} x \ln \sec x$.

Type 6: Equations reducible to linear equations. A more general case of the linear equation given in type 5 is

$$\frac{dy}{dx} + Py = Qy^n. \tag{1}$$

This is known as **Bernoulli's equation** and it can be reduced to a linear form by a substitution $z = y^{1-n}$.

$$z = y^{1-n} \Rightarrow \frac{dz}{dx} = (1-n)y^{-n}\frac{dy}{dx} \Rightarrow \frac{dy}{dx} = \frac{y^n}{1-n}\frac{dz}{dx}.$$

Substituting in equation (1) gives

$$\frac{y^n}{1-n}\frac{dz}{dx} + Py = Qy^n,$$

$$\therefore \qquad \frac{1}{1-n}\frac{dz}{dx} + Py^{1-n} = Q.$$

$$\frac{1}{1-n}\frac{dz}{dx} + Pz = Q.$$

This is clearly a linear differential equation since $1 - n$ is a constant and hence it can be solved by use of the integrating factor.

Example 8 Solve the equation $\dfrac{dy}{dx} + \dfrac{y}{x} = 2x^3y^4$.

Use the substitution $z = y^{-3}$ (i.e. $n = 4$).

Hence $\dfrac{dz}{dx} = -3y^{-4}\dfrac{dy}{dx} \Rightarrow \dfrac{dy}{dx} = \dfrac{-y^4}{3}\dfrac{dz}{dx}$

Thus the given equation becomes

$$\frac{-y^4}{3}\frac{dz}{dx} + \frac{y}{x} = 2x^3y^4.$$

Dividing by y^4:

$$\frac{-1}{3}\frac{dz}{dx} + \frac{1}{x}y^{-3} = 2x^3,$$

or $$\frac{dz}{dx} - \frac{3z}{x} = -6x^3.$$

The integrating factor is $e^{\int -3/x\,dx} = e^{-3\ln x} = e^{\ln x^{-3}} = x^{-3}$. Multiplying through by x^{-3} we have:

$$x^{-3}\frac{dz}{dx} - 3x^{-4}z = -6,$$

i.e. $$\frac{d}{dx}(x^{-3}z) = -6.$$

Integrating with respect to x gives $x^{-3}z = -6x + c$,

i.e. $z = -6x^4 + cx^3$.

But $z = y^{-3}$, $\therefore$ $y^{-3} = -6x^4 + cx^3$.

The general solution is $y^3 = \dfrac{1}{(cx^3 - 6x^4)}$.

Second-order equations

Type 7: Equations of the form $\dfrac{d^2y}{dx^2} = f(x)$. These can be solved by direct integration and will clearly give rise to two arbitrary constants.

Example 9 Solve the equation $\dfrac{d^2y}{dx^2} = \cos x$.

By integration with respect to x:

$$\frac{dy}{dx} = \sin x + A \quad \text{and} \quad y = -\cos x + Ax + B.$$

Type 8: Equations of the form $\dfrac{d^2y}{dx^2} = f(y)$. To solve equations of this form we use a substitution. Let

$$p = \frac{dy}{dx} \Rightarrow \frac{d^2y}{dx^2} = \frac{dp}{dx} = \frac{dp}{dy}\frac{dy}{dx} = p\frac{dp}{dy}.$$

Hence the equation can be written as

$$p\frac{dp}{dy}=f(y).$$

Integrating with respect to y gives $\frac{1}{2}p^2 = \int f(y)\,dy + c$. This has reduced the equation to a first-order equation which can be solved by one of the methods previously described.

Example 10 Solve the equation $\dfrac{d^2y}{dx^2} = -2y$ given that when

$$x = 0, \quad y = 2 \quad \text{and} \quad \frac{dy}{dx} = 0.$$

Let $p = \dfrac{dy}{dx} \Rightarrow \dfrac{d^2y}{dx^2} = \dfrac{dp}{dx} = p\dfrac{dp}{dy}$ as above.

The equation becomes $p\dfrac{dp}{dy} = -2y.$

Integrating with respect to y gives $\frac{1}{2}p^2 = -y^2 + c.$ (1)

Now when $y = 2$, $\dfrac{dy}{dx} = 0$, hence $c = 4$.

Since $p = \dfrac{dy}{dx}$, equation (1) is $\dfrac{1}{2}\left(\dfrac{dy}{dx}\right)^2 = 4 - y^2.$

Thus $$\frac{1}{\sqrt{2}}\frac{dy}{dx} = \sqrt{4 - y^2}.$$

Separating the variables and integrating gives:

$$\int \frac{1}{\sqrt{4 - y^2}}\,dy = \int \sqrt{2}\,dx,$$

or $$\sin^{-1}\left(\frac{y}{2}\right) = \sqrt{2}x + c_1.$$

$\therefore$ $$y = 2\sin(\sqrt{2}\,x + c_1).$$

When $y = 2$, $x = 0 \Rightarrow c_1 = \dfrac{\pi}{2}$.

$\therefore$ $$y = 2\sin\left(\sqrt{2}\,x + \frac{\pi}{2}\right) = 2\cos\sqrt{2}\,x.$$

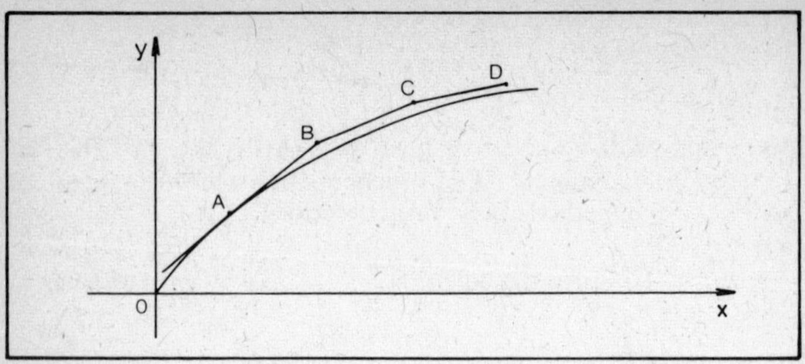

Figure 63

Numerical solution of differential equations

The equations so far have yielded exact solutions but there are many differential equations for which such solutions cannot be found. In these cases we use a numerical approximation process which produces the coordinates of points which **nearly** lie on the solution curve. This process is sometimes called a **step-by-step solution**.

Consider figure 63 which shows a solution curve to a given differential equation. We shall try to find points on this curve starting at a known point A. If, from A, we move along the tangent at A a displacement δx in the x direction then the increment in y, $\delta y \simeq \frac{dy}{dx} \times \delta x$.

This will define point B. We now move from B along a line whose direction is found by substituting the coordinates of B into the differential equation. Note that this is only approximately the same line as the tangent at B. Thus we shall determine the coordinates of C and by repetition the coordinates of D and so on. The following example will illustrate the technique.

Example 11 Solve the equation $\frac{dy}{dx} = 2x$ for $x = 0{\cdot}0(0{\cdot}1)0{\cdot}5$ given that $y = 4$ when $x = 0$.

The notation $x = 0{\cdot}0(0{\cdot}1)0{\cdot}5$ means that x takes the values from 0·0 to 0·5 in steps of 0·1.

The working is best done in tabular form

Interval	δx	$\frac{dy}{dx}$	δy	x	y
				0	4
$0{\cdot}0 \le x \le 0{\cdot}1$	0·1	0	0		
				0·1	4
$0{\cdot}1 \le x \le 0{\cdot}2$	0·1	0·2	0·02		
				0·2	4·02
$0{\cdot}2 \le x \le 0{\cdot}3$	0·1	0·4	0·04		
				0·3	4·06
$0{\cdot}3 \le x \le 0{\cdot}4$	0·1	0·6	0·06		
				0·4	4·12
$0{\cdot}4 \le x \le 0{\cdot}5$	0·1	0·8	0·08		
				0·5	4·20

By integration, the solution is $y = x^2 + c$ and the initial condition gives $c = 4$. The analytical solution is $y = x^2 + 4$. This would give $y = 4{\cdot}25$ when $x = 0{\cdot}5$. Clearly the accuracy of the numerical process depends upon the closeness of the tangent to the curve and the size of the increment δx. We can improve the accuracy of the method by using a quadratic approximation instead of a linear one. So instead of using $\delta y \simeq (dy/dx)\ \delta x = f'(x)\ \delta x$ we make use of the second term of the Taylor series, i.e. $\delta y \simeq f'(x)\,\delta x + \frac{1}{2}f''(x)(\delta x)^2$.

Example 12 Solve the differential equation $\frac{dy}{dx} = x + y$ for 0·0(0·4)2·0 given that when $x = 0$, $y = 1$.

We shall need an initial value for $\frac{d^2y}{dx^2} = 1 + \frac{dy}{dx}$.

$f'(x)$	$f''(x)$	δx	$f'(x)\,\delta x$	$\frac{1}{2}f''(x)(\delta x)^2$	δy	x	y
						0	1
1	2	0·4	0·4	0·16	0·56		
						0·4	1·56
1·96	2·96	0·4	0·784	0·2368	1·0208		
						0·8	2·5808
3·3808	4·3808	0·4	1·3523	0·3505	1·7028		
						1·2	4·2836
5·4836	6·4836	0·4	2·1934	0·5187	2·7121		
						1·6	6·9957
8·5957	9·5957	0·4	3·4383	0·7677	4·2060		
						2·0	11·2017

This equation can be solved by using the integrating factor, i.e. $dy/dx - y = x$. The integrating factor is $e^{\int -1\,dx} = e^{-x}$.

$$\therefore \quad e^{-x}\frac{dy}{dx} - e^{-x}y = e^{-x}x \Rightarrow \frac{d}{dx}(e^{-x}y) = xe^{-x}.$$

Integrating $ye^{-x} = \int xe^{-x}\,dx = -xe^{-x} + \int e^{-x}\,dx$ (by parts)

$$= -xe^{-x} - e^{-x} + C.$$

The general solution is $y = -x - 1 + Ce^x$.
But $x = 0$ when $y = 1$, $\therefore$ $C = 2$ and $y = 2e^x - x - 1$.

Comparing the values obtained from the numerical method with those obtained by the analytical method we can see the error involved. This can be seen from the table below, where values are given correct to two decimal places.

x	0·4	0·8	1·2	1·6	2·0
Numerical y	1·56	2·58	4·28	7·00	11·20
Analytical y	1·58	2·65	4·44	7·31	11·78

Key terms

First order $\frac{dy}{dx} = f(x)$ can be solved by direct integration.

$\frac{dy}{dx} = f(y) \Rightarrow \int \frac{1}{f(y)}\,dy = \int dx$ which can be integrated directly.

Variables separable These equations can be written in the form $f(y)\frac{dy}{dx} + g(x) = 0 \Rightarrow \int f(y)\,dy + \int g(x)\,dx = C$.

Exact equations A differential equation is exact if it is formed from the equation of the curve by direct differentiation.

Linear equations $\frac{dy}{dx} + Py = Q$ is called linear if P and Q are constants or functions of x only. They are solved by multiplying by $e^{\int P\,dx}$ which is called the integrating factor.

Second order

$\frac{d^2y}{dx^2} = f(x)$ is solved by direct integration.

$\frac{d^2y}{dx^2} = f(y)$ is solved by using the substitution $p = \frac{dy}{dx}$.

Chapter 11
Vectors

Many physical quantities, for example, velocities, accelerations and forces are vector quantities; they have direction as well as magnitude. It is not only necessary to know how large a force is, but also to know in which direction the force acts and even through which point it acts. Vector methods are so powerful that there are many applications in geometry and complex numbers and they comprise a part of pure mathematics on their own.

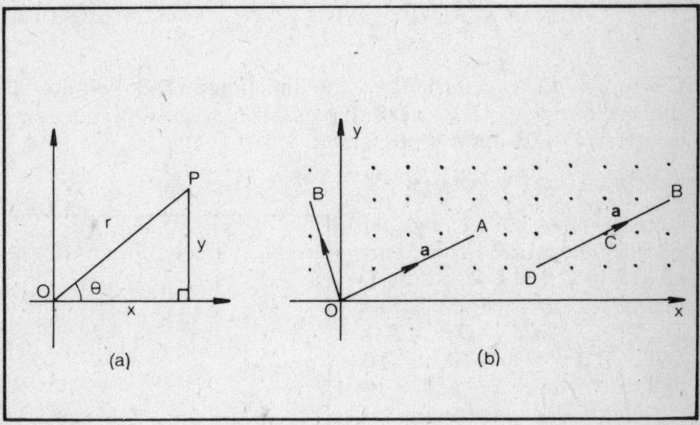

Figure 64

Two-dimensional vectors can be represented with the aid of cartesian or polar coordinates. In figure 64(a) the displacement OP can be represented by the 2 × 1 matrix $\begin{pmatrix} x \\ y \end{pmatrix}$ which is known as a column vector. **OP** could also be represented by its polar coordinates $[r, \theta]$ but this representation is only convenient when we multiply complex numbers (see chapter 9).

Two-dimensional vectors can be represented by displacements which can be represented by column vectors. In figure 64(b)

$$\mathbf{OA} = \begin{pmatrix} 4 \\ 2 \end{pmatrix}, \quad \mathbf{OB} = \begin{pmatrix} -1 \\ 3 \end{pmatrix}, \quad \mathbf{DB} = \begin{pmatrix} 4 \\ 2 \end{pmatrix}.$$

There are many notations used for vectors and it is wise to be aware of most of them. Vectors are denoted by bold face type **OA** or as directed line segments $\overrightarrow{OA}$ or sometimes by a single letter **a**. When writing them they must be underlined to distinguish them from numbers (which only have magnitude). **OA** and **OB** are the **position vectors** of the points A and B with respect to the origin O. In general, the vector **OA** = **DB**, but care must be taken if these vectors represent forces (as they act through different points).

Multiplication by a scalar (number)

In figure 64(b)

$$\mathbf{DB} = \begin{pmatrix}4\\2\end{pmatrix} = 2\begin{pmatrix}2\\1\end{pmatrix} = 2\mathbf{DC}.$$

DB represents a vector in the same direction as **DC** with double the magnitude of **DC**. k**DB** represents a vector in the same direction as **DB** but k times larger.

Addition of vectors

Vectors are added by the parallelogram law. From figure 65 **OA** + **OB** = **OD**, where D completes the parallelogram OADB. Since **OB** = **AD**, **OA** + **OB** = **OA** + **AD** = **OD** and vectors can be added by drawing them nose to tail.

$$\mathbf{OA} = \begin{pmatrix}3\\1\end{pmatrix}, \mathbf{OB} = \begin{pmatrix}1\\2\end{pmatrix},$$

$$\mathbf{OD} = \begin{pmatrix}4\\3\end{pmatrix}.$$

$$\begin{pmatrix}3\\1\end{pmatrix} + \begin{pmatrix}1\\2\end{pmatrix} = \begin{pmatrix}4\\3\end{pmatrix}.$$

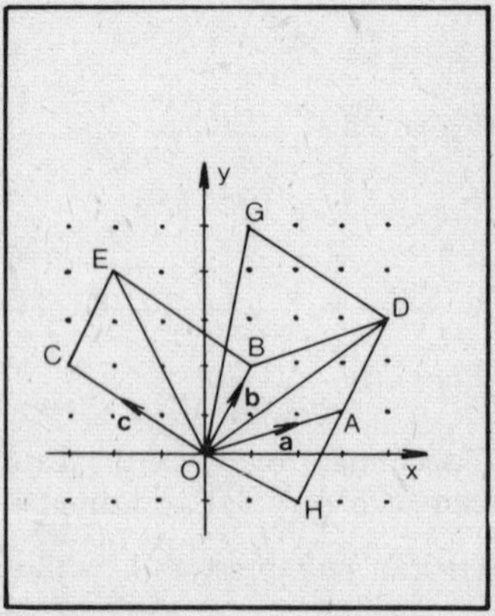

Figure 65

If **OA** = **a** and **OB** = **b** **OD** = **a** + **b** = **OB** + **BD** = **b** + **a**, so addition of vectors is **commutative**.

If **OC** = **c**, (**a** + **b**) + **c** = **OD** + **c** = **OD** + **DG** = **OG**.

a + (**b** + **c**) = **OA** + **OE** = **OA** + **AG** = **OG**.

(**a** + **b**) + **c** = **a** + (**b** + **c**) and vector addition is **associative**.

Subtraction of vectors

$$\mathbf{OB} = \mathbf{b} = \begin{pmatrix}1\\2\end{pmatrix}. \quad \mathbf{BO} = -\mathbf{b} = -\begin{pmatrix}1\\2\end{pmatrix} = \begin{pmatrix}-1\\-2\end{pmatrix}.$$

$$\mathbf{a} - \mathbf{b} = \mathbf{OA} - \mathbf{OB} = \mathbf{OA} + \mathbf{BO} = \mathbf{OA} + \mathbf{AH} = \mathbf{OH},$$

$$\mathbf{a} - \mathbf{b} = \begin{pmatrix}3\\1\end{pmatrix} - \begin{pmatrix}1\\2\end{pmatrix} = \begin{pmatrix}2\\-1\end{pmatrix}.$$

$\mathbf{a} - \mathbf{b} = -\mathbf{b} + \mathbf{a} = \mathbf{BO} + \mathbf{DA} = \mathbf{BA}$ and similarly $\mathbf{AB} = \mathbf{b} - \mathbf{a}$.

These results have been derived for two-dimensional vectors which are easy to represent with the aid of coordinates in the plane. Many of the results derived for the vectors in two dimensions are true in three dimensions. It is difficult to illustrate three-dimensional vectors in two-dimensions and you may find difficulty in visualising three-dimensional properties of figures, lines and planes. Consequently the study of vectors can appear theoretical, when it should be practically illustrated.

Ratio theorem

The vectors **a** and **b** define a plane containing O, A and B (figure 66). Any linear combination of **a** and **b** (e.g. $3\mathbf{a} + 2\mathbf{b}$) will define a vector lying in the same plane of OAB.

If P divides AB in the ratio $\lambda : \mu$ so that $\dfrac{AP}{PB} = \dfrac{\lambda}{\mu}$, then

$$\mathbf{OP} = \mathbf{p} = \mathbf{a} + \mathbf{AP}$$

$$= \mathbf{a} + \frac{\lambda}{\lambda + \mu}\mathbf{AB}$$

$$= \mathbf{a} + \frac{\lambda}{\lambda + \mu}(\mathbf{b} - \mathbf{a})$$

$$= \left(1 - \frac{\lambda}{\lambda + \mu}\right)\mathbf{a} + \frac{\lambda}{\lambda + \mu}\mathbf{b}.$$

So $\quad \mathbf{p} = \dfrac{\mu}{\lambda + \mu}\mathbf{a} + \dfrac{\lambda}{\lambda + \mu}\mathbf{b}.$

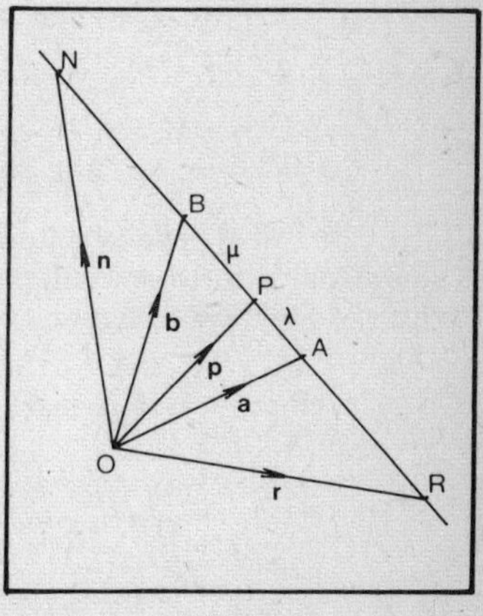

Figure 66

Ratio theorem P divides AB in the ratio $s:t \Leftrightarrow \mathbf{p} = t\mathbf{a} + s\mathbf{b}$, where $s + t = 1$. In particular if M is the mid-point of AB, $\mathbf{OM} = \mathbf{m} = \frac{1}{2}\mathbf{a} + \frac{1}{2}\mathbf{b}$. If Q is the point of trisection nearer A, $\frac{AQ}{QB} = \frac{1}{2}$ so $\mathbf{OQ} = \mathbf{q} = \frac{2}{3}\mathbf{a} + \frac{1}{3}\mathbf{b}$.

If BA = AR, then AR : RB = $-1:2$ so $\mathbf{r} = 2\mathbf{a} - \mathbf{b}$.
If AB = BN, then AN : NB = $2:-1$ so $\mathbf{n} = -\mathbf{a} + 2\mathbf{b}$.
Points on the line AB but outside AB (like R and N) divide AB in a negative ratio. If this is not clearly understood it is easier to say A is the mid-point of BR

$$\Rightarrow \mathbf{a} = \tfrac{1}{2}\mathbf{r} + \tfrac{1}{2}\mathbf{b} \Rightarrow \mathbf{r} = 2\mathbf{a} - \mathbf{b}.$$

B is the mid-point of AN

$$\Rightarrow \mathbf{b} = \tfrac{1}{2}\mathbf{a} + \tfrac{1}{2}\mathbf{n} \Rightarrow \mathbf{n} = 2\mathbf{b} - \mathbf{a} = -\mathbf{a} + 2\mathbf{b}.$$

Alternative approach Any point P on the line AB satisfies

$$\mathbf{p} = \mathbf{a} + s\mathbf{AB} = \mathbf{a} + s(\mathbf{b} - \mathbf{a}) \quad \text{where} \quad \mathbf{AP} = s\mathbf{AB} \Rightarrow s = \frac{AP}{AB}$$

(the ratio of AP : AB is $s:1$).

$$\mathbf{p} = \mathbf{a} + s(\mathbf{b} - \mathbf{a}) = \mathbf{a} + s\mathbf{b} - s\mathbf{a} = (1 - s)\mathbf{a} + s\mathbf{b}.$$

$\mathbf{p} = (1 - s)\mathbf{a} + s\mathbf{b}$ is the vector equation of the line AB.

$$s = 0 \Rightarrow \mathbf{p} = \mathbf{a},\ s = 1 \Rightarrow \mathbf{p} = \mathbf{b},\ s = \tfrac{1}{2} \Rightarrow \mathbf{p} = \tfrac{1}{2}\mathbf{a} + \tfrac{1}{2}\mathbf{b}.$$
$$s = 2 \Rightarrow \mathbf{p} = -\mathbf{a} + 2\mathbf{b},\ s = -1 \Rightarrow \mathbf{p} = 2\mathbf{a} - \mathbf{b}.$$

Each value of s defines a point on the line AB and as s takes values from $-\infty$ to $+\infty$ all points on the line AB are described. With the notation of figure 66

$$\mathbf{AP} = \frac{\lambda}{\lambda + \mu}\mathbf{AB} \Rightarrow \mathbf{p} = \left(1 - \frac{\lambda}{\lambda + \mu}\right)\mathbf{a} + \frac{\lambda}{\lambda + \mu}\mathbf{b}$$
$$= \frac{\mu}{\lambda + \mu}\mathbf{a} + \frac{\lambda}{\lambda + \mu}\mathbf{b}$$

and P divides AB in the ratio $\lambda : \mu$.

$$\mathbf{OB} = \mathbf{b} = \begin{pmatrix} 5 \\ 2 \end{pmatrix} = \begin{pmatrix} 5 \\ 0 \end{pmatrix} + \begin{pmatrix} 0 \\ 2 \end{pmatrix} = 5\begin{pmatrix} 1 \\ 0 \end{pmatrix} + 2\begin{pmatrix} 0 \\ 1 \end{pmatrix} = 5\mathbf{i} + 2\mathbf{j}$$

where $\mathbf{i} = \begin{pmatrix} 1 \\ 0 \end{pmatrix}$, $\mathbf{j} = \begin{pmatrix} 0 \\ 1 \end{pmatrix}$. $\mathbf{i}$ is the unit vector (length 1) in the direction of the x-axis and $\mathbf{j}$ is the unit vector (length 1) in the direction of the y-axis (figure 67).

In three dimensions $\mathbf{i} = \begin{pmatrix}1\\0\\0\end{pmatrix}$ $\mathbf{j} = \begin{pmatrix}0\\1\\0\end{pmatrix}$, $\mathbf{k} = \begin{pmatrix}0\\0\\1\end{pmatrix}$ and $\mathbf{k}$ is the unit vector in the direction of the z-axis. (Some examining boards use the notation **i**, **j**, **k**, instead of column vectors.)

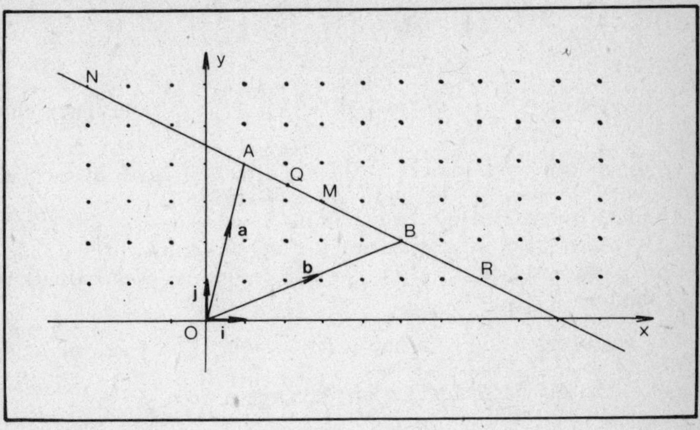

Figure 67

From figure 67, M is the mid-point of AB

$$\mathbf{OM} = \tfrac{1}{2}\mathbf{a} + \tfrac{1}{2}\mathbf{b} = \tfrac{1}{2}(\mathbf{a} + \mathbf{b}) = \frac{1}{2}\left\{\begin{pmatrix}1\\4\end{pmatrix} + \begin{pmatrix}5\\2\end{pmatrix}\right\} = \frac{1}{2}\begin{pmatrix}6\\6\end{pmatrix}$$

$$= \begin{pmatrix}3\\3\end{pmatrix} = 3\mathbf{i} + 3\mathbf{j}.$$

M has coordinates (3, 3).
The point Q dividing AB in the ratio 1 : 3 is given by

$$\mathbf{OQ} = \tfrac{3}{4}\mathbf{a} + \tfrac{1}{4}\mathbf{b} = \tfrac{1}{4}(3\mathbf{a} + \mathbf{b}) = \frac{1}{4}\left\{\begin{pmatrix}3\\12\end{pmatrix} + \begin{pmatrix}5\\2\end{pmatrix}\right\} = \frac{1}{4}\begin{pmatrix}8\\14\end{pmatrix} = \begin{pmatrix}2\\3\frac{1}{2}\end{pmatrix}.$$

Q is $(2, 3\frac{1}{2})$.

To find N : A is the mid-point of BN, so $\mathbf{a} = \tfrac{1}{2}\mathbf{ON} + \tfrac{1}{2}\mathbf{b}$

$$\mathbf{ON} = 2\mathbf{a} - \mathbf{b} = \begin{pmatrix}2\\8\end{pmatrix} - \begin{pmatrix}5\\2\end{pmatrix} = \begin{pmatrix}-3\\6\end{pmatrix}.$$ N is $(-3, 6)$.

To find R : B divides AR in the ratio 2 : 1, so $\mathbf{b} = \tfrac{2}{3}\mathbf{r} + \tfrac{1}{3}\mathbf{a}$, $3\mathbf{b} = 2\mathbf{r} + \mathbf{a}$.

$$\mathbf{r} = \tfrac{3}{2}\mathbf{b} - \tfrac{1}{2}\mathbf{a} = \tfrac{1}{2}(3\mathbf{b} - \mathbf{a}) = \frac{1}{2}\left\{\begin{pmatrix}15\\6\end{pmatrix} - \begin{pmatrix}1\\4\end{pmatrix}\right\} = \frac{1}{2}\begin{pmatrix}14\\2\end{pmatrix} - \begin{pmatrix}7\\1\end{pmatrix}.$$

$$\mathbf{OM} = \begin{pmatrix}1\\4\end{pmatrix} + \begin{pmatrix}2\\-1\end{pmatrix}, \quad \mathbf{OB} = \begin{pmatrix}1\\4\end{pmatrix} + 2\begin{pmatrix}2\\-1\end{pmatrix},$$

$$\mathbf{OR} = \begin{pmatrix}1\\4\end{pmatrix} + 3\begin{pmatrix}2\\-1\end{pmatrix}, \quad \mathbf{ON} = \begin{pmatrix}1\\4\end{pmatrix} + -2\begin{pmatrix}2\\-1\end{pmatrix}.$$

All points on AB satisfy $\mathbf{r} = \begin{pmatrix}x\\y\end{pmatrix} = \begin{pmatrix}1\\4\end{pmatrix} + t\begin{pmatrix}2\\-1\end{pmatrix}$ and this is the vector equation of the line; $\begin{pmatrix}2\\-1\end{pmatrix}$ is the direction vector of the line. The line is specified by the equations $x = 1 + 2t$, $y = 4 - t$ and each value given to t specifies a different point on the line. Eliminating t gives $x + 2y = 9$ which is the cartesian equation of the line.

The vector equation of a line is not unique. $\begin{pmatrix}-2\\1\end{pmatrix}$ can be used as a direction vector for the line AB (figure 67).

$$\mathbf{OA} = \begin{pmatrix}5\\2\end{pmatrix} + 2\begin{pmatrix}-2\\1\end{pmatrix}, \quad \mathbf{OB} = \begin{pmatrix}5\\2\end{pmatrix} + 0\begin{pmatrix}-2\\1\end{pmatrix},$$

$$\mathbf{ON} = \begin{pmatrix}5\\2\end{pmatrix} + 4\begin{pmatrix}-2\\1\end{pmatrix}.$$

This can be summarized as

$$\mathbf{r} = \begin{pmatrix}x\\y\end{pmatrix} = \begin{pmatrix}5\\2\end{pmatrix} + s\begin{pmatrix}-2\\1\end{pmatrix} \quad \text{or} \quad x = 5 - 2s, \quad y = 2 + s.$$

Eliminating s gives $x + 2y = 9$, the same equation as before in its cartesian form.

Comparing $\begin{pmatrix}x\\y\end{pmatrix} = \begin{pmatrix}1\\4\end{pmatrix} + t\begin{pmatrix}2\\-1\end{pmatrix}$ with $\begin{pmatrix}x\\y\end{pmatrix} = \begin{pmatrix}5\\2\end{pmatrix} + s\begin{pmatrix}-2\\1\end{pmatrix}$,

the point A is given by $t = 0$ and $s = 2$,
B is given by $t = 1$ and $s = 0$,
N is given by $t = -2$ and $s = 4$.

In general $\mathbf{r} = \mathbf{a} + t\mathbf{d}$, where $\mathbf{a}$ is the position vector of any point on the line and $\mathbf{d}$ represents a direction vector for the line. $\mathbf{d}$ must be a multiple of $\begin{pmatrix}2\\-1\end{pmatrix}$ for the line AB in figure 67.

Vectors in three dimensions

Vectors are specified relative to mutually perpendicular axes Ox, Oy and Oz.

If the point P has coordinates (1, 2, 3), (see figure 69, page 201), then

$$\mathbf{OP} = \begin{pmatrix}1\\2\\3\end{pmatrix} = \begin{pmatrix}1\\0\\0\end{pmatrix} + \begin{pmatrix}0\\2\\0\end{pmatrix} + \begin{pmatrix}0\\0\\3\end{pmatrix} = \mathbf{i} + 2\mathbf{j} + 3\mathbf{k}.$$

In three dimensions

$$\mathbf{i} = \begin{pmatrix}1\\0\\0\end{pmatrix}, \quad \mathbf{j} = \begin{pmatrix}0\\1\\0\end{pmatrix}, \quad \mathbf{k} = \begin{pmatrix}0\\0\\1\end{pmatrix}.$$

Vector equation of a line in three dimensions

Points on the line CJ (figure 68) can be specified by adding multiples of **JC** to the vector **OJ**.

$$\mathbf{r} = \begin{pmatrix}x\\y\\z\end{pmatrix} = \mathbf{OJ} + t\mathbf{JC}$$

$$= \begin{pmatrix}0\\1\\0\end{pmatrix} + t\begin{pmatrix}1\\-1\\1\end{pmatrix}$$

or
$$x = 0 + t, \quad y = 1 - t, \quad z = 0 + t,$$

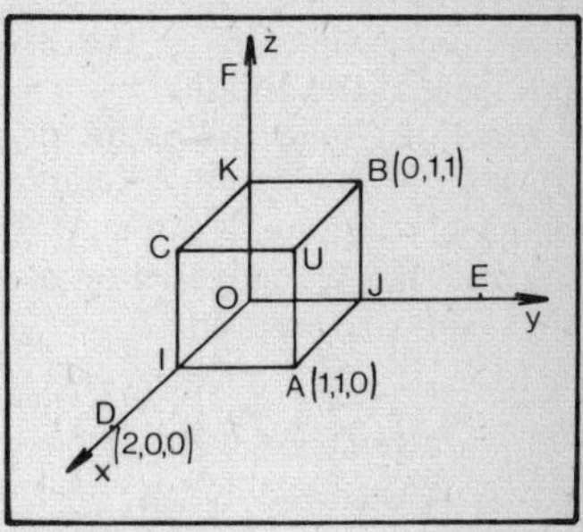

Figure 68

where J(0, 1, 0) is a point on the line and $\mathbf{i} - \mathbf{j} + \mathbf{k}$ is the direction vector.

The equation of the line through the point P(1, 2, 3) in the direction $\begin{pmatrix}4\\5\\6\end{pmatrix}$ will be $\begin{pmatrix}x\\y\\z\end{pmatrix} = \begin{pmatrix}1\\2\\3\end{pmatrix} + t\begin{pmatrix}4\\5\\6\end{pmatrix}$ or $x = 1 + 4t$, $y = 2 + 5t$, $z = 3 + 6t$.

These equations can be rearranged in the form $t = \dfrac{x-1}{4} = \dfrac{y-2}{5} = \dfrac{z-3}{6}$, which are equivalent to two equations $5x - 4y = -3$ and $6y - 5z = -3$.

Intersection of two lines in three dimensions

From figure 68 we have seen that the line CJ has equation

$$\mathbf{r} = \begin{pmatrix} x \\ y \\ z \end{pmatrix} = \begin{pmatrix} 0 \\ 1 \\ 0 \end{pmatrix} + t\begin{pmatrix} 1 \\ -1 \\ 1 \end{pmatrix}.$$

The line AK has direction vector $\mathbf{AK} = -\mathbf{i} - \mathbf{j} + \mathbf{k}$. Line AK has equation

$$\mathbf{r} = \begin{pmatrix} x \\ y \\ z \end{pmatrix} = \begin{pmatrix} 1 \\ 1 \\ 0 \end{pmatrix} + s\begin{pmatrix} -1 \\ -1 \\ 1 \end{pmatrix} \quad \text{or} \quad \begin{aligned} x &= 1 - s, \\ y &= 1 - s, \\ z &= s. \end{aligned}$$

To find the intersection of CJ with AK, equate the x, y and z values

$$\begin{aligned} x &= 1 - s = 0 + t, \\ y &= 1 - s = 1 - t, \\ z &= t \qquad = s. \end{aligned}$$

Three equations with 2 unknowns! The last two equations give $t = s$ and using this in the first equation $s = \frac{1}{2} = t$. CJ and AK intersect at the point where $s = \frac{1}{2}$, i.e. $(x, y, z) = (\frac{1}{2}, \frac{1}{2}, \frac{1}{2})$. Obviously $t = \frac{1}{2}$ must give the same point and it can be seen from the diagram that CJ and AK intersect at the centre of the unit cube, where the diagonals intersect. If two lines have the same direction vector, they are parallel or they coincide for all points. The line CD could be written

$$\begin{pmatrix} x \\ y \\ z \end{pmatrix} = \begin{pmatrix} 1 \\ 0 \\ 1 \end{pmatrix} + t\begin{pmatrix} 0 \\ 1 \\ 0 \end{pmatrix} \quad \text{or} \quad \begin{pmatrix} x \\ y \\ z \end{pmatrix} = \begin{pmatrix} 1 \\ 1 \\ 1 \end{pmatrix} + s\begin{pmatrix} 0 \\ 1 \\ 0 \end{pmatrix}$$

by first using C as the point on the line where $t = 0$ and secondly using U as the point on the line where $s = 0$. In fact $t = s + 1$ and the lines are coincident. The line CU has equation $\begin{pmatrix} x \\ y \\ z \end{pmatrix} = \begin{pmatrix} 1 \\ 0 \\ 1 \end{pmatrix} + t\begin{pmatrix} 0 \\ 1 \\ 0 \end{pmatrix}$ and IA has equation $\begin{pmatrix} x \\ y \\ z \end{pmatrix} = \begin{pmatrix} 1 \\ 0 \\ 0 \end{pmatrix} + s\begin{pmatrix} 0 \\ 1 \\ 0 \end{pmatrix}$.

They intersect when

$$\begin{aligned} x &= 1 = 1, \\ y &= t = s, \\ z &= 1 = 0. \end{aligned}$$

Since $1 \neq 0$ there is no solution to these equations and therefore they do not intersect. CJ has equation $\begin{pmatrix} x \\ y \\ z \end{pmatrix} = \begin{pmatrix} 0 \\ 1 \\ 0 \end{pmatrix} + t\begin{pmatrix} 1 \\ -1 \\ 1 \end{pmatrix}$, and IA has equation $\begin{pmatrix} x \\ y \\ z \end{pmatrix} = \begin{pmatrix} 1 \\ 0 \\ 0 \end{pmatrix} + s\begin{pmatrix} 0 \\ 1 \\ 0 \end{pmatrix}$.

They intersect when $x = t = 1$,
$y = 1 - t = s$,
$z = t = 0$.

The first and third equations are inconsistent and so have no solution. In fact the lines do not intersect, but are not parallel. They are **skew** lines.

Four cases arise for the intersection of lines in three dimensions:

(i) the lines meet in a single point, e.g. CJ and AK;
(ii) the lines are parallel, e.g. JK and AC;
(iii) the lines are coincident;
(iv) the lines are skew, e.g. CK and IA.

Direction cosines of a vector

$$\mathbf{OU} = \begin{pmatrix}1\\1\\1\end{pmatrix} \text{ has length } \lambda = \sqrt{\text{OI}^2 + \text{IA}^2 + \text{AU}^2} = \sqrt{1^2 + 1^2 + 1^2} = \sqrt{3}.$$

If OU makes an angle α with the x-axis, then in the right-angled triangle OIU, $\cos \alpha = \frac{\text{OI}}{\text{OU}} = \frac{1}{\sqrt{3}}$. Similarly if OU makes an angle β with the y-axis and γ with the z-axis, then $\cos \beta = \frac{1}{\sqrt{3}} = \cos \gamma$ and OU is equally inclined to all three axes.

For the vector $\mathbf{OP} = \begin{pmatrix}1\\2\\3\end{pmatrix}$.

$$\text{Length OP} = \sqrt{\text{ON}^2 + \text{NP}^2} = \sqrt{\text{OI}^2 + \text{IN}^2 + \text{NP}^2} = \sqrt{1^2 + 2^2 + 3^2} = \sqrt{14}.$$

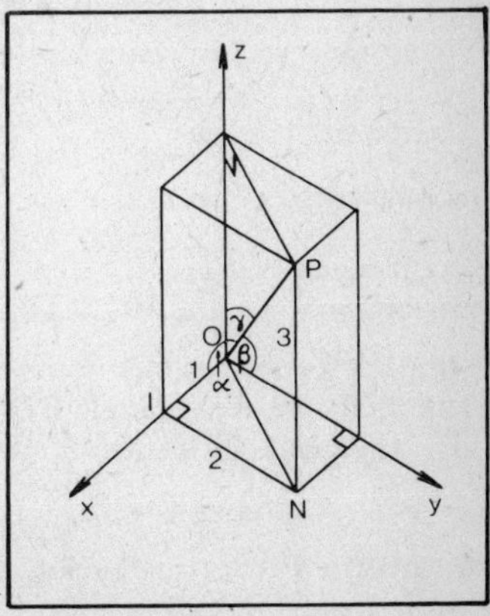

Figure 69

If **OP** makes angles α, β, and γ with the axes Ox, Oy and Oz, then $\alpha = \angle\text{IOP}$, $\beta = \angle\text{PO}y$, $\gamma = \angle\text{PO}z$. In ΔOIP (right-angled at I)

$$\cos \alpha = \frac{\text{OI}}{\text{OP}} = \frac{1}{\sqrt{14}}.$$

Similarly $\cos\beta = \dfrac{2}{\sqrt{14}}$ and $\cos\gamma = \dfrac{3}{\sqrt{14}}$.

The values $\dfrac{1}{\sqrt{14}}, \dfrac{2}{\sqrt{14}}, \dfrac{3}{\sqrt{14}}$ are the **direction cosines** of the vector **OP** which specify its direction with respect to the axes Ox, Oy and Oz. In general, the length OP where P is (a, b, c) is $l = \sqrt{a^2 + b^2 + c^2}$ and its direction cosines $\dfrac{a}{l}, \dfrac{b}{l}, \dfrac{c}{l}$.

Centroid of a triangle

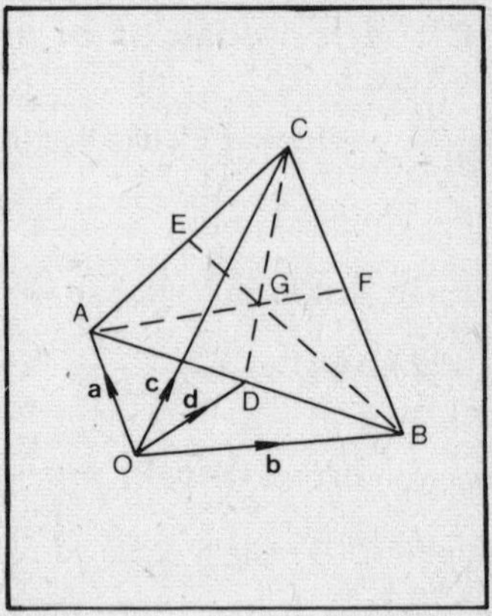

Figure 70

If **OA** = **a**, **OB** = **b**, etc. and D, E, F are the mid-points of AB, AC, BC respectively, $\mathbf{OD} = \mathbf{d} = \frac{1}{2}\mathbf{a} + \frac{1}{2}\mathbf{b}$. If CG = sCD, the equation of CD is $\mathbf{r} = \frac{1}{2}s(\mathbf{a} + \mathbf{b}) + (1 - s)\mathbf{c}$.

Similarly if BG = tBE, the equation of BE is $\mathbf{r} = \frac{1}{2}t(\mathbf{a} + \mathbf{c}) + (1 - t)\mathbf{b}$. BE and CD intersect at G where

$$\tfrac{1}{2}s(\mathbf{a} + \mathbf{b}) + (1 - s)\mathbf{c} = \tfrac{1}{2}t(\mathbf{a} + \mathbf{c}) + (1 - t)\mathbf{b}.$$

Since the vector representations must be equal, $s = t$ (coefficients of **a**), $s = 1 - t$ (coefficients of **b**), $1 - s = \dfrac{t}{2}$ (coefficients of **c**).
Hence $s = \frac{2}{3} = t$ and BG = $\frac{2}{3}$BE or G divides BE in the ratio 2 : 1.

Similarly G divides CD in the ratio 2 : 1 and AF in the ratio 2 : 1. The medians of a triangle CD, BE, AF are concurrent at G and G is the centroid (centre of mass) of the triangle.

$$\mathbf{OG} = \mathbf{g} = \tfrac{1}{3}\mathbf{a} + \tfrac{1}{3}\mathbf{b} + \tfrac{1}{3}\mathbf{c} = \tfrac{1}{3}(\mathbf{a} + \mathbf{b} + \mathbf{c}).$$

Example 1 Referring back to figure 67 on page 197, the centroid of the triangle OAB is given by

$$\mathbf{OG} = \frac{1}{3}\left\{\begin{pmatrix}0\\0\end{pmatrix} + \begin{pmatrix}1\\4\end{pmatrix} + \begin{pmatrix}5\\2\end{pmatrix}\right\} = \frac{1}{3}\begin{pmatrix}6\\6\end{pmatrix} = \begin{pmatrix}2\\2\end{pmatrix}.$$

It can be seen that G(2, 2) divides OM in the ratio 2 : 1 and similarly for the other medians from A and B.

Example 2 Referring back to figure 68 on page 199, the centroid of triangle IJK is given by

$$\mathbf{OG} = \frac{1}{3}\left\{\begin{pmatrix}1\\0\\0\end{pmatrix} + \begin{pmatrix}0\\1\\0\end{pmatrix} + \begin{pmatrix}0\\0\\1\end{pmatrix}\right\} = \frac{1}{3}\begin{pmatrix}1\\1\\1\end{pmatrix} = \begin{pmatrix}\frac{1}{3}\\\frac{1}{3}\\\frac{1}{3}\end{pmatrix}$$

Similarly the centroid of $\triangle$ABC has coordinates $(\frac{2}{3}, \frac{2}{3}, \frac{2}{3})$.

Centroid of a tetrahedron

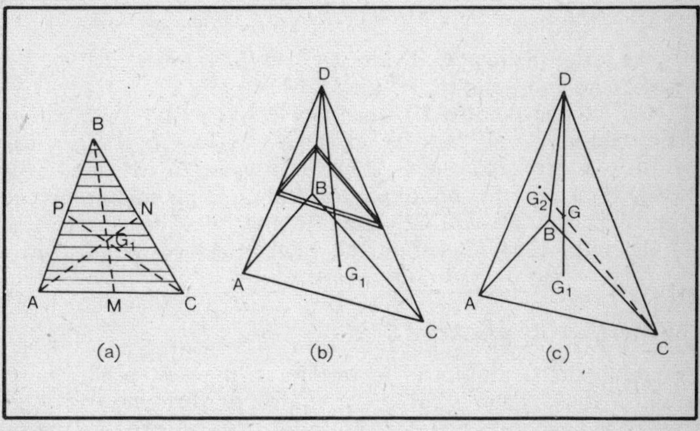

Figure 71

The centroid of a triangle is at the intersection of the medians. This is also the centre of mass (centre of gravity). Imagine the triangle divided into narrow strips parallel to AC (figure 71a). The centre of mass of each strip is at its centre, so the centre of mass of the triangle lies on the locus of the mid-points of the strips, which is the median BM. Similarly, the centre of mass of the whole triangle lies on all three medians and is therefore at G_1 the centroid. Imagine the tetrahedron divided into narrow triangles parallel to the base ABC. The centre of mass of each triangle is the centroid of the triangle. The centre of mass of the tetrahedron lies on the locus of these centroids and by the properties of enlargement, or similar figures, these centroids lie on the line joining D to G_1 the centroid of $\triangle$ABC. Similarly the centre of mass G lies on the line joining each vertex to the centroid of the opposite face.

Equation of DG_1 is $\mathbf{r} = \mathbf{d} + s\{\mathbf{d} - \frac{1}{3}(\mathbf{a} + \mathbf{b} + \mathbf{c})\}$ using $\mathbf{OA} = \mathbf{a}$, $\mathbf{OB} = \mathbf{b}$, etc. with origin O. Equation of CG_2 is $\mathbf{r} = \mathbf{c} + t(\mathbf{c} - \frac{1}{3}(\mathbf{a} + \mathbf{b} + \mathbf{d})\}$.

At G, the intersection of these lines, $\mathbf{d} + s\{\mathbf{d} - \frac{1}{3}(\mathbf{a} + \mathbf{b} + \mathbf{c})\}$ $= \mathbf{c} + t\{\mathbf{c} - \frac{1}{3}(\mathbf{a} + \mathbf{b} + \mathbf{d})\}$ from which $1 + s = -\frac{t}{3}$, $-\frac{s}{3} = -\frac{t}{3}$, $-\frac{1}{3}s = 1 + t$.

$$s = t \text{ gives } s = t = -\tfrac{3}{4}, \text{ so } \mathbf{r} = \mathbf{c} - \tfrac{3}{4}\{\mathbf{c} - \tfrac{1}{3}(\mathbf{a} + \mathbf{b} + \mathbf{d})\}$$
$$= \tfrac{1}{4}\mathbf{c} + \tfrac{1}{4}\mathbf{a} + \tfrac{1}{4}\mathbf{b} + \tfrac{1}{4}\mathbf{d}.$$

Since $\mathbf{OG}_1 = \frac{1}{3}(\mathbf{a} + \mathbf{b} + \mathbf{c})$, $\mathbf{OG} = \frac{1}{4}\mathbf{d} + \frac{3}{4} \times \frac{1}{3}(\mathbf{a} + \mathbf{b} + \mathbf{c})$ $= \frac{1}{4}\mathbf{d} + \frac{3}{4}\mathbf{OG}_1$.

By the ratio theorem, G divides DG_1 in the ratio of 3 : 1.
If M is the mid-point of AC then $\mathbf{OM} = \frac{1}{2}(\mathbf{a} + \mathbf{c})$.
If N is the mid-point of BD then $\mathbf{ON} = \frac{1}{2}(\mathbf{b} + \mathbf{d})$.
The mid-point of MN is given by $\frac{1}{2}\{\frac{1}{2}(\mathbf{a} + \mathbf{c}) + \frac{1}{2}(\mathbf{b} + \mathbf{d})\}$ $= \frac{1}{4}(\mathbf{a} + \mathbf{b} + \mathbf{c} + \mathbf{d})$. So G is the mid-point of MN and will similarly lie at the mid-point of lines joining mid-points of opposite edges AB and CD and AD and BC. These properties of the tetrahedron illustrate the geometrical power of vector methods, especially the ratio theorem.

Multiplying vectors

Vectors can be multiplied in two ways, the first producing a number (scalar) and is called the **scalar or dot product** and the second produces a vector and is called the **vector or cross product**. Both products have important applications, especially in mechanics. The study of their properties gives valuable insight into the algebraic structure of vectors and mathematical structures in general. Two-dimensional vectors are easy to represent and understand, while three-dimensional vectors are difficult to draw on plane paper and three-dimensional ideas difficult to convey. Algebraic derivations are included to help the understanding of three-dimensional work. The many geometrical applications of vector methods are emphasized whenever possible. This also helps understanding and conveys an idea of the power and simplicity of working with vectors.

Scalar product

The scalar product (or dot product) of two vectors **a** and **b** is denoted by $\mathbf{a} \cdot \mathbf{b} = ab \cos \theta$, where a and b are the lengths of the vectors **a** and **b**, and θ is the angle between them, measured positively with an anticlockwise rotation from OA to OB (figure 72).

If the angle is measured from OB to OA as $-\theta$, since $\cos(-\theta) = \cos\theta$, the scalar product takes the same value.

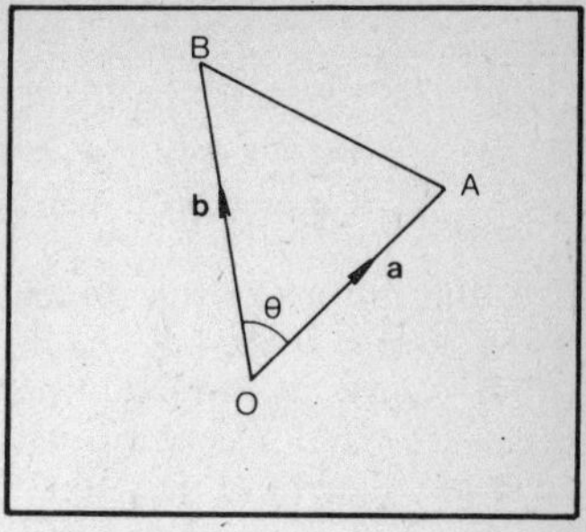

Figure 72

Example 3 If $a = 2$, $b = 3$ and $\theta = 60°$ $\left(\frac{\pi}{3} \text{ radians}\right)$,
$\mathbf{a} \cdot \mathbf{b} = 2 \times 3 \times \cos 60° = 3$.

Example 4 If $a = 2$, $b = 3$ and $\theta = 120°$,
$\mathbf{a} \cdot \mathbf{b} = 2 \times 3 \times \cos 120° = -3$.

Example 5 $(2\mathbf{a}) \cdot (3\mathbf{b}) = 2a \times 3b \times \cos\theta = 6ab\cos\theta$ $\Rightarrow (2\mathbf{a}) \cdot (3\mathbf{b}) = 6(\mathbf{a} \cdot \mathbf{b})$. From figure 72, $\mathbf{b} \cdot \mathbf{a} = ba\cos(360 - \theta)$ measuring the angle positively. Since $\cos(360 - \theta) = \cos\theta$, $\mathbf{b} \cdot \mathbf{a} = ba\cos\theta = \mathbf{a} \cdot \mathbf{b}$, so the scalar product of two vectors is commutative. An operation (denoted by *) combining two elements p and q is commutative if $p{*}q = q{*}p$ (see chapter 13, Groups).

For the vectors $\mathbf{i} = \begin{pmatrix}1\\0\\0\end{pmatrix}$, $\mathbf{j} = \begin{pmatrix}0\\1\\0\end{pmatrix}$ and $\mathbf{k} = \begin{pmatrix}0\\0\\1\end{pmatrix}$

since the angle between any two is 90°

$$\mathbf{i} \cdot \mathbf{j} = \mathbf{j} \cdot \mathbf{k} = \mathbf{i} \cdot \mathbf{k} = 0, \qquad (1)$$

and
$$\mathbf{i} \cdot \mathbf{i} = 1 \times 1 \times \cos 0° = 1.$$

So
$$\mathbf{i} \cdot \mathbf{i} = \mathbf{j} \cdot \mathbf{j} = \mathbf{k} \cdot \mathbf{k} = 1. \qquad (2)$$

If the angle between two (non-zero) vectors **a** and **b** is 90° then $\mathbf{a} \cdot \mathbf{b} = ab\cos 90° = 0$ and this property provides an easy way of telling whether two vectors are at right angles.

Length of a vector $\mathbf{a} \cdot \mathbf{a} = a \times a \times \cos 0° = a^2$, the square of the length of **a**.

$$\mathbf{a} = \begin{pmatrix}a_1\\a_2\\a_3\end{pmatrix}, \mathbf{a} \cdot \mathbf{a} = \begin{pmatrix}a_1\\a_2\\a_3\end{pmatrix} \cdot \begin{pmatrix}a_1\\a_2\\a_3\end{pmatrix} \Rightarrow a^2 = a_1^2 + a_2^2 + a_3^2.$$

Scalar product in component form

Can we find an expression for **a . b** if $\mathbf{a} = \begin{pmatrix} a_1 \\ a_2 \\ a_3 \end{pmatrix}$ and $\mathbf{b} = \begin{pmatrix} b_1 \\ b_2 \\ b_3 \end{pmatrix}$?

Writing $\mathbf{a} = a_1\mathbf{i} + a_2\mathbf{j} + a_3\mathbf{k}$ and $\mathbf{b} = b_1\mathbf{i} + b_2\mathbf{j} + b_3\mathbf{k}$,

$$\mathbf{a} \,.\, \mathbf{b} = (a_1\mathbf{i} + a_2\mathbf{j} + a_3\mathbf{k}) \,.\, (b_1\mathbf{i} + b_2\mathbf{j} + b_3\mathbf{k}).$$

To multiply out the brackets we need to establish that $\mathbf{a} \,.\, (\mathbf{b} + \mathbf{c}) = \mathbf{a} \,.\, \mathbf{b} + \mathbf{a} \,.\, \mathbf{c}$, i.e. that the scalar product is distributive over vector addition. The next section will show that this is true and it enables the brackets to be multiplied out just as in the algebra of numbers.

$$\begin{aligned}\mathbf{a} \,.\, \mathbf{b} &= a_1b_1\mathbf{i}\,.\,\mathbf{i} + a_1b_2\mathbf{i}\,.\,\mathbf{j} + a_1b_3\mathbf{i}\,.\,\mathbf{k} + a_2b_1\mathbf{j}\,.\,\mathbf{i} + a_2b_2\mathbf{j} \\ &\quad + a_2b_3\mathbf{j}\,.\,\mathbf{k} + a_3b_1\mathbf{k}\,.\,\mathbf{i} + a_3b_2\mathbf{k}\,.\,\mathbf{j} + a_3b_3\mathbf{k}\,.\,\mathbf{k} \\ &= a_1b_1 + a_2b_2 + a_3b_3\end{aligned}$$

using $\mathbf{i}\,.\,\mathbf{j} = \mathbf{j}\,.\,\mathbf{k} = \mathbf{k}\,.\,\mathbf{i} = 0$ from (1), and $\mathbf{i}\,.\,\mathbf{i} = \mathbf{j}\,.\,\mathbf{j} = \mathbf{k}\,.\,\mathbf{k} = 1$ from (2).

Scalar product is distributive over vector addition a . (b + c) = a . b + a . c

Let **OA** = **a** be along the *x*-axis.
Let **OB** = **b** in the *xy*-plane.

Then $\mathbf{a}\,.\,\mathbf{b} = ab\cos\theta = a\text{OF}$ where OF is the projection of **OB** on to the *x*-axis (figure 73).
Similarly
$\mathbf{a}\,.\,(\mathbf{b}+\mathbf{c}) = a\text{OD}\cos\angle\text{DOG}$, where OG is the projection of $\mathbf{b}+\mathbf{c} = \mathbf{OD}$ on to the *x*-axis.
So $\mathbf{a}\,.\,(\mathbf{b}+\mathbf{c}) = a\text{OG}$.
Let $\mathbf{c} = \mathbf{OC} = \mathbf{BD}$.
D is not in the *xy*-plane.

$$\begin{aligned}\mathbf{a}\,.\,\mathbf{c} &= a\text{OC}\cos\angle\text{COA} \\ &= a\text{BD}\cos\angle\text{DBH} \\ &= a\text{BH}.\end{aligned}$$

Since $\mathbf{OG} = \mathbf{OF} + \mathbf{BH}$
$a\text{OG} = a\text{OF} + a\text{BH}$.
Thus $\mathbf{a}\,.\,(\mathbf{b}+\mathbf{c}) = \mathbf{a}\,.\,\mathbf{b} + \mathbf{a}\,.\,\mathbf{c}$.

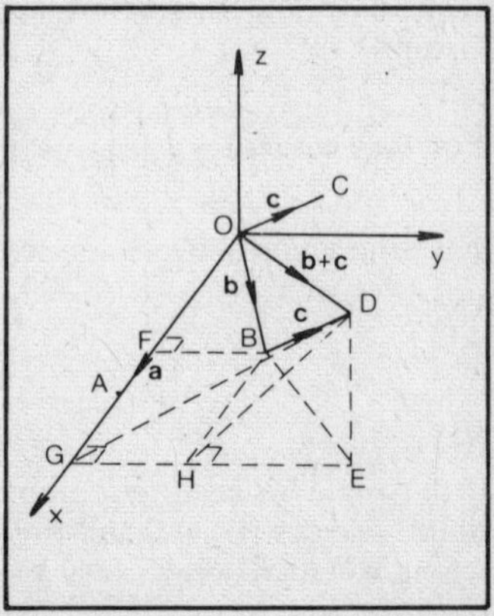

Figure 73

The diagram in figure 73 is difficult to draw and visualize and you may find the following algebraic proof easier to understand.

Alternative definition of scalar product

If $\mathbf{a} = \begin{pmatrix} a_1 \\ a_2 \\ a_3 \end{pmatrix}$ and $\mathbf{b} = \begin{pmatrix} b_1 \\ b_2 \\ b_3 \end{pmatrix}$, then $\mathbf{a} \cdot \mathbf{b} = a_1 b_1 + a_2 b_2 + a_3 b_3$.

Distributive property, i.e. $\mathbf{a} \cdot (\mathbf{b} + \mathbf{c}) = \mathbf{a} \cdot \mathbf{b} + \mathbf{a} \cdot \mathbf{c}$.

If $\mathbf{c} = \begin{pmatrix} c_1 \\ c_2 \\ c_3 \end{pmatrix}$ then $\mathbf{b} + \mathbf{c} = \begin{pmatrix} b_1 + c_1 \\ b_2 + c_2 \\ b_3 + c_3 \end{pmatrix}$.

$$\mathbf{a} \cdot (\mathbf{b} + \mathbf{c}) = \begin{pmatrix} a_1 \\ a_2 \\ a_3 \end{pmatrix} \begin{pmatrix} b_1 + c_1 \\ b_2 + c_2 \\ b_3 + c_3 \end{pmatrix} = a_1(b_1 + c_1) + a_2(b_2 + c_2) + a_3(b_3 + c_3)$$

$$= a_1 b_1 + a_2 b_2 + a_3 b_3 + a_1 c_1 + a_2 c_2 + a_3 c_3$$
$$= \mathbf{a} \cdot \mathbf{b} + \mathbf{a} \cdot \mathbf{c},$$

which seems simple enough. However, if we start with this definition, we shall need to show that it is equivalent to our previous definition, that is, that $\mathbf{a} \cdot \mathbf{b} = a_1 b_1 + a_2 b_2 + a_3 c_3 = ab \cos \theta$.

With the algebraic definition,

$$\mathbf{a} \cdot \mathbf{a} = a_1^2 + a_2^2 + a_3^2 = a^2,$$

the square of the length. Referring to figure 72 and using the cosine rule in triangle OAB

$$AB^2 = OA^2 + OB^2 - 2OA\ OB \cos \theta$$
$$(\mathbf{b} - \mathbf{a}) \cdot (\mathbf{b} - \mathbf{a}) = a^2 + b^2 - 2ab \cos \theta$$
$$\mathbf{b} \cdot \mathbf{b} - \mathbf{b} \cdot \mathbf{a} - \mathbf{a} \cdot \mathbf{b} + \mathbf{a} \cdot \mathbf{a} = a^2 + b^2 - 2ab \cos \theta.$$

From the algebraic definition we know $\mathbf{a} \cdot \mathbf{a} = a^2$ and $\mathbf{b} \cdot \mathbf{b} = b^2$ and it is easy to show that $\mathbf{b} \cdot \mathbf{a} = \mathbf{a} \cdot \mathbf{b} = a_1 b_1 + a_2 b_2 + a_3 b_3$. So

$$b^2 - 2(\mathbf{a} \cdot \mathbf{b}) + a^2 = a^2 + b^2 - 2ab \cos \theta \Rightarrow \mathbf{a} \cdot \mathbf{b} = ab \cos \theta.$$

and the two definitions are equivalent.

Angle between two vectors

When two vectors are given in component form it is easy to find their scalar product.

Example 6 If $\mathbf{a} = \begin{pmatrix} 1 \\ 2 \\ 3 \end{pmatrix}$ and $\mathbf{b} = \begin{pmatrix} 2 \\ 3 \\ 4 \end{pmatrix}$

then $\mathbf{a} \cdot \mathbf{b} = 1 \times 2 + 2 \times 3 + 3 \times 4 = 20$.

Since the length of **a** is $a = \sqrt{1^2 + 2^2 + 3^2} = \sqrt{14}$ and $b = \sqrt{2^2 + 3^2 + 4^2} = \sqrt{29}$, $\quad ab \cos\theta = \sqrt{14}\sqrt{29} \cos\theta = 20$ (from the component definition)

$$\cos\theta = \frac{20}{\sqrt{14}\sqrt{29}} = 0{\cdot}99258 \Rightarrow \theta = 6{\cdot}98°.$$

Example 7 $\mathbf{a} = \begin{pmatrix} 1 \\ 2 \\ 2 \end{pmatrix}$ $\mathbf{b} = \begin{pmatrix} 2 \\ 1 \\ -2 \end{pmatrix} \Rightarrow \mathbf{a} \,.\, \mathbf{b} = 2 + 2 - 4 = 0$ and **a** and **b** are perpendicular.

Example 8 Referring to figure 68 on page 199 find the angle between (i) a diagonal and a plane face of the cube and (ii) the angle between two diagonals (a diagonal joins opposite vertices).

(i) The angle between the diagonal OU and the plane face OIAJ is the angle between the vectors $\mathbf{OU} = \mathbf{i} + \mathbf{j} + \mathbf{k}$ and $\mathbf{OA} = \mathbf{i} + \mathbf{j}$.

$$\text{OU} = \sqrt{1^2 + 1^2 + 1^2} = \sqrt{3}, \qquad \text{OA} = \sqrt{1^2 + 1^2 + 0^2} = \sqrt{2},$$

$$\mathbf{OU} \,.\, \mathbf{OA} = 2 \qquad \mathbf{OU} \,.\, \mathbf{OA} = \sqrt{3}\sqrt{2} \cos \angle\text{AOU} = 2$$

$$\cos \angle\text{AOU} = \frac{2}{\sqrt{3}\sqrt{2}} = \frac{2}{\sqrt{6}} = \frac{2\sqrt{6}}{6} = \frac{\sqrt{6}}{3} = 0{\cdot}8165.$$

Thus $\angle\text{AOU} = 35{\cdot}3°.$

(ii) The angle α between two diagonals is equal to the angle between **OU** and **IB**.

$$\mathbf{OU} = \mathbf{i} + \mathbf{j} + \mathbf{k}, \quad \mathbf{IB} = -\mathbf{i} + \mathbf{j} + \mathbf{k}, \quad \text{OU} = \text{IB} = \sqrt{3}.$$

$$\mathbf{OU} \,.\, \mathbf{IB} = \sqrt{3}\sqrt{3} \cos\alpha = \begin{pmatrix} 1 \\ 1 \\ 1 \end{pmatrix}\begin{pmatrix} -1 \\ 1 \\ 1 \end{pmatrix}, \quad \cos\alpha = \frac{1}{\sqrt{3}\sqrt{3}} = \frac{1}{3}$$

$\Rightarrow \alpha = 70{\cdot}5°.$

In general, the angle between two vectors **a** and **b** is given by

$$\cos\theta = \frac{\mathbf{a} \,.\, \mathbf{b}}{ab} = \frac{a_1 b_1 + a_2 b_2 + a_3 b_3}{\sqrt{a_1^2 + a_2^2 + a_3^2}\sqrt{b_1^2 + b_2^2 + b_3^2}}$$

if $\mathbf{a} = \begin{pmatrix} a_1 \\ a_2 \\ a_3 \end{pmatrix}$ and $\mathbf{b} = \begin{pmatrix} b_1 \\ b_2 \\ b_3 \end{pmatrix}$.

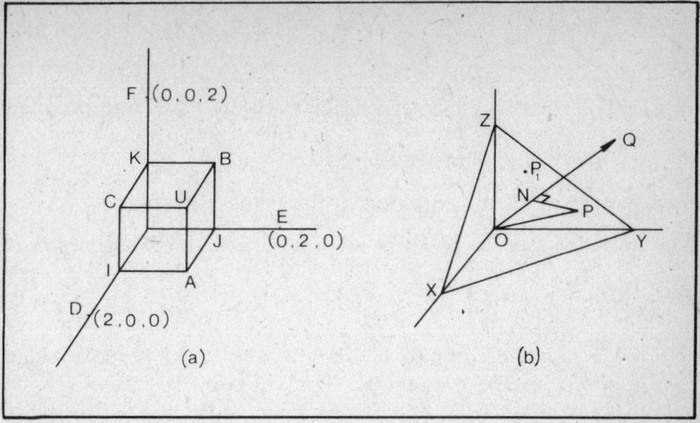

Figure 74

Equation of a plane in three dimensions

The points IJK (figure 74a) define a plane which includes the mid-points of JK, IJ, IK, i.e. $(0, \frac{1}{2}, \frac{1}{2})$, $(\frac{1}{2}, \frac{1}{2}, 0)$, $(\frac{1}{2}, 0, \frac{1}{2})$ respectively. The centroid of triangle IJK $(\frac{1}{3}, \frac{1}{3}, \frac{1}{3})$ must also lie in the plane. This suggests the equation of the plane is $x + y + z = 1$.
Similarly the plane through A(1, 1, 0), B(0, 1, 1) and C(1, 0, 1) includes the points D(2, 0, 0), E(0, 2, 0) and F(0, 0, 2). All the points satisfy the equation $x + y + z = 2$. A little knowledge of the geometry and symmetry of the cube will enable you to see that these two planes are parallel.
Consider the plane passing through XYZ (figure 74b). This plane is defined by specifying a vector **OQ** perpendicular to the plane and a point lying on the plane $P_1(x_1, y_1, z_1)$. Let $\mathbf{OQ} = \begin{pmatrix} a \\ b \\ c \end{pmatrix}$ which cuts the plane at N (where $ON = h$). For any point $P(x, y, z)$ on the plane, **NP** is perpendicular to **OQ**

$$\Rightarrow \mathbf{NP} \cdot \mathbf{OQ} = 0$$
$$\Rightarrow (\mathbf{OP} - \mathbf{ON}) \cdot \mathbf{OQ} = 0$$
$$\mathbf{OP} \cdot \mathbf{OQ} = \mathbf{ON} \cdot \mathbf{OQ} \Rightarrow ax + by + cz = hOQ = d,$$

which is the equation of the plane XYZ. d can be calculated if P_1 is known and $d = ax_1 + by_1 + cz_1$.
$d = hOQ = h\sqrt{a^2 + b^2 + c^2}$, so the distance of the plane from the

origin is $h = \dfrac{d}{\sqrt{a^2 + b^2 + c^2}}$. $d = 0 \Rightarrow$ plane passes through the origin. The vector $\begin{pmatrix} a \\ b \\ c \end{pmatrix}$ is perpendicular to the plane and is called the **normal vector** of the plane.

Example 8 Find the equation of the plane IJK (figure 74a).

The normal vector is **OU** = **i** + **j** + **k**. Check this by seeing that

$$\mathbf{OU} \,.\, \mathbf{IJ} = \begin{pmatrix} 1 \\ 1 \\ 1 \end{pmatrix}\begin{pmatrix} -1 \\ 1 \\ 0 \end{pmatrix} = 0 \quad \text{and} \quad \mathbf{OU} \,.\, \mathbf{IK} = \begin{pmatrix} 1 \\ 1 \\ 1 \end{pmatrix}\begin{pmatrix} -1 \\ 0 \\ 1 \end{pmatrix} = 0,$$

so **OU** is perpendicular to **IJ** and **IK** and so is perpendicular to all points in the plane IJK. The equation of plane IJK is $x + y + z = d$. I(1, 0, 0) lies on the plane, so $d = 1 + 0 + 0 = 1$ and $x + y + z = 1$. Similarly **OU** is the normal vector of plane ABC, so the equation of plane ABC is $x + y + z = k = 2 + 0 + 0 = 2$, since A(2, 0, 0) lies on plane ABC.

The two planes are parallel as they have the same normal vector.

Example 9 Find the equation of the plane cutting the x-axis at (1, 0, 0) the y-axis at (0, 2, 0) and the z-axis at (0, 0, 3).

The equation of the plane is $ax + by + cz = d$. Point (1, 0, 0) lies on the plane $\Rightarrow a = d$ and similarly $b = \frac{1}{2}d$ and $c = \frac{1}{3}d$, so $dx + \frac{1}{2}dy + \frac{1}{3}dz = d$ or $6x + 3y + 2z = 6$.

Example 10 Find the equation of the plane passing through the three points P(1, 1, 1), Q(1, 2, 3) and R(3, 2, 1).

The normal vector $\begin{pmatrix} a \\ b \\ c \end{pmatrix}$ of the plane must be perpendicular to **PQ** and to **PR**.

$$\begin{pmatrix} a \\ b \\ c \end{pmatrix}\begin{pmatrix} 0 \\ 1 \\ 2 \end{pmatrix} = 0 \Rightarrow b + 2c = 0, \quad b = -2c. \qquad \begin{pmatrix} a \\ b \\ c \end{pmatrix}\begin{pmatrix} 2 \\ 1 \\ 0 \end{pmatrix} = 0 \Rightarrow 2a + b = 0 \Rightarrow b = -2a.$$

Since the normal vector can have any magnitude, choose $a = 1$ then $b = -2$, $c = 1$. The normal vector is **i** − 2**j** + **k** and the equation of the plane is $x - 2y + z = d$. Since P lies on the plane, $d = 1 - 2 + 1 = 0$. The equation of plane PQR is $x - 2y + z = 0$.

Distances from the origin. All points on **OU** have coordinates which are multiples of (1, 1, 1).

OU meets the plane IJK $(x + y + z = 1)$ at $G = (\frac{1}{3}, \frac{1}{3}, \frac{1}{3})$ and **OU** meets the plane ABC at $H = (\frac{2}{3}, \frac{2}{3}, \frac{2}{3})$.

$$OG = \sqrt{(\tfrac{1}{3})^2 + (\tfrac{1}{3})^2 + (\tfrac{1}{3})^2} = \sqrt{\tfrac{1}{3}} = \frac{1}{\sqrt{3}}$$

and $$OH = \sqrt{(\tfrac{2}{3})^2 + (\tfrac{2}{3})^2 + (\tfrac{2}{3})^2} = \frac{2}{\sqrt{3}}.$$

$OU = \sqrt{3} = \dfrac{3}{\sqrt{3}}$ so G and H are the points of trisection of OU.

This fact is reasonably obvious and can be proved using the ratio theorem.

Distance of a point $P_1(x_1, y_1, z_1)$ from the plane $ax + by + cz = d$

With the notation in figure 75,

$\mathbf{OP_1 . OQ}$

$= (\mathbf{OM} + \mathbf{MP_1}) . \mathbf{OQ}$
$= \mathbf{OM . OQ} + \mathbf{MP_1 . OQ}.$

$\mathbf{OM . OQ} = d$, since M is on the plane and all points (x, y, z) satisfy $\begin{pmatrix} a \\ b \\ c \end{pmatrix} \begin{pmatrix} x \\ y \\ z \end{pmatrix} = d.$

$\mathbf{MP_1 . OQ} = MP_1\, OQ \cos 0°$, since MP_1 is parallel to OQ.

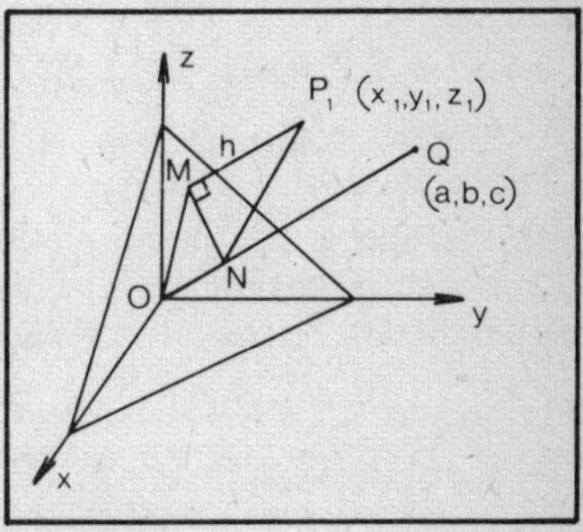

Figure 75

$$\mathbf{OP_1 . OQ} = \begin{pmatrix} x_1 \\ y_1 \\ z_1 \end{pmatrix} \cdot \begin{pmatrix} a \\ b \\ c \end{pmatrix} = d + MP_1\, OQ.$$

$$ax_1 + by_1 + cz_1 = d + MP_1 \sqrt{a^2 + b^2 + c^2}$$

$$h = MP_1 = \frac{ax_1 + by_1 + cz_1 - d}{\sqrt{a^2 + b^2 + c^2}}.$$

Example 11 Find the distance of the point (1, 1, 1) from the plane $x + 2y + 3z = 6$.

$h = \dfrac{ax_1 + by_1 + cz_1 - d}{\sqrt{a^2 + b^2 + c^2}}$ gives $h = \dfrac{1 + 2 + 3 - 6}{\sqrt{1^2 + 2^2 + 3^2}} = 0$ which means that (1, 1, 1) lies on the plane $x + 2y + 3z = 6$ which can easily be seen since it satisfies the equation of the plane.

Intersection of a line with a plane

Three cases arise:

(i) the line is parallel to the plane,

(ii) the line lies in the plane,

(iii) the line intersects the plane in a single point.

Example 12 Find the intersection of the line $x = 1 + 3t$, $y = 2 + 4t$, $z = 3 + 5t$ with the plane $x + 2y + 3z = 6$.

Substitute the x, y, z values of the line into the equation of the plane. This will give the value of t which specifies the point on the line and the plane

$$1 + 3t + 2(2 + 4t) + 3(3 + 5t) = 6 \Rightarrow 26t + 14 = 6 \Rightarrow t = -\tfrac{4}{13}.$$

The point of intersection is $x = 1 - \frac{12}{13} = \frac{1}{13}$, $y = 2 - \frac{16}{13} = \frac{10}{13}$, $z = 3 - \frac{20}{13} = \frac{19}{13}$.

The line meets the plane at the point $(\frac{1}{13}, \frac{10}{13}, \frac{19}{13})$, i.e. case (iii).
If the line is parallel to the plane, its direction vector will be perpendicular to the normal vector of the plane. Consider the line $x = 1 + 3t$, $y = 2 + 4t$, $z = 3 + 5t$ and the plane $3x + 4y - 5z = 10$.

The direction vector of the line is $\begin{pmatrix} 3 \\ 4 \\ 5 \end{pmatrix}$. The normal vector of the plane is $\begin{pmatrix} 3 \\ 4 \\ -5 \end{pmatrix}$. To find t:

$$\begin{aligned} 3(1 + 3t) + 4(2 + 4t) - 5(3 + 5t) &= 10 \\ \Rightarrow 3 + 9t + 8 + 16t - 15 - 25t &= 10 \\ \Rightarrow \qquad 11 - 15 &= 10, \end{aligned}$$

which is not true. Our assumption that the line meets the plane is false, so the line must be parallel to the plane.
Consider the same line with the plane $3x + 4y - 5z = -4$. Finding t leads to $11 - 15 = -4$ which is always true. In this case all values of t satisfy the equation of the plane and the line lies in the plane.

Key terms

Notation If the coordinates of A are (1, 2, 3) then

$$\mathbf{OA} = \overrightarrow{OA} = \mathbf{i} + 2\mathbf{j} + 3\mathbf{k} = \begin{pmatrix}1\\2\\3\end{pmatrix}.$$

Addition and subtraction $\mathbf{OA} = \begin{pmatrix}a_1\\a_2\\a_3\end{pmatrix}$, $\mathbf{OB} = \begin{pmatrix}b_1\\b_2\\b_3\end{pmatrix}$,

$$\mathbf{OA} + \mathbf{OB} = \begin{pmatrix}a_1+b_1\\a_2+b_2\\a_3+b_3\end{pmatrix} \quad \mathbf{AB} = \mathbf{OB} - \mathbf{OA} = \begin{pmatrix}b_1-a_1\\b_2-a_2\\b_3-a_3\end{pmatrix}.$$

Ratio theorem P divides AB in ratio

$$p:q \Rightarrow \mathbf{OP} = \frac{q}{p+q}\mathbf{OA} + \frac{p}{p+q}\mathbf{OB}.$$

If $\mathbf{OA} = \mathbf{a}$ and $\mathbf{OB} = \mathbf{b}$, then the mid-point of $\mathbf{AB}$ is $\frac{1}{2}\mathbf{a} + \frac{1}{2}\mathbf{b}$ and the points of trisection of $\mathbf{AB}$ are $\frac{2}{3}\mathbf{a} + \frac{1}{3}\mathbf{b}$ (nearer A) and $\frac{1}{3}\mathbf{a} + \frac{2}{3}\mathbf{b}$ (nearer B).

Vector equation of a line $\begin{pmatrix}x\\y\\z\end{pmatrix} = \begin{pmatrix}1\\2\\3\end{pmatrix} + t\begin{pmatrix}3\\4\\5\end{pmatrix}$

represents a straight line through the point (1, 2, 3) in the direction of the vector $3\mathbf{i} + 4\mathbf{j} + 5\mathbf{k}$.

Scalar product of $\mathbf{a} = \begin{pmatrix}a_1\\a_2\\a_3\end{pmatrix}$ and $\mathbf{b} = \begin{pmatrix}b_1\\b_2\\b_3\end{pmatrix}$ is

$\mathbf{a} \cdot \mathbf{b} = ab\cos\theta = a_1b_1 + a_2b_2 + a_3b_3$.

Equation of a plane is $ax + by + cz = d$, where $\begin{pmatrix}a\\b\\c\end{pmatrix}$ is the normal vector of the plane.

Distance of the point (x_1, y_1, z_1) to the plane $ax + by + cz = d$ is $\dfrac{ax_1 + by_1 + cz_1 - d}{\sqrt{a^2 + b^2 + c^2}}$.

Chapter 12 Probability

A fourteen-year-old boy was asked to distinguish between 'weather' and 'climate'. After a moment's thought he came up with his answer. 'Climate is what should happen and weather is what does happen.' He meant that climate is a theoretical forecast based on years of observation of the weather. The same is true of probability and statistics. Probability is a theoretical prediction of events which resembles closely the statistical patterns and results which are actually observed.

For example, the probability of a coin landing heads is the same as a coin landing tails, because in the long run it is observed that half the time coins land heads and half the time coins land tails. Our probability theory is based on the statistics of many observed results.

Measuring probability

If we throw a coin 100 times we expect to observe 50 'heads' and 50 'tails'. This may not actually happen, but if the coin is fair (not biased) then as the number of throws increases, the proportion of heads and tails each gets closer to one half. In the event of throwing the coin, 'heads' and 'tails' are two equally likely outcomes, so the probability of a head is one out of two, or 1 in 2, or $\frac{1}{2}$. Similarly, the probability of a tail is also $\frac{1}{2}$. In throwing a 'die' there are 6 equally likely outcomes (if the die is fair). The probability of a six with one throw is 1 in 6 or $\frac{1}{6}$.

Example 1 The probability of cutting a shuffled pack of cards and getting the ace of spades is $\frac{1}{52}$ since all 52 cards have an equal chance of being cut.

Example 2 The probability of cutting any spade is $\frac{13}{52} = \frac{1}{4}$ because there are 13 spades in the pack.

The probability of an event is measured in fractions between 0 and 1 and is cancelled down to its lowest terms.

The probability of an event A

$$= \frac{\text{number of outcomes producing A}}{\text{number of possible outcomes}}.$$

Example 3 The probability of a person's birthday being in January is $\frac{1}{12}$ since there are 12 months in the year. More accurately, it is $\frac{31}{365}$ since there are more days in some months than others, or even $\dfrac{31}{365\frac{1}{4}}$.

If an event B has probability 0 it means it will never happen. For example, the probability of gaining a score of seven with one throw of a die. If an event C has probability 1 this means the event is certain to happen.

Success and failure

If the probability of throwing a 6 is $\frac{1}{6}$, the probability of not throwing a 6 is $\frac{5}{6}$.

Adding probabilities

In one throw of a die, the probability of a six, $p(6) = \frac{1}{6}$ and $p(5) = \frac{1}{6}$. The probability of a 5 or 6, $p(5 \text{ or } 6) = \frac{1}{6} + \frac{1}{6} = \frac{1}{3}$. But p(even number) $\frac{3}{6} = \frac{1}{2}$ and $p(\text{even or } 6) = \frac{1}{2}$ and not $\frac{1}{2} + \frac{1}{6}$, since the event of throwing an even number includes a six.

If A and B are mutually exclusive outcomes

$$p(A) + p(B) = p(A \text{ or } B) = p(A \cup B).$$

In this case A and B cannot both happen at once which means that $p(A \text{ and } B) = p(A \cap B) = 0$.

The events E(even number) and S(a six) do not exclude each other.

In this case $p(E \text{ or } S) = \frac{1}{2} = p(E) + p(S) - p(E \text{ and } S)$

$$= \tfrac{1}{2} \quad + \tfrac{1}{6} \quad - \tfrac{1}{6}.$$

This result is true in general for 2 events A and B.

$$p(A \text{ or } B) = p(A) + p(B) - p(A \text{ and } B),$$

or

$$p(A \cup B) = p(A) + p(B) - p(A \cap B).$$

This result is similar to that for the intersection and union of 2 sets.

Example 4 Find the probability of drawing from a pack of cards: (i) a red card (ii) a picture card (iii) a red card or a picture card.

$$p(\text{red}) = \tfrac{26}{52} = \tfrac{1}{2}.$$

A picture card is a jack, queen or king, and p(picture) $= \frac{12}{52} = \frac{3}{13}$.

$$\text{p(red or picture)} = \text{p(red)} + \text{p(picture)} - \text{p(red and picture)}$$
$$= \frac{1}{2} + \frac{3}{13} - \frac{6}{52} = \frac{26+12-6}{52} = \frac{32}{52} = \frac{8}{13}.$$

Combined events

What is the probability of throwing a head and a tail with 2 coins? Writing H for head and T for tail the possible outcomes are HH, HT, TH, TT the second and third being different since a head on the first coin and a tail on the second, is different from a tail on the first and a head on the second.

$$\text{p(H and T)} = \tfrac{2}{4} = \tfrac{1}{2}.$$

This information can be shown pictorially on a tree diagram (figure 76) where 4 outcomes are represented as branches. Starting from S there are 4 routes along the branches providing the 4 outcomes. The probabilities of each branch are multiplied along the branches to give the probability of each double event. Thus

$$\text{p(HH)} = \tfrac{1}{2} \times \tfrac{1}{2} = \tfrac{1}{4}.$$

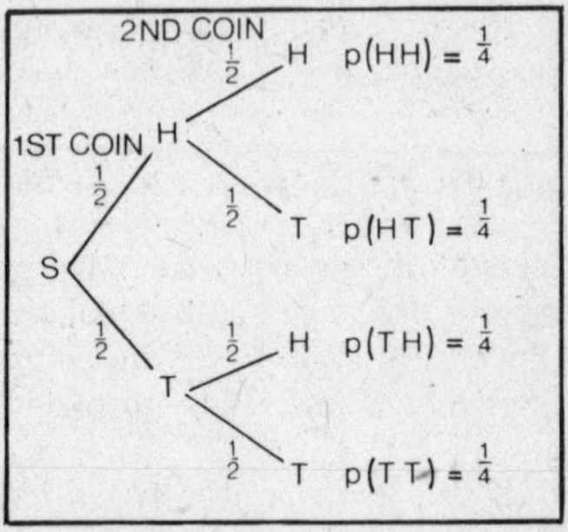

Figure 76

Throwing two dice

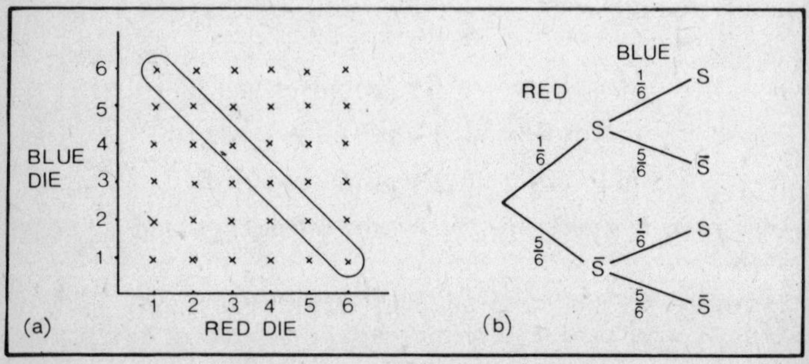

Figure 77

If two dice are thrown together, there are 36 possible outcomes like (3, 6)(6, 3)(1, 2), etc. The ordered pairs can best be illustrated with a 'coordinate' diagram (figure 77a) with the first coordinate the score on the red die and the second the blue die score. Each point represents an outcome and (3, 2) is different from (2, 3).

To find the probability of a total score of 7, we look for the outcomes which produce 7, i.e. (6, 1)(5, 2)(4, 3)(3, 4)(2, 5) and (1, 6). There are six, so $p(7) = \frac{6}{36} = \frac{1}{6}$. Figure 77(a) shows the six possible outcomes enclosed in a set which looks similar to the 'line $x + y = 7$'.

Similarly $\text{p(six on red die)} = \frac{6}{36} = \frac{1}{6}$.

$\text{p(six on blue die)} = \frac{6}{36} = \frac{1}{6}$.

$$\begin{aligned}\text{p(six on either die)} &= \text{p(six on red)} + \text{p(six on blue)} \\ &\quad - \text{p(six on both)} \\ &= \tfrac{1}{6} + \tfrac{1}{6} - \tfrac{1}{36} = \tfrac{11}{36}.\end{aligned}$$

If we are only looking for sixes, the tree diagram in figure 77(b) gives a simplified version.

$$\begin{aligned}\text{p(red 6 and blue 6)} &= \tfrac{1}{6} \times \tfrac{1}{6} = \tfrac{1}{36} \\ \text{p(red 6 and no blue 6)} &= \tfrac{1}{6} \times \tfrac{5}{6} = \tfrac{5}{36} \\ \text{p(no red 6 and blue 6)} &= \tfrac{5}{6} \times \tfrac{1}{6} = \tfrac{5}{36} \\ \text{p(no red 6 and no blue 6)} &= \tfrac{5}{6} \times \tfrac{5}{6} = \tfrac{25}{36}\end{aligned}$$

These 4 outcomes exhaust all the possibilities and their sum = 1.

The tree diagram in figure 77(b) could be drawn with the score on the blue die as the first event, and the results would be the same, since the outcome on one die does not depend on the other die. In some cases the two events are dependent. In other words, the outcome of the second event depends on the result of the first event. In some cases, two events which may appear simultaneous like the choosing of 2 cards from a pack, have to be treated as one event following another, as they are dependent, the result of the first affecting the result of the second.

Drawing with replacement

Consider drawing 2 balls from a bag containing 5 green balls and 3 yellow balls, all eight being of equal size and weight so that each ball has an equal probability of being drawn. There are 4 possible results: (i) green followed by green, (ii) green and yellow, (iii) yellow and green and (iv) yellow and yellow. Figure 78(a) shows the tree diagram for the case when the first ball is replaced after the first drawing.

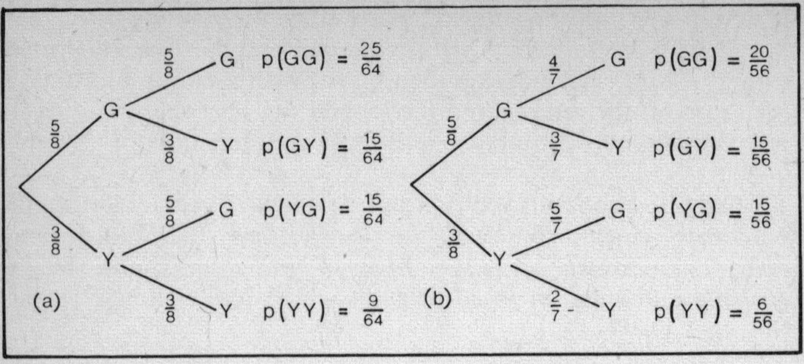

Figure 78

Drawing without replacement

Figure 78(b) shows the tree diagram for the case when the first ball is not replaced. In this case the probability of drawing a green ball on the second drawing depends on whether a green or yellow was drawn first. If a green ball was drawn first, there remain 4 green and 3 yellow, and the probability of a green on the second draw is $\frac{4}{7}$. The probability of 2 green balls is $\frac{5}{8} \times \frac{4}{7} = \frac{20}{56} = \frac{5}{14}$. Check that in both tree diagrams the probabilities of the four possible outcomes add up to 1. The probability $\frac{4}{7}$ is called the **conditional probability**. It is the probability of a green on the second draw given that a green has occurred on the first draw and denoted by $p(G|G)$. The probability $\frac{3}{7}$ of a yellow on the second draw is the probability of 'yellow given green', i.e. $p(Y|G) = \frac{3}{7}$.

The probability of G and Y is found by multiplying along the branches and this gives us a formal definition of the conditional probability, $p(Y|G)$.

$$p(G \text{ and } Y) = \tfrac{5}{8} \times \tfrac{3}{7} = p(G) \times p(Y|G) \Rightarrow p(Y|G) = \frac{p(G \text{ and } Y)}{p(G)} = \frac{p(G \cap Y)}{p(G)}.$$

Similarly $$p(G|Y) = \frac{p(G \cap Y)}{p(Y)}.$$

The third 'branch' gives $p(G|Y) = \dfrac{p(G \text{ and } Y)}{p(Y)} = \dfrac{15/56}{3/8} = \dfrac{5}{7}$.

Since the events G and Y are exhaustive (they include all

possibilities of drawing one ball), Y can be written as ~G or $\bar{G}$ standing for not green.

Then $$p(Y) + p(G) = 1 \Rightarrow p(\bar{G}) = 1 - p(G).$$

The last two examples of drawing balls from a bag can be thought of as two distinct events happening one after the other in two separate 'trials'. Sometimes two distinct events can be defined within one 'trial'.

Example 5 Let the 'trial' be throwing the blue and red die together and let the event F be a five on the blue die and the event S a total score of seven. Figure 79(a) shows the tree diagram with the first 2 branches representing F and $\bar{F}$.

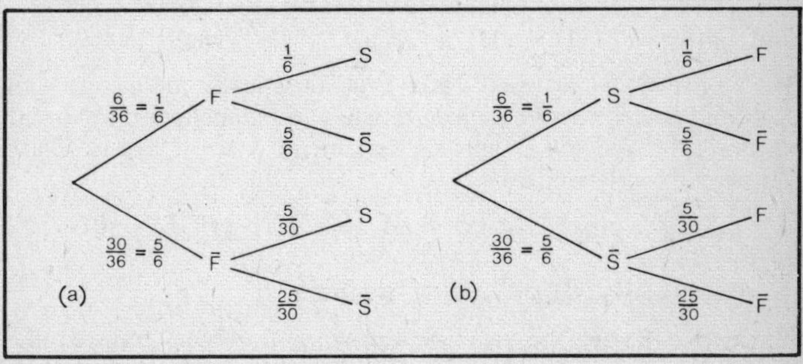

Figure 79

Five on the blue die can occur in 6 ways (1, 5) (2, 5) (3, 5) (4, 5) (5, 5) and (6, 5), out of 36; so $p(F) = \frac{6}{36} = \frac{1}{6}$. Out of these 6 ways, one gives a total of seven, so $p(S|F) = \frac{1}{6}$ and $p(\bar{S}|F) = \frac{5}{6}$. Out of the 30 ways of getting $\bar{F}$, 5 have a total of 7 and 25 do not.

$$p(S|\bar{F}) = \frac{5}{30} = \frac{1}{6} \text{ and } p(\bar{S}|\bar{F}) = \frac{5}{6}.$$

Since $p(S|F) = \frac{1}{6} = p(S|\bar{F})$, we say that the event S is **statistically independent** of the event F, i.e. its probability does not depend on whether F has happened or not. Figure 79(b) shows the same situation with S and $\bar{S}$ on the first branches of the tree diagram. Out of 6 ways of getting S, one gives F and 5 give $\bar{F}$; so $p(F|S) = \frac{1}{6}$ and $p(\bar{F}|S) = \frac{5}{6}$.

Out of 30 outcomes that produce $\bar{S}$, 5 give F and 25 give $\bar{F}$, so $p(F|\bar{S})$ and F is statistically independent of S. The event S in a general way depends on F since if F occurs, the outcomes then producing S are different from those producing S if $\bar{F}$ occurs.

However, we can say S is statistically independent of F if the probabilities of $S|F$ and $S|\bar{F}$ have the same value.

The question now arises, 'if S is statistically independent of F, is F statistically independent of S?'

Firstly, let $p(F) = f$ so that $p(\bar{F}) = 1 - f$, and let $p(S|F) = s \Rightarrow p(\bar{S}|F) = 1 - s$. From figure 79(a), S can happen in two ways, i.e. $F \cap S$ or $\bar{F} \cap S$. So $p(S) = p(F \cap S) + p(\bar{F} \cap S)$.

If S is statistically independent of F, then $p(S|F) = s$.

$\therefore \quad p(S) = p(F \cap S) + p(\bar{F} \cap S) = fs + (1-f)s = s.$

From figure 92(b), $p(S \cap F) = fs$ and $p(S) = s \Rightarrow p(F|S) = f$,

and $p(\bar{S} \cap F) = (1-s)f$ and $p(\bar{S}) = 1 - s \Rightarrow p(F|\bar{S}) = f$.

Hence $p(F|S) = f = p(F|\bar{S})$ and F is statistically independent of S. F and S can now be called statistically independent, now that we have proved the reciprocal statement. Also, if F and S are statistically independent

$$p(S|F) = p(S|\bar{F}) = p(S) \quad \text{and} \quad p(F|S) = p(F|\bar{S}) = p(F).$$

Statistically dependent events

Example 6 Let the trial be throwing the blue and red die together and consider the event T to be a total score greater than 10, which can be gained in 3 ways (5, 6) (6, 5) and (6, 6). The event F stands for a 5 on the blue die which can arise in 6 ways (1, 5) (2, 5) (3, 5) (4, 5) (5, 5) and (6, 5). Figure 80(a) shows the tree diagram writing T on the first branches. Out of the 3 ways for T, one gives F and 2 do not. So $p(F|T) = \frac{1}{3}$ and $p(\bar{F}|T) = \frac{2}{3}$.

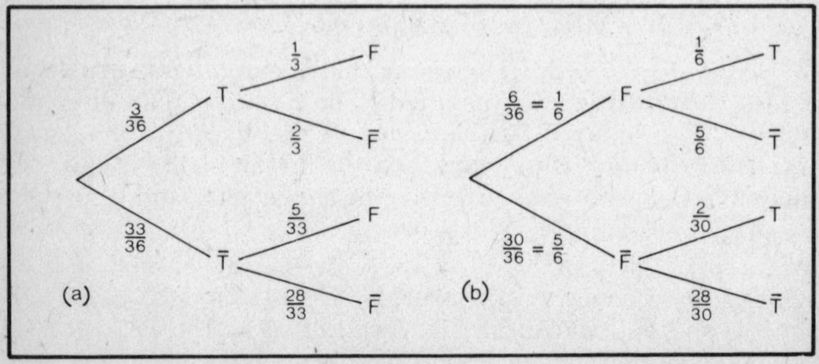

Figure 80

Out of the 33 ways for $\bar{T}$, 5 give F and 28 do not.

So $p(F|\bar{T}) = \frac{5}{33}$ and $p(\bar{F}|\bar{T}) = \frac{28}{33}$. F is therefore statistically dependent on T.

Figure 80(b) shows the tree diagram with F on the first branches. Out of the 6 outcomes producing F, one produces T and 5 produce $\bar{T}$. $p(T|F) = \frac{1}{6}$ and $p(\bar{T}|F) = \frac{5}{6}$.
Out of the 30 outcomes producing $\bar{F}$, 2 produce T and 28 produce $\bar{T}$.

$$p(T|\bar{F}) = \tfrac{2}{30} = \tfrac{1}{15} \quad \text{and} \quad p(\bar{T}|\bar{F}) = \tfrac{28}{30} = \tfrac{14}{15}.$$

T is not statistically independent of F.
As a check from figure 80(a)

$$\begin{aligned} p(T \cap F) + p(\bar{T} \cap F) &= \tfrac{3}{36} \times \tfrac{1}{3} + \tfrac{33}{36} \times \tfrac{5}{33} = \tfrac{1}{36} + \tfrac{5}{36} = \tfrac{6}{36} \\ &= \tfrac{1}{6} = p(F), \end{aligned}$$

and from figure 80(b)

$$\begin{aligned} p(F \cap T) + p(\bar{F} \cap T) &= \tfrac{1}{6} \times \tfrac{1}{6} + \tfrac{5}{6} \times \tfrac{1}{15} = \tfrac{1}{36} + \tfrac{2}{36} \\ &= \tfrac{3}{36} = \tfrac{1}{12} = p(T). \end{aligned}$$

Tree diagrams give an easily understandable way of representing the two events in examples 5 and 6. Another way is to make a **contingency table** which includes all the information in a table of probabilities.

Contingency tables

In figure 81(a) the first entry represents $p(S \cap F) = \frac{1}{36}$ and the second entry (top row) represents $p(S \cap \bar{F}) = \frac{5}{36}$. These two added give $\frac{1}{6}$, which is p(S). The sum of the second row $\frac{5}{6} = p(\bar{S})$ and

$\cap$	F	$\bar{F}$	
S	$\frac{1}{36}$	$\frac{5}{36}$	$\frac{1}{6} = p(S)$
$\bar{S}$	$\frac{5}{36}$	$\frac{25}{36}$	$\frac{5}{6} = p(\bar{S})$
	$\frac{1}{6}$	$\frac{5}{6}$	1

(a)

$\cap$	T	$\bar{T}$	
F	$\frac{1}{36}$	$\frac{5}{36}$	$\frac{1}{6} = p(F)$
$\bar{F}$	$\frac{2}{36}$	$\frac{28}{36}$	$\frac{5}{6} = p(\bar{F})$
	$\frac{1}{12}$	$\frac{11}{12}$	1
	$= p(T)$	$= p(\bar{T})$	

(b)

Figure 81

similarly the sum of the first column will be p(F) and the second column p($\bar{F}$). For conditional probabilities, p(F|S) and p($\bar{F}$|S) are found by comparing the top row entries $\frac{1}{36}$ and $\frac{5}{36}$, i.e. p(F|S) and p($\bar{F}$|S) are in the ratio 1 : 5 and must be $\frac{1}{6}$ and $\frac{5}{6}$. The probabilities of F and $\bar{F}$ given $\bar{S}$ are in the same ratio which leads to statistical independence.

In figure 81(b) for example 6, $p(F \cap T) = \frac{1}{36}$; $p(F \cap \bar{T}) = \frac{5}{36}$; $p(\bar{F} \cap T) = \frac{2}{36}$ and the rest can be calculated. The rows are not in the same ratio and the events will not be statistically independent.

Example 7 Draw up the contingency table for $p(A) = \frac{3}{7}$, $p(B) = \frac{2}{5}$ and $p(A \text{ and } B) = \frac{11}{35}$. Are the events statistically independent?

$\cap$	p(A)	p($\bar{A}$)	
p(B)	$\frac{11}{35}$	$\frac{3}{35}$	$\frac{2}{5}$
p($\bar{B}$)	$\frac{4}{35}$	$\frac{17}{35}$	$\frac{3}{5}$
	$\frac{3}{7}$	$\frac{4}{7}$	1

$$p(A|B) = \frac{p(A \cap B)}{p(B)} = \frac{11/35}{2/5} = \frac{11}{14}$$

$$p(A|\bar{B}) = \frac{p(A \cap \bar{B})}{p(\bar{B})} = \frac{4/35}{3/5} = \frac{4}{21}$$

$p(A|B) \neq p(A|\bar{B}) \Rightarrow$ A and B are not statistically independent. It has been shown that if A is statistically independent of B, then $p(A|B) = p(A|\bar{B}) = p(A)$ and in this case $p(B|A) = p(B|\bar{A}) = p(B)$ and B is statistically independent of A. In this case the events are independent of each other. Can anything be said about the conditional probability B|A if we only know the conditional probabilities A|B and A|$\bar{B}$? The probability of B must be known to complete the tree diagram with event B preceding A, then the tree diagram must be reversed to show event A preceding B. In some situations it is convenient to measure p(A|B) and p(A|$\bar{B}$) and yet desirable to know p(B|A) and p(B|$\bar{A}$). This reversing of the tree diagram leads to a result known as **Bayes theorem**, but it is easier and more straightforward to reverse the tree diagram.

Example 8 Although I carry my raincoat $\frac{3}{4}$ of the time I always seem to be getting wet. It rains on $\frac{2}{3}$ of the days when I do not take it, but only $\frac{1}{3}$ of the days when I do take it. How true is it that I am always getting wet?

Let the events be R for 'it rains' and C for 'I take my raincoat'. Then $p(C) = \frac{3}{4}$, $p(R|C) = \frac{1}{3}$, $p(R|\bar{C}) = \frac{2}{3}$. This information is summarized in figure 95(a), from which we deduce that

$$p(C \cap R) = \tfrac{3}{4} \times \tfrac{1}{3} = \tfrac{1}{4} \quad \text{and} \quad p(\bar{C} \cap R) = \tfrac{1}{4} \times \tfrac{2}{3} = \tfrac{1}{6}.$$

Hence $p(R) = p(C \cap R) + p(\bar{C} \cap R) = \frac{1}{4} + \frac{1}{6} = \frac{5}{12}$.

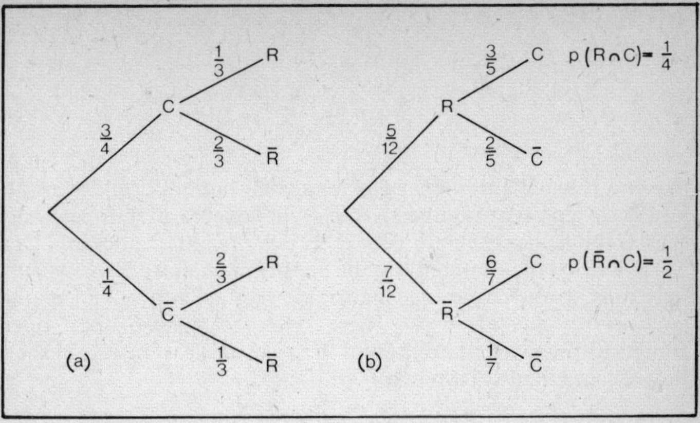

Figure 82

This enables the reversed tree diagram figure 82(b) to be constructed since

$$p(C|R) = \frac{p(C \cap R)}{p(R)} = \frac{1/4}{5/12} = \frac{3}{5}$$

and $$p(C|\bar{R}) = \frac{p(C \cap \bar{R})}{p(\bar{R})} = \frac{1/2}{7/12} = \frac{6}{7}.$$

I only get wet when it rains and I do not take my raincoat.

This is the event $R \cap \bar{C}$. When it rains I have my raincoat $\frac{3}{5}$ of the time since $p(C|R) = \frac{3}{5}$. Since $p(R \cap \bar{C}) = \frac{5}{12} \times \frac{2}{5} = \frac{1}{6}$. I only get wet $\frac{1}{6}$ of the time, but this is a high proportion $(\frac{2}{3})$ of the small fraction $(\frac{1}{4})$ of the times I do not take my raincoat.

Key terms

$p(A \text{ or } B) = p(A) + p(B) - p(A \cap B)$.

Conditional probability

$$p(A|B) = \frac{p(A \cap B)}{p(B)}.$$

The event A is statistically independent of the event B if

$$p(A|B) = p(A|\bar{B}) \qquad \{= p(A)\}.$$

A is statistically independent of B

$$\Leftrightarrow \text{B is statistically independent of A.}$$

Chapter 13
Groups

Algebra is usually thought of as long complicated formulae using letters to represent numbers. This is the algebra of numbers and the numbers obey certain rules. The study of mathematics involves much more than numbers and the mathematician works with sets, functions, transformations, matrices, vectors and other mathematical tools and ideas. These other 'elements' obey rules similar to those for numbers and the rules need to be established before mathematical work can be done.

Algebra of numbers

The most elementary numbers are the **natural** or **counting numbers** $N = \{1, 2, 3, 4, \ldots\}$. These form an infinite set and if we add two members of the set together we always obtain another member of the set. N is said to be closed under the operation of addition. N is not closed under subtraction since $3 - 4 = -1$ which is not a member of the set N. N is closed under multiplication, but not closed under division. The operation of subtraction on the members of N produces new elements like $0, -1, -2$, etc. and these together with N form a new set whose members are called the **integers**. Integers are denoted by $Z = \{\ldots, -3, -2, -1, 0, 1, 2, \ldots\}$. The operation of division on Z produces numbers like $\frac{1}{2}, \frac{2}{3}$ which are **fractions**. The fractions belong to a set of numbers called the **rational numbers**. The set of rational numbers, Q, consists of numbers of the form $\frac{a}{b}$, where $a, b \in Z$ and $b \neq 0$. So the rational numbers include positive and negative whole numbers and fractions. Even recurring decimals belong to the set of rational numbers, since they can be changed into rational numbers by the following process. If $x = 0{\cdot}\dot{1}23\,12\dot{3}$, then $1000x = 123{\cdot}123\,123$, $1000x - x = 123$, since the recurring decimal parts are equal and $999x = 123 \Leftrightarrow x = \frac{123}{999} = \frac{41}{333}$ and x is rational.

The rational numbers form the first set of numbers which are closed under the four rules of arithmetic.

Addition

$\frac{p}{q}+\frac{r}{s}=\frac{ps+qr}{qs}$ which is rational, $q \neq 0; s \neq 0 \Rightarrow qs \neq 0.$

Subtraction $\frac{p}{q}-\frac{r}{s}=\frac{ps-qr}{qs}$ which is rational.

Multiplication $\frac{p}{q}\times\frac{r}{s}=\frac{pr}{qs}$ which is rational.

Division $\frac{p}{q}\div\frac{r}{s}=\frac{ps}{qr}$ which is rational, provided $r \neq 0, q \neq 0.$

The number system has now been extended to include all rational numbers both positive and negative and might be thought to be complete. All the rational numbers appear on the number line (figure 83) but there are others as well.

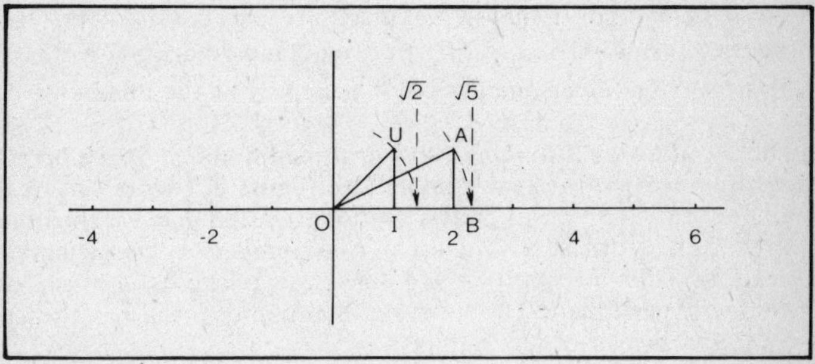

Figure 83

For instance the number 0·1010010001... is not a recurring decimal. It has a pattern but cannot be converted into the ratio of two natural numbers. Similarly $\sqrt{2}$ and $\sqrt{5}$ are not rational numbers. We can calculate the value of $\sqrt{5}$ to any prescribed accuracy ($\sqrt{5} = 2{\cdot}23606798$ to 8 decimal places), but its decimal equivalent does not recur.

Proof that $\sqrt{5}$ is irrational

Assume $\sqrt{5}$ is rational and produce a contradiction. (You may not be familiar with this kind of proof, but if a logically sound argument produces a false conclusion then the original assumption must have been wrong.)

$\sqrt{5}$ rational $\Rightarrow \sqrt{5}$ can be written as p/q, where p and q belong to N (whole numbers), $q \neq 0$ and p and q are coprime (the fraction p/q is in its lowest form; i.e. p and q have no common factors).

$$\sqrt{5} = \frac{p}{q} \Rightarrow 5 = \frac{p^2}{q^2} \Leftrightarrow 5q^2 = p^2 \Rightarrow p^2 \text{ is divisible by 5.}$$

$p \times p$ has a factor of $5 \Rightarrow p$ has a factor of $5 \Rightarrow p = 5r$, where r is a whole number.

$$5q^2 = p^2 \Leftrightarrow 5q^2 = 25r^2 \Leftrightarrow q^2 = 5r^2 \Rightarrow q^2 \text{ is divisible by 5.}$$
$$\Rightarrow q \text{ is divisible by 5.}$$

p and q have both been shown to contain a factor 5 and this contradicts the fact that they were coprime. This is the contradiction proving the original assumption was false, that is, $\sqrt{5}$ must be irrational. Other irrational numbers occur in mathematics like $\sqrt{2}$, π and e. These numbers are on the number line (figure 83) since OA $= \sqrt{5}$ (Pythagoras) and rotating OA down to meet the number line gives OB to represent the number $\sqrt{5}$.

The set of rational numbers and irrational numbers make up all the numbers on the real number line (figure 83), since any real number which is not rational is irrational, and this set is called the set of **real numbers**. The set of real numbers is closed under the four rules of number and for most purposes this set of numbers is adequate. The set of rational numbers will be denoted by Q and the set of real numbers by R.

Rules in the algebra of numbers

When numbers are combined under the operation of the four rules, certain properties are obvious and are taken for granted.

For example $3 + 4 = 4 + 3 = 7$,
$3 \times 4 = 4 \times 3 = 12$.

The operations of addition and multiplication on the set of real numbers are commutative. Subtraction is not commutative since $3 - 4 \neq 4 - 3$. Division is not commutative since $3 \div 4 \neq 4 \div 3$. When three numbers are combined, why do we talk of 'adding three numbers together' but not 'subtracting three numbers together'?

$3 + 4 + 5$ either means $(3 + 4) + 5$ or $3 + (4 + 5)$ the brackets indicating which numbers are added first. Since $(3 + 4) + 5 = 3 + (4 + 5) = 12$, the order of addition does not matter and addition of numbers is said to be associative. Similarly, the operation of multiplication on the set of real numbers is also associative since $(3 \times 4) \times 5 = 3 \times (4 \times 5) = 60$. However, subtraction is not associative since $5 - (4 - 3) \neq (5 - 4) - 3$ and neither is division since $12 \div (6 \div 2) \neq (12 \div 6) \div 2$.

The distributive laws

When two or more operations are being performed, certain laws and conventions need to be established. For example, what does $3 \times 4 + 5$ mean? $3 \times 4 + 5$ should be punctuated by using brackets. The language of mathematics and its conventions must be clearly understood before progress can be made. It is important to appreciate the basic rules even though they may only be conventions. $3 \times 4 + 5$ either stands for $(3 \times 4) + 5 = 17$ or $3 \times (4 + 5) = 27$ and realistically the brackets must be left in. However, it has been agreed that the brackets are left out in the first case so that $3 \times 4 + 5$ means $(3 \times 4) + 5 = 17$ and $3 + 4 \times 5$ means $3 + (4 \times 5) = 23$. In other words, multiplication is done first. Also $3 - 4 \times 5 = 3 - (4 \times 5) = -17$ and $6 \div 2 - 3 = (6 \div 2) - 3 = 0$. Multiplication and division are done before addition and subtraction. Consider $3 - 4 + 5$ which could mean $(3 - 4) + 5 = 4$ or $3 - (4 + 5) = -6$. In the first case the brackets are omitted and in the second they are inserted.

Similarly $12 \div 6 \times 2$ could mean $(12 \div 6) \times 2 = 4$ or $12 \div (6 \times 2) = 1$. The brackets are retained only in the second case so that $24 \div 8 \times 2$ means $(24 \div 8) \times 2 = 6$.

All these conventions may seem puzzling, but they form the language of mathematics and language without punctuation can be meaningless.

$3 \times (4 + 5)$ can be written without brackets although the brackets are essential to convey the proper meaning.
$3 \times (4 + 5) = 3 \times 4 + 3 \times 5 = (3 \times 4) + (3 \times 5)$, the brackets being inserted to emphasize this distributive law.
Multiplication is distributive over addition for the set of real numbers.
Multiplication is also distributive over subtraction which can be summarized formally as
$a \times (b - c) = (a \times b) - (a \times c)$.

Addition is not distributive over multiplication. If it were $3 + (4 \times 5) = 23$ would equal $(3 + 4) \times (3 + 5) = 56$.
Multiplication is distributive over addition from the right hand side $(4 + 5) \times 3 = (4 \times 3) + (5 \times 3) = 27$.
Care must be taken with division.
$(6 + 9) \div 3 = (6 \div 3) + (9 \div 3) = 5$ is true, but
$3 \div (6 + 9) \neq (3 \div 6) + (3 \div 9)$.

Many of these rules are assimilated carefully over a long learning period and are taken for granted. Each time a new algebra is presented, these rules of combination and procedure should be tested, otherwise wrong assumptions and mistakes are made.

Algebra of sets

Sets are made up of elements which may be numbers, letters, functions, cars or anything.
If set $V = \{a, e, i, o, u\}$ and $F = \{a, b, c, d, e\}$ the union of V and F, written $V \cup F = \{a, b, c, d, e, i, o, u\}$, i.e. the set containing elements which are in V or in F or both.
The intersection of V and F written $V \cap F = \{a, e\}$, i.e. the set containing the elements which are in both V and F.

The picture in figure 84 is a Venn diagram illustrating the situation clearly. $V \cap F$ is the overlapping region containing a and e. The universal set for this situation could be the letters of the alphabet. The complement of V, denoted by V', is the set of elements which are not in V. $V' = \{\text{consonants}\}$
$F' = \{f, g, h, i, \ldots, x, y, z\}$.

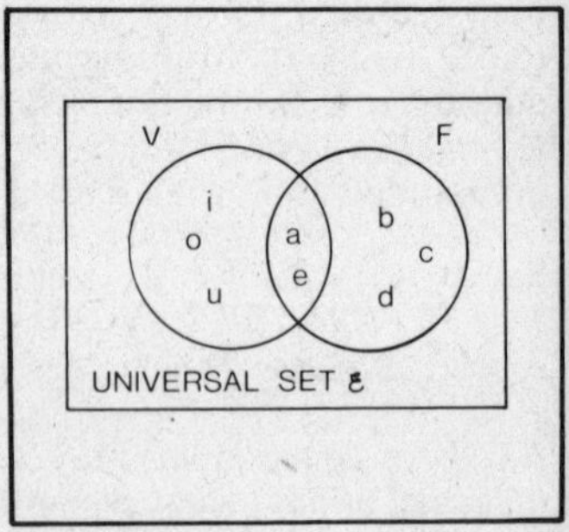

Figure 84

In figure 85(a) $V \cap F$ is completely unshaded, V' is vertically shaded and F' is horizontally shaded.
In figure 85(b), the region shaded horizontally is in V and outside F, so that it is in V and in F', hence it is in $V \cap F'$.
In figure 85(b) the region shaded vertically is in F and outside V, so that it is in F and in V' and is therefore in $F \cap V'$.

Commutativity $V \cap F = \{a, e\} = F \cap V$, since they both describe those elements in both V and F.

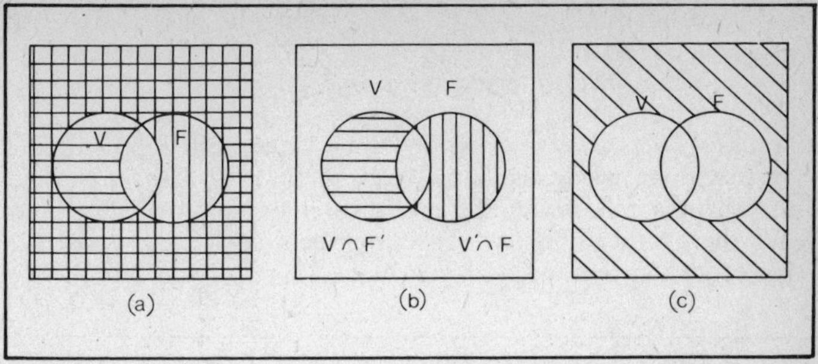

Figure 85

Similarly $V \cup F = F \cup V$ and both the union and the intersection of sets of commutative.
The region unshaded in figure 85(c) represents $V \cup F$. The region shaded is therefore the complement of this, i.e. $(V \cup F)'$.
Referring to figure 85(a), this same region is double shaded and is therefore in both V' and F', thus $(V \cup F)' = V' \cap F'$. This is one of **De Morgan's laws**. It can easily be verified by writing out the individual elements, but is quite easily illustrated by the diagrams.

In figure 85(a) the region unshaded is $V \cap F$ and the region outside it is $(V \cap F)'$. This region is shaded either vertically or horizontally and therefore is $V' \cup F'$. $(V \cap F)' = V' \cup F'$. This is the second of De Morgan's laws. Again this can be verified by listing the elements of each set.

Associativity

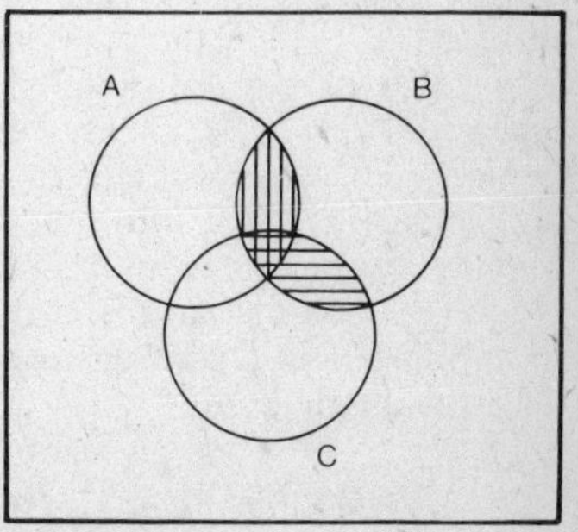

Figure 86

Figure 86 shows the Venn diagram for the possible situation with three sets. An element could be in A or in A', B or B', C or C', which gives eight possible combinations which correspond with the eight regions in figure 86. These eight regions are listed in figure 87 (page 230). The region $A \cap B$ is vertically shaded. Its intersection with C is the middle region (doubly shaded). $B \cap C$

is shaded horizontally and its intersection with A is the central region. $(A \cap B) \cap C = A \cap (B \cap C)$, since both represent the middle region; thus intersection is associative.

Similarly $(A \cup B) \cup C = A \cup (B \cup C)$ since both represent the seven regions inside the three circles. Union is associative and we can now talk about the intersection or union of three sets since there is no ambiguity. (Three numbers can be added together since addition is associative.)

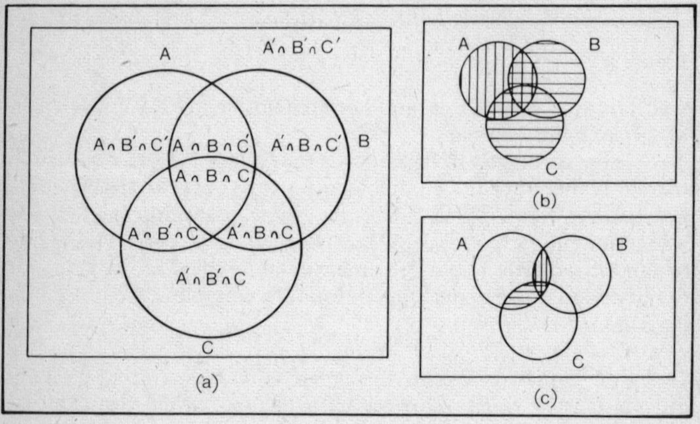

Figure 87

Figure 87(a) shows the eight possible regions that can result with three sets (7 within the circles and the outside region). Since the union of two sets will involve at least two regions, these eight individual regions can only be described with the aid of intersections. The central region is in all three sets (as we have seen) and so is $A \cap B \cap C$ (no brackets since associative). The region directly above this is in A and B (and so $A \cap B$) but not in C (so in C'). This region is $A \cap B \cap C'$. The elements in C only are in A' and B' and C and this region is $A' \cap B' \cap C$. The complete list is in the diagram.

Distributivity

$A \cap (B \cup C) = (A \cap B) \cup (A \cap C)$.
In figure 87(b), $B \cup C$ is shaded horizontally and A vertically, so $A \cap (B \cup C)$ is doubly shaded.

In figure 87(c) $A \cap B$ is shaded vertically and $A \cap C$ horizontally, so the region with any shading at all represents $(A \cap B) \cup (A \cap C)$ and this is the same as $A \cap (B \cup C)$. So

$$A \cap (B \cup C) = (A \cap B) \cup (A \cap C).$$

$$A \cup (B \cap C) = (A \cup B) \cap (A \cup C).$$

By shading the appropriate diagrams, the above rule can be shown to be true. For the operations of union and intersection of sets we have: (i) union is distributive over intersection, (ii) intersection is distributive over union. The algebraic laws of closure, commutativity, associativity and distributivity must be established before the usual operations are performed.

Finite or modular arithmetic

The system of modular arithmetic may seem a little contrived, but its extension has important applications in number theory and its simpler examples provide valuable insight into algebraic laws and structure. The set of integers can be partitioned into five sets (modulo 5) if they leave the same remainder after division by 5. The integers 1, 6, 11, ... leave a remainder 1 after dividing by 5. This set is denoted by 1. Similarly the set denoted by 2 contains

2, 7, 12, 17, ... and also -3, -8, ...

$0 \equiv 0, \pm 5, \pm 10, \ldots$ $\quad$ $1 \equiv \ldots, -9, -4, 1, 6, 11, 16, \ldots$

$2 \equiv \ldots, -8, -3, 2, 7, 12, \ldots$ $\quad$ $3 \equiv \ldots, -7, -2, 3, 8, 13, \ldots$

$4 \equiv \ldots, -6, -1, 4, 9, 14, \ldots$

Alternatively, the sets can be defined as sequences.
The members of the set denoted by 0 can be expressed as $5n$, where n is an nteger. The members of the set denoted by 1 can be expressed as $5n + 1$, where n is an integer.

The members of the set denoted by 2 can be expressed as $5n + 2$, where n is an integer and similarly for 3 and 4. All the integers are now partitioned into the 5 sets 0, 1, 2, 3, 4 and each integer belongs to only one set. Addition is defined in the usual way with the sum reduced modulo 5 to one of the five sets 0, 1, 2, 3, 4. $3 + 4 = 7 \equiv 2$ (modulo 5) so $3 + 4 \equiv 2$. Any members of 3 and 4 would give the same result, since $5n + 3 + 5k + 4 = 5n + 5k + 5 + 2 \equiv 2$ (mod 5). This system is also known as clock arithmetic and illustrated with a clock-face with only 5 numbers (figure 88a). The operation of addition (mod 5) can be summarized with the aid of a combination table (figure 88b). $1 + 2 \equiv 3$ is in the 2nd row, 3rd column. $2 + 1 \equiv 3$ is in the 3rd row, 2nd column. $3 + 4 \equiv 2$ (4th row, 5th column).

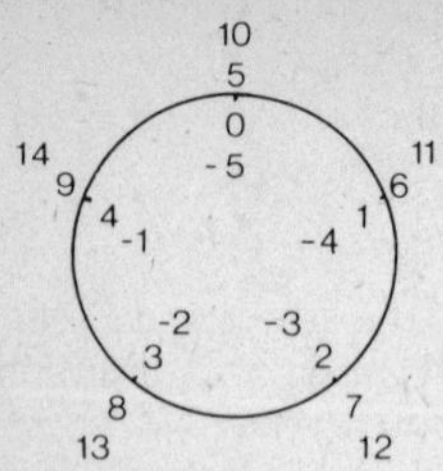

(a)

⊕	0	1	2	3	4
0	0	1	2	3	4
1	1	2	3	4	0
2	2	3	4	0	1
3	3	4	0	1	2
4	4	0	1	2	3

(b)

⊗	0	1	2	3	4
0	0	0	0	0	0
1	0	1	2	3	4
2	0	2	4	1	3
3	0	3	1	4	2
4	0	4	3	2	1

(c)

Figure 88

The set $\{0, 1, 2, 3, 4\}$ under addition (modulo 5) forms a closed set, the combination of any two giving another member in the set. When 0 is added to any element a, the result is a. 0 is the **identity element** for addition.

$3 + 4 = 4 + 3 = 2$ and this is true for every pair of elements and so addition is commutative. This is easily recognized from the table, since $3 + 4$ and $4 + 3$ are symmetrically placed on either side of the main diagonal and the table is symmetrical about the main diagonal.

$x + 4 = 2$ is solved from the table by looking in the last column for the number 2. Since $3 + 4 = 2$, $x = 3$ is the solution. A more formal solution follows:

$$
\begin{aligned}
x + 4 = 2 &\Rightarrow (x + 4) + 1 = 2 + 1 \\
&\Rightarrow x + (4 + 1) = 3 \quad \text{if the operation is associative} \\
&\Rightarrow x + 0 = 3 \quad \text{since } 4 + 1 = 0 \\
&\Rightarrow x = 3 \quad \text{since } x + 0 = x \text{ and 0 is the identity.}
\end{aligned}
$$

Any equation of the form $x + a = b$ (mod 5) can be solved provided that the operation of addition modulo 5 is associative and that there exists an element which combines with a to give 0. 1 is called the **inverse element** of 4, since $4 + 1 = 0$. 4 is also the inverse of 1, so 1 and 4 form an inverse pair. Similarly 2 and 3 form an inverse pair and 0 is its own inverse (self-inverse). To show that the operation is associative we must show that $(a + b) + c = a + (b + c)$ for all values of a, b and c Write $a = 5m + a$ $(0 \leq a \leq 4)$, $b = 5k + b$ $(0 \leq b \leq 4)$, $c = 5n + c$ $(0 \leq c \leq 4)$.

$$a + (b + c) \equiv 5m + a + (5k + b + 5n + c) = 5(m + k + n) + a + b + c = a + b + c \pmod 5.$$
$$(a + b) + c \equiv (5m + a + 5k + b) + 5n + c = 5(m + k + n) + a + b + c = a + b + c \pmod 5.$$

So $$a + (b + c) = (a + b) + c.$$

A set which is (i) **closed** for a certain operation (ii) **associative** (iii) contains an **identity element** and (iv) possesses **inverse** elements for all its members is called a **group** under that operation. The group properties enable equations with that operation to be solved uniquely with members from that set. The set Z of integers forms a group under addition, with 0 as the identity and the inverse of a being $-a$. There are an infinite number of elements in the group, so the group has infinite order. The **order of a group** is the number of elements in the group. Figure 88(b) shows the set of integers (modulo 5) under the operation of multiplication (modulo 5). 0 multiplied with any element gives 0, and is called the zero member of the set. If 0 is omitted the set $Z_5^* = \{1, 2, 3, 4\}$ forms a group under multiplication (modulo 5). The suffix 5 indicates modulo 5 and * means that 0 is omitted.

$3 \times 3 = 9 \equiv 4$ so $3 \times 3 \equiv 4 \pmod 5$ and similarly for the other elements. The set is closed and this can be seen by the fact that only 1, 2, 3 and 4 occur in the multiplication table.

To verify the associative rule, with the notation as above,

$$\begin{aligned}(a \times b) \times c &\equiv \{(5m + a)(5k + b)\}(5n + c)\\ &= (25km + 5mb + 5ka + ab)(5n + c)\\ &= 5(25kmn + 5mbn + 5kan + 5kmc + mbc + kac + nab)\\ &\quad + abc\\ &\equiv abc \pmod 5.\end{aligned}$$

$a \times (b \times c)$ will give a similar expression which will be equivalent to $abc \pmod 5$.

The associative rule holds for all triples. The identity element is 1; $a \times 1 = 1 \times a = a$ for all elements a. The elements 2 and 3 form an inverse pair, and 1 and 4 are self-inverse.

The set forms a group, under multiplication (modulo 5), of order 4 (4 elements). The operation is commutative.
A commutative group is called **Abelian**.

The equation $3 \times x \equiv 2$ has the solution $x \equiv 4$ (from the table).

More formally $\quad 3 \times x \equiv 2$

$$\begin{aligned}
&\Rightarrow 2 \times (3 \times x) \equiv 2 \times 2 \\
&\Rightarrow (2 \times 3) \times x \equiv 4 \text{ (associative rule)} \\
&\Rightarrow \quad 1 \times x \equiv 4 \\
&\Rightarrow \quad\quad x = 4 \text{ (identity is 1).}
\end{aligned}$$

⊕	0	1	2	3	4	5
0	0	1	2	3	4	5
1	1	2	3	4	5	0
2	2	3	4	5	0	1
3	3	4	5	0	1	2
4	4	5	0	1	2	3
5	5	0	1	2	3	4

(a)

⊗	1	2	3	4	5
1	1	2	3	4	5
2	2	4	0	2	4
3	3	0	3	0	3
4	4	2	0	4	2
5	5	4	3	2	1

(b)

Figure 89

Any equation of this form would have a unique solution from the set {1, 2, 3, 4}. Figure 89(a) shows the combination table for the set {0, 1, 2, 3, 4, 5} under addition (modulo 6). The table is similar to the addition table (modulo 5) with 0 as the identity element, but with 6 elements instead of 5. 1 and 5 form an inverse pair as do 2 and 4. 0 and 3 are self-inverse. The equations $x + a \equiv b$ have unique solutions for all a, b in the set $Z_5 = \{0, 1, 2, 3, 4, 5\}$ and the set forms an Abelian group under the operation of addition (modulo 6).

Figure 89(b) shows the combination table for $Z_6^* = \{1, 2, 3, 4, 5\}$ under the operation of multiplication modulo 6. The set is **not** closed under this operation, since $2 \times 3 = 0$, which is not a member of the set. The operation is still associative and the identity element is still 1. The elements 1 and 5 are self-inverse but 2, 3 and 4 do **not** have inverses. Z_6^* does **not** form a group under multiplication.

The equation $3 \times x \equiv 2$ has no solutions in this set, while the equation $3 \times x \equiv 1$ has three solutions $x = 1$, 3 or 5. Equations of the form $a \times x \equiv b$ can have no solution, or 1, 2 or 3 solutions. This type of equation does not necessarily have a unique solution. Even if $Z_6 = \{0, 1, 2, 3, 4, 5\}$ under multiplication were considered, while this set under multiplication is closed, associative and has 1 as the identity, not all elements have an inverse and so Z_6 does not form a group under multiplication.

All sets Z_n ($n = 2, 3, 4, \ldots$) form groups under addition (modulo n) and the combination tables are similar to those in figure 89(a) and figure 88(b).

It is interesting to consider the sets Z_n^* under multiplication modulo n. Figure 90(a) shows the combination table for Z_7^* under multiplication modulo 7. Z_7^* is closed, associative with identity 1. Inverse pairs are 2 and 4, 3 and 5; 1 and 6 are self-inverse. Z_7^* forms a group of order 6.

⊗	1	2	3	4	5	6
1	1	2	3	4	5	6
2	2	4	6	1	3	5
3	3	6	2	5	1	4
4	4	1	5	2	6	3
5	5	3	1	6	4	2
6	6	5	4	3	2	1

(a)

⊗	1	2	3	4	5	6	7	8
1	1	2	3	4	5	6	7	8
2	2	4	6	8	1	3	5	7
3	3	6	0	3	6	0	3	6
4	4	8	3	7	2	6	1	5
5	5	1	6	2	7	3	8	4
6	6	3	0	6	3	0	6	3
7	7	5	3	1	8	6	4	2
8	8	7	6	5	4	3	2	1

(b)

Figure 90

Figure 90(b) shows the combination table for Z_9^* under multiplication (modulo 9). It is not closed and 3 and 6 have no inverses. At first sight it might be thought that Z_n^* forms a group under multiplication (modulo n) if n is an odd number, but figure 90(b) disproves this theory. In fact Z_n^* forms a group if p is prime, since in this case there will be no zeros in the combination table. The tables for Z_p^* (p prime) would all be Latin squares, that is, each row or column contains each element once only. The four group conditions are satisfied. In Z_9^*, if 3 and 6 are omitted the remaining 6 elements form a group. The sets Z_n of integers (modulo n) under addition all form cyclic groups. In figure 89(a), in each row the elements stay in the same cyclic order with 0, 1, ... appearing at the end. The multiplication tables provide more interesting structures which will be considered later.

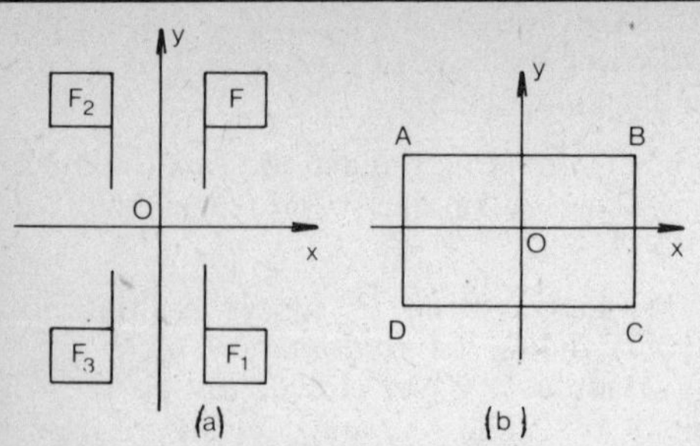

	I	X	Y	H
I	I	X	Y	H
X	X	I	H	Y
Y	Y	H	I	X
H	H	Y	X	I

(c)

Figure 91

Symmetry operations

Figure 91(a) shows an object, F, with 3 images F_1, F_2, F_3. F_1 is the image of F after reflection in the x-axis. Denoting this transformation by X, we have $X(F) = F_1$. If Y denotes reflection in the y-axis, then $Y(F) = F_2$, and if H denotes a half-term (rotation of 180°) about 0, the origin (0, 0), then $H(F) = F_3$.

$$Y(F_1) = F_3, \quad Y\{X(F)\} = F_3 = H(F) \Rightarrow YX = H.$$

X followed by Y is equivalent to H. Similarly XY = H, since

$$XY(F) = X\{Y(F)\} = X(F_2) = F_3 = H(F).$$

The effect of XY (Y followed by X) is the same as H.

$$XH(F) = X\{H(F)\} = X(F_3) = F_2 = Y(F) \Rightarrow XH = Y.$$

$$HH(F) = H\{H(F)\} = H(F_3) = F. \; HH = I,$$

where I denotes the identity transformation, i.e. I(F) = F.
These results for the combination of I, X, Y and H can be summarized in the form of a table (figure 91c). The result of XY is placed in the second row, third column and the result of YX in the third row, second column. So the heading column represents the transformation which is written first in each product, while the heading row represents the transformation which is performed first, since XY means Y followed by X. This convention applies to all combination tables, but unfortunately in many of our examples like transformations, functions and matrices, in the product AB, B refers to the operation which is performed first.

Figure 91(b) shows a rectangle ABCD which has 2 lines of symmetry Ox and Oy and rotational symmetry of order 2 (180°)

about O. Denoting reflection in Ox by X, reflection in Oy by Y and a half-turn about O by H, the application of these transformations to the rectangle will leave it occupying its same outline with ABC and D occupying different positions X(ABCD) = DCBA (labelling the vertices clockwise from the top left-hand vertex).

YX(ABCD) = Y(DCBA) = CDAB = H(ABCD). The combination of these transformations gives the same table of results (figure 91c). These transformations are called the **symmetry operations** of the rectangle and they form a group of order 4 under the operation of successive application. (This is sometimes called 'followed by' since XY means 'Y followed by X'.) The set {I, X, Y, H} is closed. This can be seen from the table (figure 91c). To demonstrate the associative rule, consider (AB)C and A(BC) where A, B and C are isometries. (AB)C means {C fb (B fb A)} where fb stands for 'followed by'. A(BC) means (C fb B) fb A. In both cases C is performed first then B and A last, and will give the same final result. This is true of all transformations in space which are one to one (i.e. one object point gives one image point), since these then have an inverse transformation defined. (If the transformation is represented by a 3×3 matrix, the one-to-one property ensures that its determinant is not zero, which implies that an inverse matrix exists).

The set {I, X, Y, H} has an identity element I, and all four elements are self-inverse, so the set forms an Abelian (commutative) group of order 4. (This particular group with all elements self-inverse is called the **Klein group** of order 4).

Symmetry group of the equilateral triangle

The equilateral triangle has three lines of symmetry and order of rotational symmetry 3 (120°). The symmetry operations are R_1 (rotation of +120° about O), R_2 (rotation of +240° about O = rotation of −120° about O) and I; L, M, N represent the three reflections in the lines of symmetry AO, BO and CO. The successive transformations are best evaluated by considering their effect on ABC, the vertices of the triangle. (Remember ML means L first, then M.)

The full group table is in figure 92(b) and four elements I, M, N, L are self-inverse, while R_1 and R_2 form an inverse pair. The

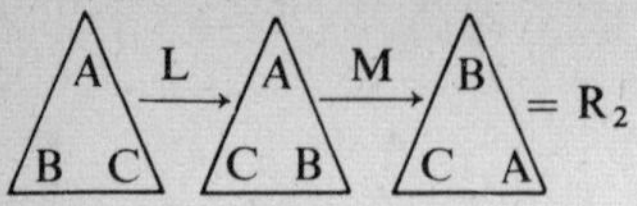

$ML = R_2$

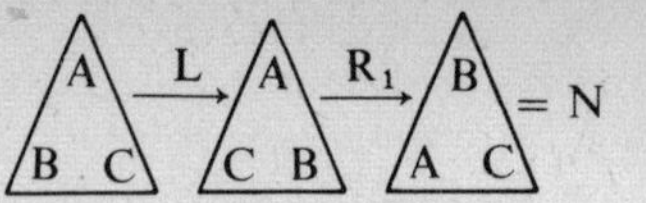

$R_1L = N$

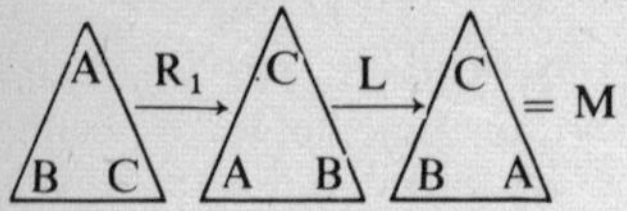

$LR_1 = M$

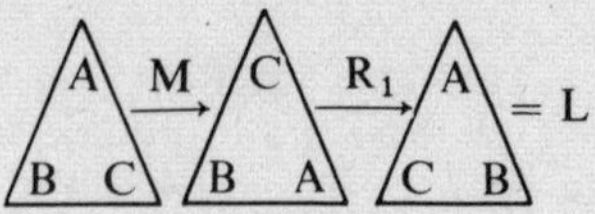

$R_1 M = L$

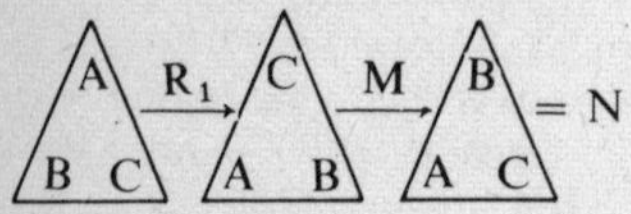

$MR_1 = N$

$MM = M^2 = I$

group is not commutative since $MR_1 \neq R_1 M$. (It is only necessary to find one pair of elements which do not commute to prove the group is not commutative, but when the group is commutative, all pairs must be commutative.) The set $\{I, R_1, R_2\}$ form a group and this is called a subgroup of the main group $\{I, R_1, R_2, L, M, N\}$. Similarly {I, M}, {I, L} and {I, N} form subgroups of order 2. A subgroup of a group is a subset of elements which form a group with the same operation as the original group. The group {I, X, Y, H} has 3 subgroups of order 2 i.e. {I, H}, {I, Y}, {I, X}.

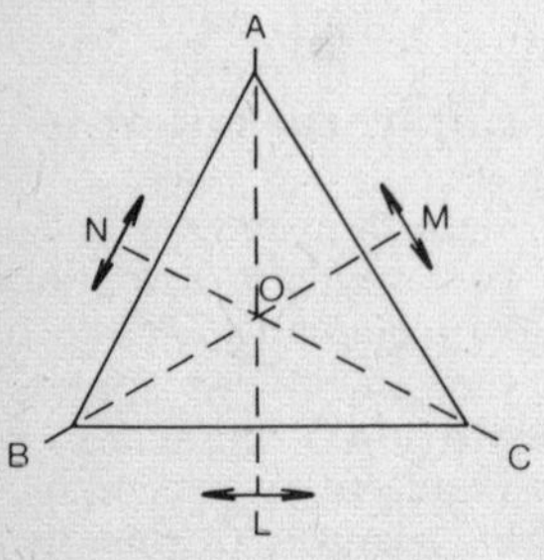

(a)

	I	R_1	R_2	L	M	N
I	I	R_1	R_2	L	M	N
R_1	R_1	R_2	I	N	L	M
R_2	R_2	I	R_1	M	N	L
L	L	M	N	I	R_1	R_2
M	M	N	L	R_2	I	R_1
N	N	L	M	R_1	R_2	I

(b)

Figure 92

Example 1 Groups of functions Show that the functions $f(x) = \frac{1}{x}$, $g(x) = -x$ and $h(x) = -\frac{1}{x}$ together with the identity function $e(x) = x$ form a group, the functions being applied successively.

$$f\,g(x) = f\{g(x)\} = f(-x) = -\frac{1}{x} = h(x),$$

$$g f(x) = g\{f(x)\} = g\left(\frac{1}{x}\right) = -\frac{1}{x} = h(x),$$

$$g\,h(x) = g\{h(x)\} = g\left(-\frac{1}{x}\right) = \frac{1}{x} = f(x),$$

$$h\,g(x) = h\{g(x)\} = h(-x) = \frac{1}{x} = f(x),$$

$$h f(x) = h\left(\frac{1}{x}\right) = -x = g(x),$$

$$f\,h(x) = f\left(-\frac{1}{x}\right) = -x = g(x),$$

$$f f(x) = f\left(\frac{1}{x}\right) = \frac{1}{(1/x)} = x$$
$$= e(x),$$

$$g\,g(x) = g(-x) = -(-x) = x$$
$$= e(x),$$

$$h\,h(x) = -\frac{1}{x} = \frac{1}{(-1/x)} = x$$
$$= e(x).$$

	e	f	g	h
e	e	f	g	h
f	f	e	h	g
g	g	h	e	f
h	h	g	f	e

Figure 93

The results are summarized in the group table in figure 93. All the four functions are self-inverse, and form the Klein group of order 4.

Key terms

Natural numbers $N = \{1, 2, 3, 4, \ldots\}$.

Integers $Z = \{\ldots, -3, -2, -1, 0, +1, +2, +3, +4, \ldots\}$.

Rationals $Q = \left\{\text{numbers of the form } \dfrac{p}{q}\right\}$ where $p, q \in Z$ and $q \neq 0$.

Reals R rationals and irrationals.

Complex numbers C: $a + bj$, where a, b are real and $j^2 = -1$.

A **set of elements** $a, b, c, \ldots$ forms a group G under the operation $*$ if (i) the set is closed: $a * b \in G$, for all a, b, (ii) $*$ is associative: $a * (b * c) = (a * b) * c$, for all a, b, c, (iii) there is an identity e such that $a * e = e * a = a$, for all a, (iv) every element a has an inverse a^{-1} such that $a * a^{-1} = a^{-1} * a = e$.

An **Abelian group** is commutative, $a * b = b * a$, for all a, b. $*$ is distributive over $\circ$ if $a * (b \circ c) = (a * b) \circ (a * c)$.

Chapter 14
Statics of a Particle

In mechanics we study the motion and equilibrium of bodies under the application of forces. The study of forces in equilibrium is known as statics, and the study of motion is called dynamics. In both cases the motion or equilibrium is governed by the forces acting on the bodies and we shall first discuss some of these forces. (All quantities will be measured in SI units, metres, kilograms and seconds, unless it is more meaningful to do otherwise.)

Forces have magnitude and direction and can be represented by vectors. The forces have a point of application and this must be considered when the resultant of forces is found. Forces are of many different kinds and arise in different situations.

Gravitational forces

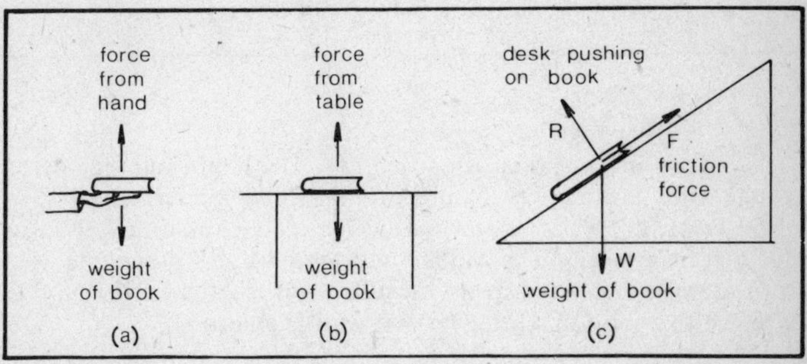

Figure 94

The weight of a body is the force with which the earth attracts it. A mass of 10 kg is attracted towards the earth with a force of 98 newtons. If a book is supported by a person's hand (figure 94a) the forces balance because the book is in equilibrium (held stationary). The weight of the book (downwards) is balanced by an equal force (upwards) from the person's hand. If the book is placed on a desk, then to balance its weight the desk must push upwards on the book (figure 94b). In figure

94c, the desk-lid is raised and to counteract the weight of the book there is a reaction force R from the desk which is perpendicular to the desk-top surface, and a friction force F which opposes the motion of the book. The book has a tendency to slide down the line of greatest slope, so F opposes this motion. If the book is in equilibrium, W, F and R will balance and their vector sum is zero, or the resultant of F and R is equal and opposite to W.

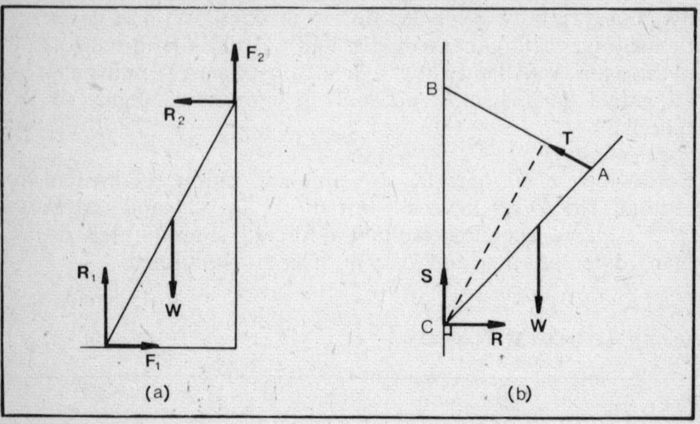

Figure 95

The forces on a ladder standing on rough ground against a rough wall are shown in figure 95(a). The ladder is called a rigid body (not a particle) whose weight acts at its centre of mass (centre of gravity). The ladder will tend to slip down the wall and along the ground away from the wall so the friction forces oppose this motion. If the vertical wall is smooth, $\mathbf{F}_2 = \mathbf{0}$.

In figure 95(b) a flag-pole is supported at an angle to a vertical wall by reaction forces $\mathbf{R}$ and $\mathbf{S}$ at the flag-pole base and by the tension $\mathbf{T}$ in the rope AB. The tension force is an example of a pulling force.

The book in figure 94 is an example of a body which can be regarded as a particle, in that its mass can be thought of as concentrated at one point (the centre of mass or centre of gravity), and the forces acting on the book are taken to act through that point.

The ladder in figure 95(a) is an example of a rigid body where the point of application of each force is very important. The reactions and frictional forces act at the ends of the ladder and are related to each other through the rigid structure of the ladder.

Equilibrium of a particle

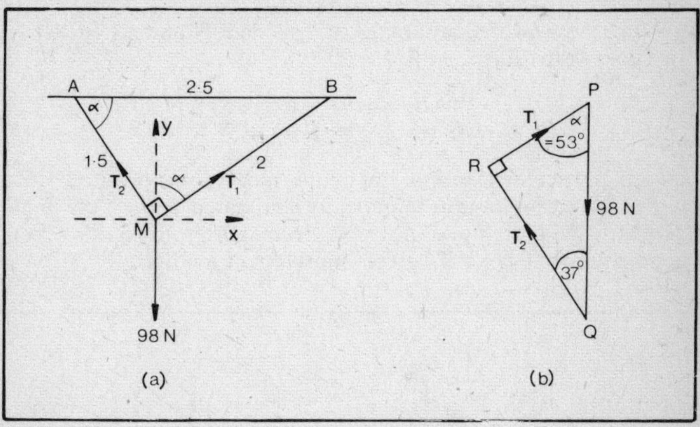

Figure 96

Figure 96(a) shows a mass M of 10 kg supported by two strings attached to two points A and B at the same level 2·5 metres apart. The strings have lengths AM = 1·5 m, BM = 2 m. If the tensions in the strings are $\mathbf{T}_1$ and $\mathbf{T}_2$, then the three forces $\mathbf{T}_1$, $\mathbf{T}_2$ and 98 N (weight) balance as the particle is in equilibrium. The vector sum of these forces is zero and they form a triangle of forces (figure 96b), the sides of the triangle representing the magnitudes of the forces.

ΔABM is right-angled at M (3,4,5 triangle).
$\angle A = \tan^{-1}\frac{4}{3} = 53{\cdot}1° = \alpha = \angle BMy$.

$\mathbf{T}_1$ and $\mathbf{T}_2$ can be resolved into **i** and **j** components along x- and y-axes with origin M.

$$\mathbf{T}_1 = T_1 \sin\alpha\,\mathbf{i} + T_1 \cos\alpha\,\mathbf{j}; \qquad \mathbf{T}_2 = -T_2 \cos\alpha\,\mathbf{i} + T_2 \sin\alpha\,\mathbf{j}.$$

The weight can be expressed as $-98\mathbf{j}$.
Since the forces balance, the components add to zero.

So $T_1 \sin\alpha - T_2 \cos\alpha = 0$ and $T_1 \cos\alpha + T_2 \sin\alpha - 98 = 0$.

These two simultaneous equations give T_1 and T_2 since α is known. The first equation comes from resolving horizontally; the components of $\mathbf{T}_1$ and $\mathbf{T}_2$ balance in the horizontal direction.
The second equation comes from resolving vertically; the forces upwards balance the forces downwards.

$$T_1 \sin\alpha - T_2 \cos\alpha = 0 \Rightarrow 4T_1 = 3T_2.$$
$$T_1 \cos\alpha + T_2 \sin\alpha = 98 \Rightarrow \tfrac{3}{5}T_1 + \tfrac{4}{5}T_2 = 98.$$

Solving these two equations gives $T_1 = 58{\cdot}8$ N and $T_2 = 78{\cdot}4$ N.
In figure 96(b) since $\angle \mathrm{PRQ} = 90°$,

$$T_1 = 98 \cos\alpha = 98 \times 0{\cdot}6 = 58{\cdot}8 \text{ N}.$$
$$T_2 = 98 \sin\alpha = 98 \times 0{\cdot}8 = 78{\cdot}4 \text{ N}.$$

In this particular example this is the more direct method since one of the variables is automatically eliminated. Even if the angle between $\mathbf{T}_1$ and $\mathbf{T}_2$ is not 90°, resolving in directions perpendicular to $\mathbf{T}_1$ and $\mathbf{T}_2$ give a more direct method.

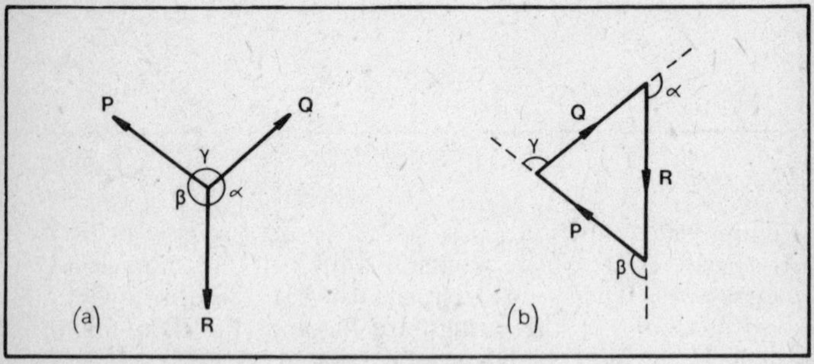

Figure 97

If three forces are in equilibrium acting through the same point (figure 97a), then the triangle of forces in figure 97(b) enables **P**, **Q** and **R** to be related by the sine rule, which gives

$$\frac{P}{\sin(180-\alpha)} = \frac{Q}{\sin(180-\beta)} = \frac{R}{\sin(180-\gamma)}$$
$$\Rightarrow \frac{P}{\sin\alpha} = \frac{Q}{\sin\beta} = \frac{R}{\sin\gamma}.$$

This result is known as **Lami's theorem**, each force being proportional to the sine of the angle between the other two forces.

The resultant of any number of forces acting through a point in a plane can be found by adding them vectorially, either by finding the resultant components in two perpendicular directions (for example the **i** and **j** directions) or by drawing them nose to tail and finding the resultant.

Example 1 A string ABCD carries a mass of 10 kg tied at B and w N at C, and is suspended from points A and D which are at the same level. AB and CD make angles of 60° and 45° with the horizontal and BC is horizontal. Find the tensions in the strings and the weight w.

Let the tensions in each part of the string be as shown in figure 98(a). The string BC pulls on the 10 kg mass at B with force T_2 so it will pull with the same force on the mass at C The mass of 10 kg is in equilibrium under the action of 3 forces T_1, T_2 and 98 N.

Resolving vertically $98 = T_1 \cos 30° \Rightarrow T_1 = 113$ N.
Resolving horizontally $T_1 \cos 60° = T_2 \Rightarrow T_2 = \frac{1}{2}T_1 = 56{\cdot}5$ N.
At C horizontally $T_2 = T_3 \cos 45° \Rightarrow T_3 = \sqrt{2}\, T_2 = 79{\cdot}9$ N.
At C vertically $w = T_3 \cos 45° \Rightarrow w = 56{\cdot}5$ N.

From now on resolving horizontally will be denoted by → and resolving vertically by ↑.

Friction

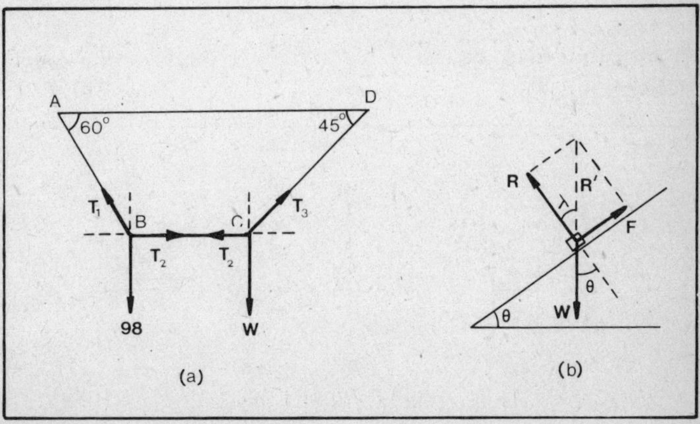

Figure 98

Consider a book resting on a plane surface inclined at an angle θ to the horizontal (figure 98b). Its weight acts vertically downwards so the resultant force of the plane surface on the book must be vertically upwards. It is more usual to regard this resultant force as made up of two forces R (perpendicular to the plane) and F (a frictional force up the plane in the line of greatest slope as the frictional force opposes the intended motion of the body).

As the inclination of the plane is increased, F will increase until the book begins to slide. The maximum value of the frictional force, called the **limiting friction**, occurs just as the book is about to move. The limiting friction bears a constant ratio to the normal reaction between the surfaces, and $F = \mu R$, where μ is called the coefficient of (static) friction. When motion takes place the dynamic friction is slightly less than this but it is usually taken to be the same. The resultant of F and R (R' in figure 98b) makes an angle $\lambda = \tan^{-1}\left(\frac{F}{R}\right)$ with the normal (perpendicular to the plane). This angle gradually increases as F increases until F reaches its maximum value μR. Then $\lambda = \tan^{-1}\left(\frac{\mu R}{R}\right)$ and $\mu = \tan \lambda$, and λ is known as the angle of friction. In fact λ is the value of θ when the body is about to slide in figure 98(b).

Example 2 Find the least force to move a particle on a rough horizontal plane.

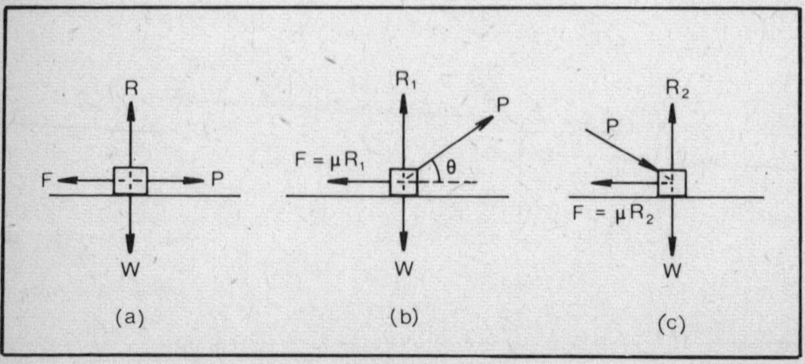

Figure 99

If the force P is horizontal (figure 99a) then it must overcome the friction F which will be equal to μR $(=\mu W)$ when the particle is about to move.

If the force P acts upwards at an angle θ to the horizontal, its vertical component $P \sin \theta$ will reduce the reaction, so $R_1 = W - P \sin \theta$. For motion to take place

$$P \cos \theta > F = \mu R_1 = \mu W - \mu P \sin \theta,$$
$$\Rightarrow P \cos \theta + \mu P \sin \theta > \mu W.$$

If λ is the angle of friction, $\mu = \tan \lambda$, so

$$P \cos \theta + P \sin \theta \tan \lambda > W \tan \lambda,$$
$$\Rightarrow P(\cos \lambda \cos \theta + \sin \theta \sin \lambda) > W \sin \lambda$$

$$P > \frac{W \sin \lambda}{\cos(\theta - \lambda)},$$

which takes its least value when $\cos(\theta - \lambda)$ is greatest, i.e. when $\theta = \lambda$.
In this case $P > W \sin \lambda$ so $P = W \sin \lambda$ is the critical value.

If P acts downwards (figure 99c) the reaction $R_2 > R$ and the friction increases. The critical value of P will increase and the new P(critical) will be greater than the old P(critical).

Example 3 Find the least force required to move a particle up a rough inclined plane when the inclination of the plane is less than the angle of friction and the friction prevents the particle sliding down under its own weight.

If P acts at an angle θ with the plane (figure 100), then

$$R = W \cos \alpha - P \sin \theta.$$

The limiting friction $F = \mu R$, thus

$$\mu R = \mu W \cos \alpha - \mu P \sin \theta.$$

P must overcome F and the resolved part of the weight down the slope, so

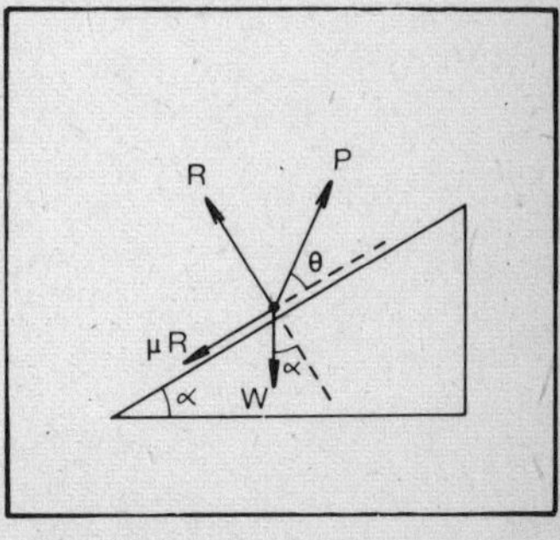

Figure 100

$$P \cos \theta > F + W \sin \alpha$$
$$= \mu W \cos \alpha - \mu P \sin \theta + W \sin \alpha$$

$P(\cos\theta + \mu\sin\theta) > W(\mu\cos\alpha + \sin\alpha)$
$P(\cos\theta + \tan\lambda\sin\theta) > W(\tan\lambda\cos\alpha + \sin\alpha)$

$$P > \frac{W\sin(\alpha + \lambda)}{\cos(\theta - \lambda)}.$$

This is least when $\theta = \lambda$ and then $P = W\sin(\alpha + \lambda)$.

The calculation is similar when the particle is sliding down the plane. When the inclination of the plane is greater than the angle of friction, the particle slides down under its own weight. By similar methods, the least force required to (i) prevent the particle sliding down and (ii) move it up the slope, can be found.

Key terms

Statics Study of bodies in equilibrium.
Dynamics Study of bodies in motion.
Resolving forces A force P acting through a point in a direction making an angle θ with a fixed direction OA can be resolved into two components $P\cos\theta$ in the direction of OA and $P\sin\theta$ in the direction perpendicular to OA.
Limiting friction $F = \mu R$.
Angle of friction λ, where $\tan\lambda = \mu$.

Chapter 15
Statics of a Rigid Body

A rigid body is regarded as one whose size cannot be neglected when dealing with the forces acting on it, and which remains undistorted by these forces. The forces may act at different points of the body and their effect is transmitted to the rest of the body by the rigidity of the body's composition. For example, in figure 95(a) on page 242, the weight of the ladder acts through its centre of mass (half-way along if the ladder is uniform) and the reactions and friction act on the ends of the ladder. In practice, the ladder sags slightly, but we must ignore this at this stage, otherwise we bring in the internal forces which hold the ladder together. We assume a rigid body retains its shape at all times. When forces do **not** act through the same point, care must be taken in finding their resultant.

Resultant of two like parallel forces

Consider two parallel forces **P** and **Q** acting in the same sense through points A and B (figure 101a). At A and B introduce two equal and opposite forces **R** and $-$**R** which balance. Then the resultant of **P** and **R** acting at A is represented by **AS** and the

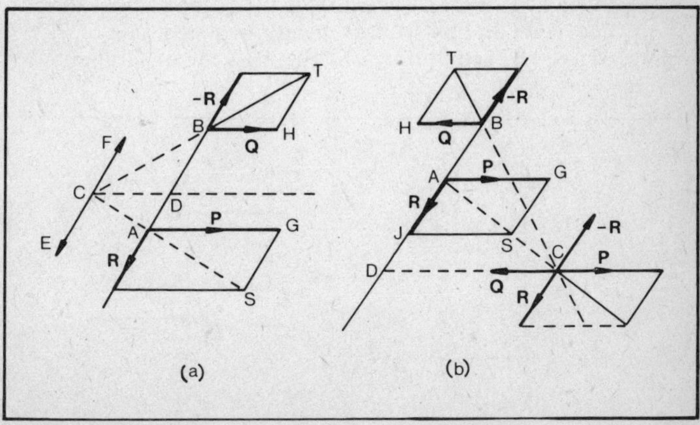

Figure 101

resultant of **Q** and **R** acting at B is represented by **BT**. AS and BT intersect at C so that **P** and **Q** are equivalent to forces **P** and **Q** along CD (parallel to BH and AG) since the forces **R** represented by **CE** and $-$**R** by **CF** balance and cancel each other out. The resultant is a force of **P** + **Q** acting along CD, where D divides AB in the ratio $Q : P$.

Resultant of two unlike parallel forces

In figure 101(b), **P** and **Q** are parallel forces which are opposite in sense. Introduce equal and opposite forces $-$**R** at B and **R** at A, so that the resultant of **P** and **R** acting at A is represented by **AS**, and the resultant of **Q** and $-$**R** at B by **BT**. AS and BT meet at C, so at C we have forces **P**, **Q** and forces **R** and $-$**R** (which cancel) to give a resultant **P** $-$ **Q** along DC.

By similar triangles $\text{AG} : \text{GS} = P : R = \text{CD} : \text{AD}$,

$$\text{BH} : \text{HT} = Q : R = \text{CD} : \text{DB}.$$

$$P \times \text{AD} = R \times \text{CD} = Q \times \text{DB} \Rightarrow \text{AD} : \text{DB} = Q : P.$$

So D divides AB externally in the ratio $Q : P$.

This method breaks down if AS and BT are parallel, which will happen when **P** and **Q** are equal and opposite. In this case there is no resultant force, but **P** and **Q** together have a turning effect which is called a **couple**. This turning effect is described by the **moment** of the couple, the magnitude of its turning effect.

Moment of a force

In figure 102(a) **AB** represents the force **P** whose **moment** about O is defined as $P \times \text{OD}$, where OD is the perpendicular distance of the line of action of **P** from O and P the magnitude of **P**. If **Q** (parallel to $-$**P**) acts through C, the moment of **Q** about O is

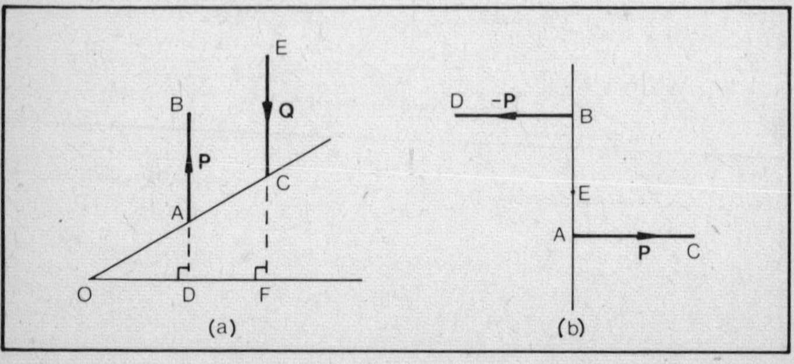

Figure 102

$Q \times$ OF. If the moment of **P** about O is regarded as positive, then the moment of **Q** about O is negative. (In three dimensions, the moment of a force is a vector in the direction of the axis of rotation of the turning effect.)

Moment of a couple

Figure 102(b) shows two equal and opposite forces **P** represented by **AC** and **–P** by **BD**. The total anticlockwise moment of these forces about A is $P \times$ AB, since **P** along AC has zero moment about A. The total moment about B is $P \times$ AB (anticlockwise). The total moment about E is

$$P \times \text{BE} + P \times \text{AE} = P \times (\text{BE} + \text{AE}) = P \times \text{AB}.$$

The total moment of these two forces about any point in the plane is $P \times$ AB and this is the moment of the couple formed by the two parallel forces **P** and **–P** represented by **AC** and **BD**. The moment of the couple is the product of one of the arms of the couple and the perpendicular distance between the lines of action of the two equal and opposite forces constituting the couple.

Centre of parallel forces

Consider a system of parallel forces $\mathbf{F}_1, \mathbf{F}_2, \mathbf{F}_3, \ldots$ acting through points $A_1, A_2, A_3, \ldots$ (figure 103a). The resultant of forces $\mathbf{F}_1$ and $\mathbf{F}_2$ is a single force $\mathbf{F}_1 + \mathbf{F}_2$ acting through $\mathbf{B}_1$, where $F_1 \times A_1B_1 = F_2 \times A_2B_1$ (the moments of $\mathbf{F}_1$ and $\mathbf{F}_2$ about B_1 balance). The resultant of forces $\mathbf{F}_1 + \mathbf{F}_2$ acting through B_1 and $\mathbf{F}_3$ through A_3 is a force $\mathbf{F}_1 + \mathbf{F}_2 + \mathbf{F}_3$ acting through B_2, where $(F_1 + F_2) \times B_1B_2 = F_3 \times A_3B_2$.

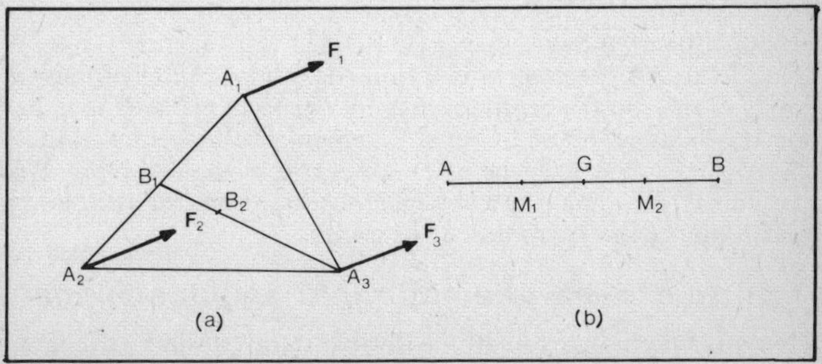

Figure 103

This process can be repeated for any number of parallel forces and the resultant force is the sum of the parallel forces. This resultant acts through a point whose position is fixed relative to $A_1, A_2, A_3, \ldots$. The result holds even if $\mathbf{F}_1, \mathbf{F}_2, \mathbf{F}_3, \ldots$ are not in the plane of $A_1, A_2, A_3, \ldots$. Compare it with the centroid of a triangle (see vectors page 202). It is also equivalent to finding the centre of mass of a number of masses $M_1, M_2, M_3 \ldots$ positioned at $A_1, A_2, A_3 \ldots$

Centre of mass (centre of gravity) of a rigid body

A rigid body is made up of many constituent masses fixed together. For a particle, the mass is considered to be concentrated at one point and the force of gravity acts through that point. When dealing with rigid bodies, the constituent masses give a total mass which produces the force of gravity acting through a particular point called the centre of gravity. For each body it is necessary to specify the centre of gravity in order to locate the line of action of the force of gravity.

Centre of mass of a uniform thin rod

Since the rod AB in figure 103(b) is uniform, equal lengths have equal masses. Divide the rod into many equal lengths either side of G, the mid-point. For each mass M_1 to the left of G there is an equal mass M_2 to the right of G. The resultant of the forces of gravity due to M_1 and M_2 acts through G and so the resultant force of all the masses making up the rod acts through G (the centre of mass).

Centre of mass of a uniform thin rectangular plate (or lamina)

Figure 104(a) shows a rectangle divided into narrow strips ST parallel to AB. Each strip is a thin rod whose centre of mass is at its centre. So the centre of mass of the rectangular lamina lies on the locus of the mid-points, EG. Similarly the centre of mass lies on HF, so it must be at O the centre of the rectangle. This argument also applies to a lamina in the shape of a parallelogram, where the centre of mass is at its centre.

Centre of mass of a uniform triangular lamina

The triangle in figure 104(b) is divided into strips ST parallel to AC. The centre of mass of each strip is at its centre, so the centre of mass of the lamina lies on BD, the locus of the mid-points

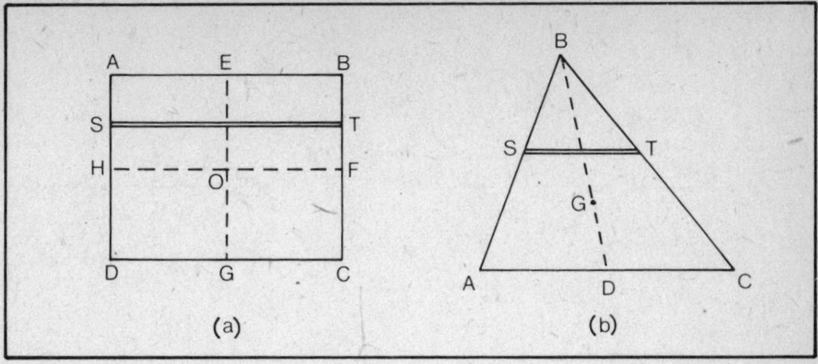

Figure 104

(BD is the median from B to AC). Similarly the centre of mass lies on the medians from A to BC and from C to AB. These meet at G, where BG : GD = 2 : 1, and G (the centre of mass) is called the **centroid** of the triangle (see vectors page 202).

The centre of mass of a uniform triangular lamina is the same as that of three equal masses placed at its vertices. Consider equal masses M_1 at A, M_2 at B, M_3 at C. The resultant force caused by masses M_1 and M_3 is a force caused by a mass $M_1 + M_2$ at D, and the resultant of masses $M_1 + M_3$ at D and M_2 at B is a force due to masses $M_1 + M_2 + M_3$ at G where

$$BG : GD = (M_1 + M_3) : M_2 = 2 : 1.$$

The centre of mass of a tetrahedron is found by dividing it into slices of triangular laminae and then relying on the geometrical configuration of the tetrahedron (see vectors page 203). The centre of mass is on a line joining the vertex to the centroid of the base at a height $\frac{1}{4}h$ above the base, where h is the height of the tetrahedron.

This result is true for any pyramid, in that the centre of mass lies at a height $\frac{1}{4}h$ from the base to the vertex lying on the line joining the vertex to the centre of mass of the base, whose position will depend on the shape of the base. The base can be divided into triangles and so the pyramid can be divided into tetrahedra each of whose centres of mass lies $\frac{1}{4}h$ from base to vertex. A cone can be regarded as a pyramid whose base has an infinite number of sides.

As confirmation of the above result, the centre of mass of a uniform solid cone can be found by a calculus method.

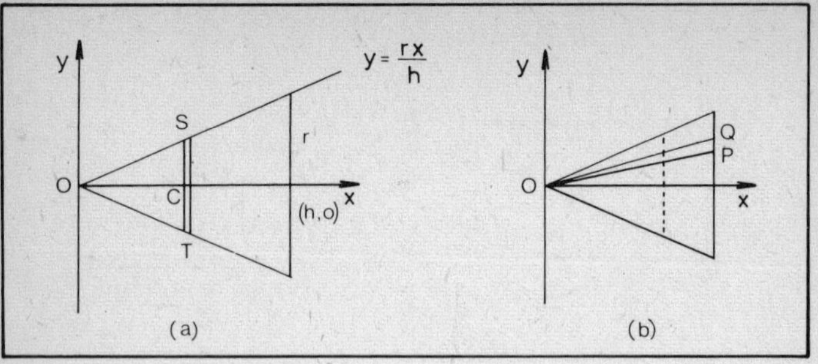

Figure 105

Centre of mass of a uniform solid cone

Let a cone of base radius r, height h and vertex O be formed by rotating the line $y = rx/h$ about the x-axis (figure 105a). By symmetry, the centre of mass lies on Ox at the point $(\bar{x}, 0)$. Divide the cone into slices SCT (parallel to the base) of height δx. Let OC $= x$ and CS $= y = rx/h$. If the mass per unit volume of the cone is m, then the mass of this circular slice is approximately $m\pi y^2\,\delta x = m\pi \dfrac{r^2x^2}{h^2}\,\delta x$.

Its moment about O is $m\pi y^2 x\,\delta x = \dfrac{m\pi r^2 x^3}{h^2}\,\delta x$.

The sum of the moments of these slices about Oy is equal to the moment of the whole body about Oy. The moment is the mass of the body multiplied by the distance of the centre of mass from Oy. This is really the definition of centre of mass. Realistically we should only talk about moments of forces and not moments of masses, and in this process g (acceleration due to gravity) should be inserted to change the masses to forces. For convenience it is usually omitted. Similarly the mass per unit volume cancels, but this is usually retained.

Total mass $\times\ \bar{x}$ = sum of the moments of all the slices,

i.e. $$\tfrac{1}{3}\pi m r^2 h\bar{x} = \lim_{\delta x \to 0} \sum \frac{m\pi r^2 x^3}{h^2}\,\delta x = \int \frac{m\pi r^2}{h^2}\,x^3\,dx = \left[\frac{m\pi r^2}{h^2}\,\frac{x^4}{4}\right]_0^h$$

$$= \frac{m\pi r^2 h^4}{4h^2} = \tfrac{1}{4}m\pi r^2 h^2.$$

Thus $\bar{x} = \frac{3}{4}h$ and the centre of mass is $\frac{1}{4}h$ from the base to the vertex.

Centre of mass of a uniform hollow cone (without base)

The curved surface of a cone (figure 105b) is divided into small narrow nearly triangular laminae OPQ by drawing lines from the vertex to points on the edge of the base which are very close together. The centres of mass of each triangular lamina is $\frac{2}{3}$ of the height from the vertex to the base of the triangle and so the centres of mass of all the triangles form a circle parallel to the plane of the base. The centre of mass of the cone will lie on the axis of the cone at the centre of this circle, which is $\frac{2}{3}$ of the height from the vertex to the base.

Centre of mass of a hollow cone with base (of the same material)

This is an example of a composite body made from the curved surface of a cone and a circular lamina base. The centre of mass of the composite body will be between $\frac{2}{3}h$ and h (the height of the cone), the larger the value of r (radius of base) the nearer it will be to the base. With the vertex at the origin and axis along the x-axis, if $G(\bar{x}, 0)$ is the centre of mass, the moment of the base about G will be equal to the moment of the hollow cone about G.

Thus mass of cone × $(\bar{x} - \frac{2}{3}h)$ = mass of base × $(h - \bar{x})$,
or (mass of cone + mass of base) × $\bar{x}$
= mass of cone × $\frac{2}{3}h$ + mass of base × h.

This is equivalent to taking moments about the y-axis and saying that the moment of the constituent bodies about O is equal to the mass of the composite body × $\bar{x}$. (This is how problems are usually done.)

If k is the mass per unit area of both cone and base,

$$(k\pi r\sqrt{r^2 + h^2} + k\pi r^2)\bar{x} = k\pi r\sqrt{(r^2 + h^2)}\tfrac{2}{3}h + k\pi r^2 h$$

$$\{\sqrt{(r^2 + h^2)} + r\}\bar{x} = \tfrac{2}{3}h\sqrt{(r^2 + h^2)} + hr.$$

If $l = \sqrt{r^2 + h^2}$ is the length of the curved surface of the cone, (called the slant height)

$$(l + r)\bar{x} = \tfrac{2}{3}hl + hr \Rightarrow \bar{x} = \frac{h(2l + 3r)}{3(l + r)}.$$

If the masses of the base and the cone are equal, $\pi rl = \pi r^2$, $\Rightarrow l = r$, which can never happen.

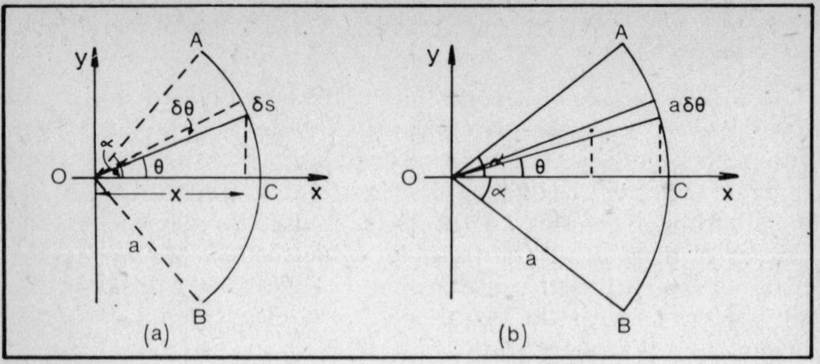

Figure 106

Centre of mass of a circular arc

Let an arc ACB of mass per unit length k and radius a enclose an angle 2α at O (figure 106a). Consider a small length $\delta s = a\,\delta\theta$ distant x from Oy. By symmetry the centre of mass lies on OC and we need consider only the portion CA since CB has the same effect.

$$\text{Mass of arc AC} \times \bar{x} = \sum \text{mass of element } \delta s \times x.$$

$$\alpha a k\bar{x} = \lim_{\delta s\to 0}\sum k\,\delta s\,x = \lim_{\delta s\to 0}\sum a\,\delta\theta\,ka\cos\theta = k\int_0^\alpha a^2\cos\theta\,d\theta.$$

$$\bar{x} = \frac{ka^2}{\alpha ak}\int_0^\alpha \cos\theta\,d\theta = \frac{a}{\alpha}\Big[\sin\theta\Big]_0^\alpha = \frac{a\sin\alpha}{\alpha}.$$

$$\text{If}\quad \alpha = \frac{\pi}{2},\quad \bar{x} = \frac{2a}{\pi};\qquad \text{if}\quad \alpha = \frac{\pi}{4},\quad \bar{x} = 2\sqrt{2}\left(\frac{a}{\pi}\right) \simeq 0{\cdot}9a.$$

Centre of mass of a circular sector

With the notation in figure 106(b), the typical element is a sector which can be considered as a triangle of height a, base $a\,\delta\theta$ whose centre of mass is distant $\tfrac{2}{3}a\cos\theta$ from Oy. For the whole figure

$$\tfrac{1}{2}a^2 2\alpha k\bar{x} = \lim_{\delta\theta\to 0}\sum \tfrac{1}{2}a\,\delta\theta\,ak\tfrac{2}{3}a\cos\theta = \int_{-\alpha}^{\alpha}\tfrac{1}{3}ka^3\cos\theta\,d\theta.$$

$$\Rightarrow \alpha a^2 k\bar{x} = \tfrac{1}{3}ka^3\Big[\sin\theta\Big]_{-\alpha}^{\alpha} \Rightarrow \bar{x} = \frac{\tfrac{1}{3}ka^3 2\sin\alpha}{\alpha a^2 k} = \frac{2a\sin\alpha}{3\alpha}.$$

If $\alpha = \dfrac{\pi}{2}$, $\bar{x} = \dfrac{4a}{3\pi} \simeq 0{\cdot}42a$. The same result is obtained using limits 0 to α and doubling.

This can be checked by a calculus method which can be extended to the general case.

In figure 107 the quadrant OAB of radius a is divided into strips of width δx and height y, distant x from OB, where $x^2 + y^2 = a^2$. The mass of this strip is $ky\,\delta x$ (taking it as a rectangle with mass per unit area k). The quadrant is symmetrical about $y = x$ so that it is only necessary to find $\bar{x}$ $(= \bar{y})$.

Moments about Oy give

$$\tfrac{1}{4}k\pi a^2\bar{x} = \lim_{\delta x \to 0} \sum kxy\,\delta x$$

$$= \int_0^a kxy\,dx$$

$$= \int_0^a k\sqrt{(a^2 - x^2)}\,x\,dx.$$

$$\tfrac{1}{4}k\pi a^2\bar{x} = \left[\tfrac{1}{3}k(a^2 - x^2)^{3/2}\right]_0^a$$

$$\Rightarrow \bar{x} = \frac{\tfrac{1}{3}ka^3}{\tfrac{1}{4}k\pi a^2} = \frac{4a}{3\pi}.$$

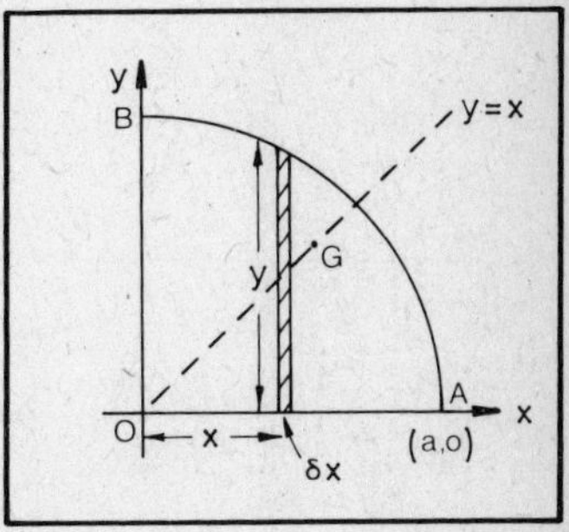

Figure 107

By symmetry $\bar{y} = \dfrac{4a}{3\pi}$, so $\text{OG} = \dfrac{4a}{3\pi} \times \sqrt{2} \simeq 0{\cdot}6a.$

From page 256, for a sector of semi-angle 45°,

$$\bar{x} = \frac{2a \sin 45°}{3\pi/4} = \frac{8a}{(3\sqrt{2})\pi} \simeq 0{\cdot}6a \text{ as above.}$$

In general $\bar{x} = \dfrac{\int xy\,dx}{\int y\,dx}$, since $\int y\,dx$ gives the total area.

Also, by taking moments about O and remembering the distance of the centre of mass of each strip from Ox is $\tfrac{1}{2}y$, we get

$$\tfrac{1}{4}k\pi a^2\bar{y} = k\int_0^a \tfrac{1}{2}y^2\,dx = \tfrac{1}{2}k\int_0^a (a^2 - x^2)\,dx$$

$$= \tfrac{1}{2}k\left[a^2x - \tfrac{1}{3}x^3\right]_0^a = \tfrac{1}{3}ka^3.$$

So $\bar{y} = \dfrac{4a}{3\pi}$ and in general $\bar{y} = \dfrac{\int \frac{1}{2}y^2\,dx}{\int y\,dx}$.

Centre of mass of a uniform solid hemisphere

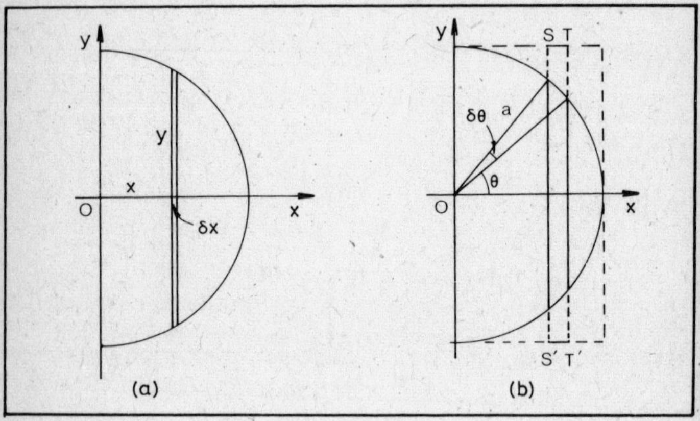

Figure 108

A hemisphere of radius a, is symmetrical about Ox and has mass per unit volume m (figure 108a). A typical narrow slice of width δx, radius y, distant x from Oy has mass $m\pi y^2\,\delta x$ (taking it to be a cylinder) and $x^2 + y^2 = a^2$. Moments about Oy give

$$\tfrac{2}{3}\pi a^3 m\bar{x} = \lim_{\delta x \to 0} \sum m\pi y^2\,\delta x\,x = m\int_0^a \pi(a^2 - x^2)x\,dx$$

$$= m\pi\left[\tfrac{1}{2}a^2x^2 - \tfrac{1}{4}x^4\right]_0^a = \tfrac{1}{4}\pi a^4 m.$$

So $$\bar{x} = \frac{3a}{8}.$$

Centre of mass of a hollow hemisphere

A hemispherical shell of mass per unit area k and radius y is divided into narrow circular bands of width δx perpendicular to Ox and distance x from Oy, where $x^2 + y^2 = a^2$ (figure 108b). The surface area of the narrow band is $2\pi a\ \delta x$ (the same as the circumscribing cylindrical band STS′T′). Moments about Oy give

$$2\pi a^2 k\bar{x} = \lim_{\delta x \to 0} \sum 2\pi akx\ \delta x = k\int_0^a 2\pi ax\ dx = k\left[\pi ax^2\right]_0^a = k\pi a^3.$$

Thus $$\bar{x} = \tfrac{1}{2}a.$$

Alternatively, the surface area of the narrow band is approximately $2\pi a \sin\theta\ \delta\theta$ distant $a\cos\theta$ from Oy.

Moments about Oy give

$$2\pi a^2 k\bar{x} = \int_0^{\frac{1}{2}\pi} 2k\pi a^3 \sin\theta \cos\theta\ d\theta = k\pi a^3 \int_0^{\frac{1}{2}\pi} \sin 2\theta\ d\theta = k\pi a^3.$$

Thus $$\bar{x} = \tfrac{1}{2}a.$$

Jointed rods

Example 1 A step ladder consists of two uniform 'legs' each of mass 10 kg which are freely jointed at the top. The ladder stands on a smooth level surface and is kept in equilibrium by a string joining the bases of the legs. Find the tension in the string and the action at the joint, if each leg is at 60° to the horizontal.

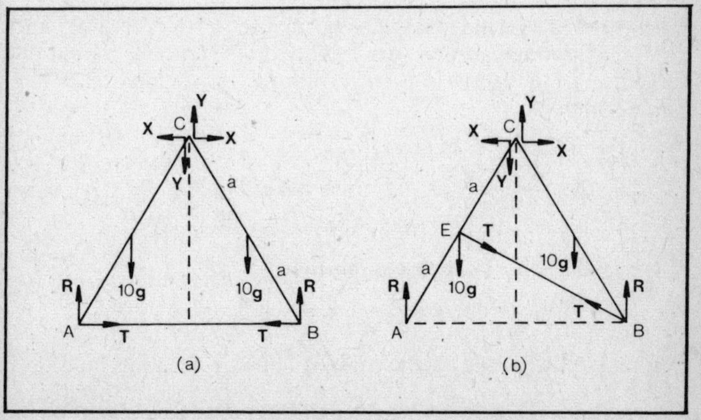

Figure 109

In figure 109(a) the forces **X** and **Y** applied to AC by BC are equal and opposite to those of AC on BC. AC ↑ $R = 10g + Y$; BC ↑ $R = 10g - Y$. Adding these equations gives

$$R = 10g = 98 \text{ N}.$$

Subtracting gives $Y = 0$. AC → gives $X = T$ and moments about B for BC gives

$$10g\, a \sin 30° = X 2a \cos 30° \Rightarrow X = T = \frac{5g}{\sqrt{3}} = 28{\cdot}3 \text{ N}.$$

Example 2 For the same ladder as in the previous example solve the problem when the string is attached from B to E in figure 109(b).

Moments about C for AC give $R_1 2a \sin 30° = 10ga \sin 30° + Ta$
$\Rightarrow R_1 = 5g + T.$

Moments about C for BC give $R_2 2a \sin 30° = 10ga \sin 30° + Ta$
$\Rightarrow R_2 = 5g + T$ and $R_1 = R_2 = R$.

AC ↑ $R = T \cos 60° + Y + 10g$, AC → $X = T \cos 30°$,
BC ↑ $R + T \cos 60° + Y = 10g$.

Adding gives $R = 10g = 98N$; subtracting gives $T = -2Y$.

$$T = 5g = 49 \text{ N}; \quad Y = -2{\cdot}5g = -24{\cdot}5 \text{ N}; \quad X = \frac{5\sqrt{3}}{2} g = 42{\cdot}5 \text{ N}.$$

Y acts downwards on BC, upwards on AC.

Example 3 A ladder is leaning against a smooth wall and standing on rough ground with coefficient of friction 0·5. Find the largest angle the ladder can make with the vertical and remain in equilibrium.

With the notation in figure 110(a)

↑ $R = W$; → $F = S$ and moments about A gives

$$Wa \cos \theta = S2a \sin \theta \Rightarrow S = \tfrac{1}{2}W \cot \theta.$$

If the ladder is in limiting equilibrium $F = \tfrac{1}{2}R$,

$$\Rightarrow F = S = \tfrac{1}{2}W \cot \theta = \tfrac{1}{2}W \Rightarrow \tan \theta = 1 \Rightarrow \theta = 45°$$

Without knowing the coefficient of friction, $\dfrac{R}{F} = 2 \tan \theta$, and thus the resultant reaction at A passes through D, the intersection of the lines of action of **S** and **W**.

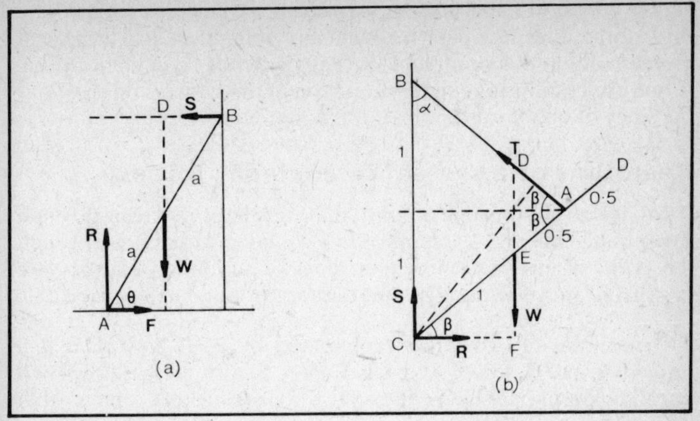

Figure 110

Example 4 In figure 110(b), the flag pole CD has a uniform mass of 30 kg and is 2 metres long. If BC = 2 m and CA = 1·5 m, find the tension in the rope (length 1·5 m) and the reaction at C.

$$\text{BA} = \text{AC} = 1{\cdot}5 \Rightarrow \angle\text{CBA} = \alpha = \angle\text{BCA} = \cos^{-1}\tfrac{2}{3} = 48{\cdot}2°$$

$$\Rightarrow \beta = 41{\cdot}8°.$$

$$\uparrow \quad T\cos 48{\cdot}2° + S = W; \qquad \rightarrow \quad T\cos 41{\cdot}8° = R.$$

Moments about C gives $T \times 1{\cdot}5 \sin 2\beta = W\cos\beta$,

$$\Rightarrow 3T\sin\beta = W, \quad \Rightarrow T = \tfrac{1}{2}W.$$
$$S = W - T\cos\alpha = W - \tfrac{2}{3} \times \tfrac{1}{2}W = \tfrac{2}{3}W;$$
$$R = \tfrac{1}{2}W\cos\beta = \sqrt{\tfrac{5}{6}}\,W.$$

Resultant force at C is $\dfrac{\sqrt{21}}{6}W$ at an angle of $\tan^{-1}\dfrac{4}{\sqrt{5}}$ to the horizontal.

$$\text{DE} = \text{EF} = \tfrac{2}{3}; \quad \text{CF} = \cos\beta = \frac{\sqrt{5}}{3} \Rightarrow \frac{\text{DF}}{\text{CF}} = \frac{4}{\sqrt{5}}.$$

So the resultant force at C passes through D, the intersection of the lines of action of **T** and **W**.

If three parallel forces are acting on a rigid body in equilibrium, then the resultant of each pair must be equal and opposite to the other force. If the three forces are not parallel, then the lines of action of two of them meet in a point. If the line of action of the third force does not pass through this point then it will have a moment about this point and the body will not be in equilibrium. A body in equilibrium under the action of three forces implies that the lines of action of the three forces are concurrent.

Resultant of a system of coplanar forces

If two forces act through a point, their resultant must act through that point. Thus the resultant of two coplanar forces acts through the point of intersection of their lines of action, unless they are parallel. If they are parallel the resultant is found as on page 249.

The resultant of two forces represented by $p\mathbf{OA}$ and $q\mathbf{OB}$ is a force $(p+q)\mathbf{OC}$, where AC : CB = $q : p$. Figure 111(a) shows two parallelograms ODAC and OEBC constructed such that AC : CB = $q : p$, that is

$$p \times AC = q \times CB. \tag{1}$$

$$\begin{aligned} p\mathbf{OA} + q\mathbf{OB} &= p(\mathbf{OD} + \mathbf{OC}) + q(\mathbf{OE} + \mathbf{OC}) \\ &= (p+q)\mathbf{OC} + p\mathbf{OD} + q\mathbf{OE} \\ &= (p+q)\mathbf{OC}, \end{aligned}$$

since $p\mathbf{OD} + q\mathbf{OE} = p\mathbf{CA} + q\mathbf{CB} = \mathbf{0}$, using the result in (1).

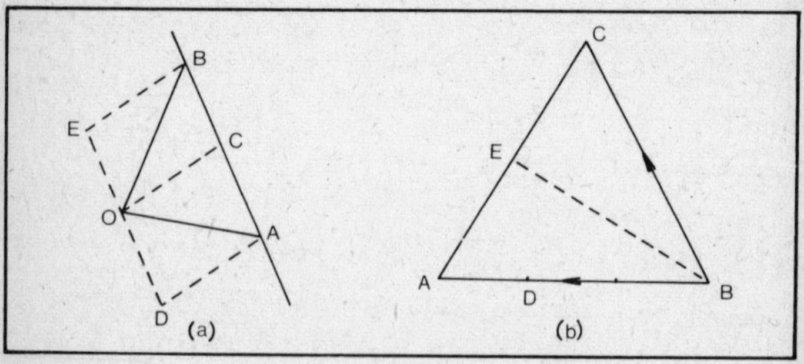

Figure 111

Example 5 Find the resultant of forces represented by **BA**, **BC**, and 2**AC** acting along the sides of triangle ABC.

BA and **BC** give a resultant of 2**BE**, where AE = EC (figure 111b). 2**AC** = 4**AE** and this force with 2**BE** give a resultant of 6**DE**, where AD : DB = 1 : 2.

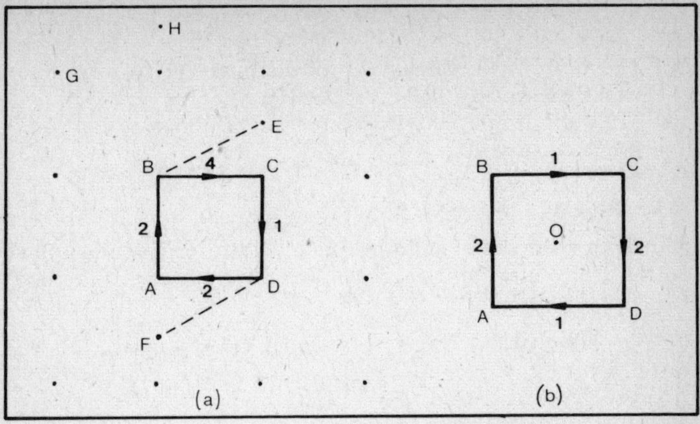

Figure 112

Example 6 Find the resultant of the systems of forces acting around the squares in figure 112.

In figure 112(a), with A as origin and D as the point $(a, 0)$, and using a scale where a distance $AB = a$ represents a force of 2, the forces along AB and BC are equivalent to a force represented by $2\mathbf{BE}$. The forces along CD and DA give a resultant of $\mathbf{DF}$. $\mathbf{DF}$ and $2\mathbf{BE}$ are parallel forces which are equivalent to $\mathbf{GH}$. The resultant of the system is a force of magnitude $\sqrt{5}$ in a direction parallel to BE through the point $G(-a, 2a)$.

An easier method uses $\mathbf{i}$ and $\mathbf{j}$ components with resultant $\mathbf{R}$, where $\mathbf{R} = 2\mathbf{j} + 4\mathbf{i} - \mathbf{j} - 2\mathbf{i} = 2\mathbf{i} + \mathbf{j}$. The magnitude of $\mathbf{R}$ is $\sqrt{5}$. Moments about A give $(\sqrt{5})h = 4a + 1a = 5a$ (clockwise), where h is the perpendicular distance from A to the line of action of $\mathbf{R}$. Since $h = (\sqrt{5})a$, $\mathbf{R}$ acts through G as AG is perpendicular to the line of action of $\mathbf{R}(=2\mathbf{i} + \mathbf{j})$.

In figure 112(b), the resultant force $\mathbf{R} = -\mathbf{i} + 2\mathbf{j} + \mathbf{i} - 2\mathbf{j} = \mathbf{0}$. However the forces do not balance, they form a couple. The total moment about O is $6 \times \frac{1}{2}a = 3a$ (clockwise). The total moment about A is also $3a$ (clockwise). This confirms the result that the moment of a couple is the same about any point in the plane.

Conditions of equilibrium of a system of coplanar forces:

(i) The sum of the components of the forces in two directions which are not parallel must be zero.

(ii) The sum of the moments of the forces about any arbitrary point must be zero.

Key terms

Centres of mass of uniform bodies (mass is evenly distributed so that the centres of mass lie on axes of symmetry, if the body has symmetry).

Body	Centre of mass	
Thin rod	centre	
Rectangular plate (lamina)	centre	
Triangular lamina	centroid (intersection of medians)	
Tetrahedron Pyramid	$\frac{3}{4}$ distance from vertex to centre of mass of base	
Solid cone	$\frac{3}{4}$ distance from vertex to centre of base	
Hollow cone	$\frac{2}{3}$ distance from vertex to centre of base	
Circular arc (angle 2θ)	$\dfrac{a \sin \theta}{\theta}$	radius a
Circular sector (angle 2θ)	$\dfrac{2a \sin \theta}{3\theta}$	radius a
Circular quadrant	$\dfrac{4a}{3\pi}$	radius a
Solid hemisphere	$\dfrac{3a}{8}$	radius a
Hollow hemisphere	$\frac{1}{2}a$	radius a

Chapter 16
Kinematics

Mechanics is the study of the motion of bodies and the forces which cause the motion. The motion is described by specifying displacement (from an origin), velocity (speed) and acceleration at any particular time t. In this chapter the description of motion, kinematics, is set out.

A particle is a body whose mass may be considered to act at a point. A lamina is a body with some mass having an appreciable area but negligible thickness.

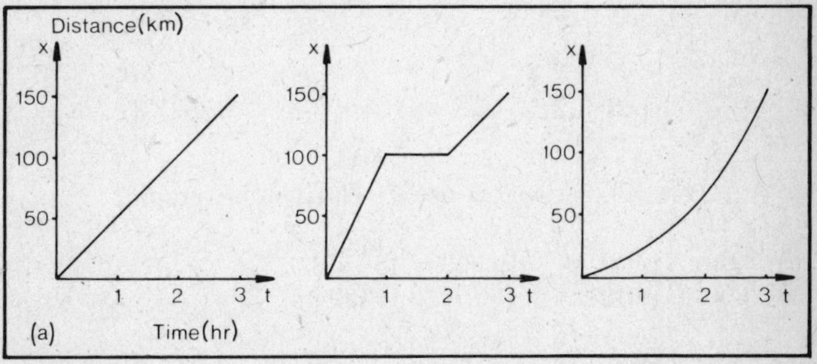

Figure 113

Figure 113 shows the distance–time graphs of three cars travelling a distance of 150 km in 3 hours. The first car in figure 113 travels at a constant speed of 50 km hr^{-1} and the graph is a straight line. The second car travels 100 km in the first hour, stops for the second hour (the graph is flat, gradient zero) and travels 50 km in the third hour. The third car's average speed is 50 km hr^{-1}, but its actual speed increases with time. If $x = \frac{50}{3} t^2$ then the speed $v = \frac{dx}{dt} = \frac{100}{3} t$, the gradient of the curve.

If x is plotted against t, the gradient at a point represents the speed at that time.

Figure 114 shows the graphs of the three cars' speeds plotted against time. In figure 114(c) the gradient is constant and is equal to $\frac{100 \text{ km hr}^{-1}}{3 \text{ hr}} = 33\frac{1}{3}$ km hr^{-2}. This is the acceleration of the car.

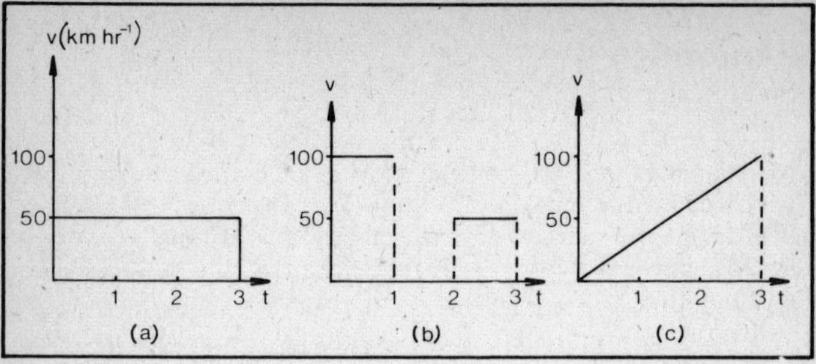

Figure 114

In general, acceleration = rate of change of speed

$$a = \frac{dv}{dt} = \frac{d^2x}{dt^2} \quad \text{also} \quad a = \frac{dv}{dt} = \frac{dv}{dx} \times \frac{dx}{dt} = v\frac{dv}{dx}.$$

$$v = \frac{dx}{dt} \Rightarrow x = \int v\, dt = \text{area under velocity–time graph.}$$

In figure 114(a) $x = 50 \times 3 = 150$ km.
In figure 114(b) $x = 100 + 50 = 150$ km.

In figure 114(c) $x = \int_0^3 33\tfrac{1}{3}t\, dt = \left[16\tfrac{2}{3}t^2\right]_0^3 = 16\tfrac{2}{3} \times 9 - 0$

$= 150$ km.

Constant (uniform) acceleration

A particle travels with constant acceleration a and starting speed u m s^{-1} (figure 115). After t seconds its speed will have increased by at m s^{-1} and if v is its speed at time t,

$$v = u + at. \qquad (1)$$

The distance x at time t is given by the area of OPQST which can be found in three ways.

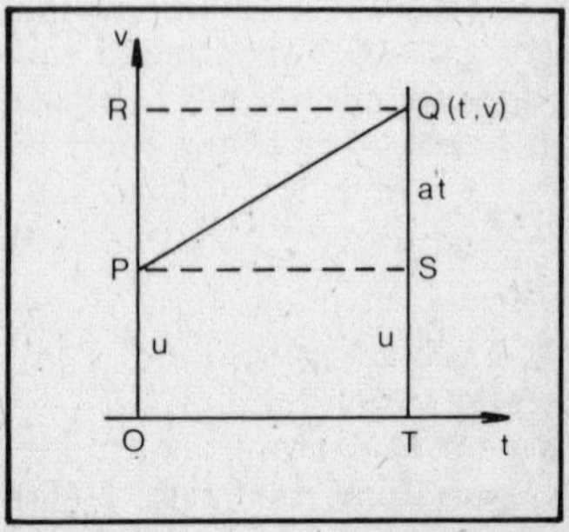

Figure 115

OPST + PSQ gives $\quad x = ut + \frac{1}{2}at^2$. (2)
ORQT − PQR gives $\quad x = vt - \frac{1}{2}at^2$. (3)
Trapezium OPQT gives $\quad x = \frac{1}{2}(u + v)t$. (4)
Squaring (1) gives $\quad v^2 = u^2 + 2uat + a^2t^2$
$\quad = u^2 + 2a(ut + \frac{1}{2}at^2)$.
Using (2) $\quad v^2 = u^2 + 2ax$. (5)

Equations (1) to (5) enable one-dimensional problems on uniform acceleration to be solved.

Example 1 A train accelerates uniformly to change its speed from 10 m s^{-1} to 20 m s^{-1} in 5 seconds. Finds its acceleration and the distance covered.

$v = u + at$ gives $20 = 10 + 5a \Rightarrow a = 2$ m s^{-2}.
$s = \frac{1}{2}(u + v)t$ gives $s = \frac{1}{2}(10 + 20) \times 5 = 75$ m.

Vertical motion under gravity

Example 2 Find the height reached by a ball thrown vertically upwards at 20 m s^{-1} and the time it is in the air.

An object unsupported falls towards the centre of the earth under the action of the gravitational force (gravity). In this problem air resistance is neglected and the acceleration due to gravity is regarded as uniform for motion close to the surface of the earth. The acceleration due to gravity is taken as 9·8 m s^{-2}, although it varies slightly at different points on the earth's surface.
If the upwards direction is regarded as positive, the acceleration $a = -9{\cdot}8$ m s^{-2}, and the starting velocity $u = 10$ m s^{-1}. At the highest point, the velocity of the ball is zero so $v = u + at$ gives $0 = 20 - 9{\cdot}8\,t \Rightarrow t = 2{\cdot}04$ seconds. When the ball returns to its original place, its displacement is zero; $x = ut + \frac{1}{2}at^2$ gives $0 = 20t - 4{\cdot}9t^2$

$$\Rightarrow t = 0 \text{ (start)} \quad \text{or} \quad t = \frac{20}{4{\cdot}9} = 4{\cdot}08 \text{ (finish)}.$$

Notice that the times for the ball to ascend and descend are the same.

Example 3 A particle moves in a straight line so that its displacement from O in metres is given by $x = t^3 - 12t + 4$ at time t seconds. Examine the motion.

$$\frac{dx}{dt} = 3t^2 - 12 \Rightarrow v = -12 \text{ m s}^{-1} \text{ and } x = 4 \text{ m when } t = 0.$$

The particle starts 4 m from O moving towards O.

$v = 0$ when $3t^2 - 12 = 0 \Rightarrow t^2 = 4 \Rightarrow t = 2$ (t positive).
$t = 2 \Rightarrow x = 8 - 24 + 4 = -12$ and the particle stops momentarily when $t = 2$ at $x = -12$ m.

For $t > 2$, $v > 0$ and the particle moves back towards 0 and continues in the positive direction building up speed.

Two-dimensional motion

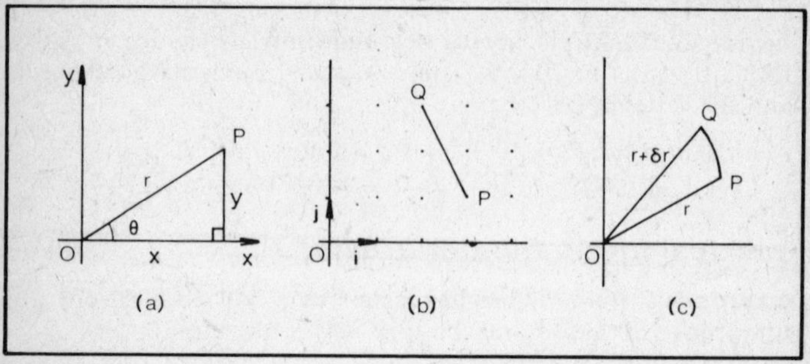

Figure 116

Reference may have to be made to chapter 11 on vectors, page 193. When a particle moves in a plane, two coordinates are needed to specify its position. These can be cartesian coordinates (x, y) or polar coordinates (r, θ) as in figure 116(a).

$$x = r\cos\theta; \quad y = r\sin\theta; \quad x^2 + y^2 = r^2; \quad \tan\theta = y/x.$$

In figure 116(b), if a particle moves from P(3, 1) to Q(2, 3) in one second, its displacement $\mathbf{PQ} = -\mathbf{i} + 2\mathbf{j} = \begin{pmatrix} -1 \\ 2 \end{pmatrix}$ which is a vector. If it moves in a straight line with constant speed, its velocity will be $-\mathbf{i} + 2\mathbf{j}$ which has a magnitude of $\sqrt{5}$ m s^{-1} in the direction of $\mathbf{PQ}$. If at time t a particle is at P (where $\mathbf{OP} = \mathbf{r}$) and a short time later, at time $t + \delta t$ it is at Q (where $\mathbf{OQ} = \mathbf{r} + \delta\mathbf{r}$) in figure 116(c), then its displacement $\mathbf{PQ} = \delta\mathbf{r}$. The displacement per unit time is $\delta\mathbf{r}/\delta t$ and this represents the average velocity in the time interval δt. Velocity is a vector quantity whose magnitude represents the speed of the particle and whose direction represents the direction in which the particle is moving. The instantaneous velocity at P is

$$\mathbf{v} = \lim_{\delta t \to 0} \frac{\delta\mathbf{r}}{\delta t} = \frac{d\mathbf{r}}{dt}$$

which is along the tangent to the curve representing the path of the particle. Similarly, the acceleration of the particle is defined as

$$\mathbf{a} = \lim_{\delta t \to 0} \frac{\delta \mathbf{v}}{\delta t} = \frac{d\mathbf{v}}{dt} = \frac{d^2\mathbf{r}}{dt^2}.$$

If the x- and y-displacements are functions of time then

$$\mathbf{r} = x\mathbf{i} + y\mathbf{j} = f(t)\mathbf{i} + g(t)\mathbf{j}, \text{ where } x = f(t), y = g(t).$$

$$\mathbf{r} + \delta\mathbf{r} = f(t+\delta t)\mathbf{i} + g(t+\delta t)\mathbf{j}$$
$$\Rightarrow \delta\mathbf{r} = \{f(t+\delta t) - f(t)\}\mathbf{i} + \{g(t+\delta t) - g(t)\}\mathbf{j}.$$

$$\mathbf{v} = \lim_{\delta t \to 0} \frac{\delta \mathbf{r}}{\delta t} = \lim_{\delta t \to 0} \left[\left\{\frac{f(t+\delta t) - f(t)}{\delta t}\right\}\mathbf{i} + \left\{\frac{g(t+\delta t) - g(t)}{\delta t}\right\}\mathbf{j}\right]$$

$$\mathbf{v} = f'(t)\mathbf{i} + g'(t)\mathbf{j}.$$

This result enables any vector expressed in terms of **i** and **j** to be differentiated. The acceleration is

$$\mathbf{a} = \frac{d\mathbf{v}}{dt} = \frac{d^2\mathbf{r}}{dt^2} = f''(t)\mathbf{i} + g''(t)\mathbf{j}.$$

If a particle moves so that its displacement after t seconds is given by $x = t - 2$, $y = (t-2)^2$, then it follows the path of $y = x^2$ (figure 117).

$$\frac{dx}{dt} = \dot{x} = 1; \quad \frac{dy}{dt} = \dot{y} = 2t - 4.$$

The notation $\dot{x}$ is used to stand for differentiation with respect to time t.

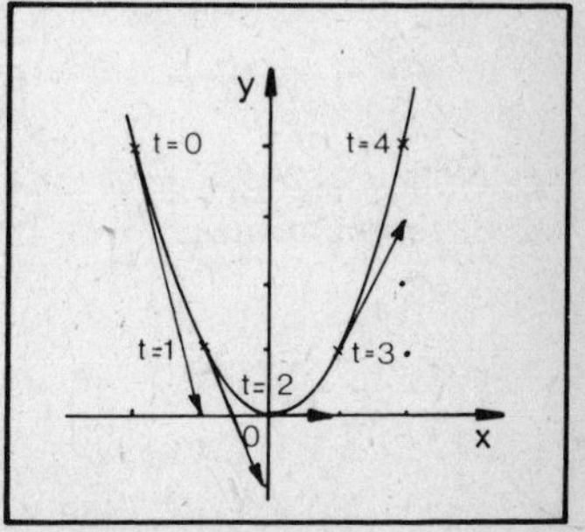

Figure 117

t	(x, y)	$\mathbf{v} = \begin{pmatrix}\dot{x}\\ \dot{y}\end{pmatrix}$
0	$(-2, 4)$	$\begin{pmatrix}1\\ -4\end{pmatrix}$
1	$(-1, 1)$	$\begin{pmatrix}1\\ -2\end{pmatrix}$
2	$(0, 0)$	$\begin{pmatrix}1\\ 0\end{pmatrix}$
3	$(1, 1)$	$\begin{pmatrix}1\\ 2\end{pmatrix}$
4	$(2, 4)$	$\begin{pmatrix}1\\ 4\end{pmatrix}$

The table of values of the displacement at different times enables the curve to be plotted.
The velocities can be drawn on the displacement curve and can be seen to be along the tangent at each point.
The acceleration $\mathbf{a} = 2\mathbf{j}$ is in the y-direction and is constant for all values of t.

The cycloid

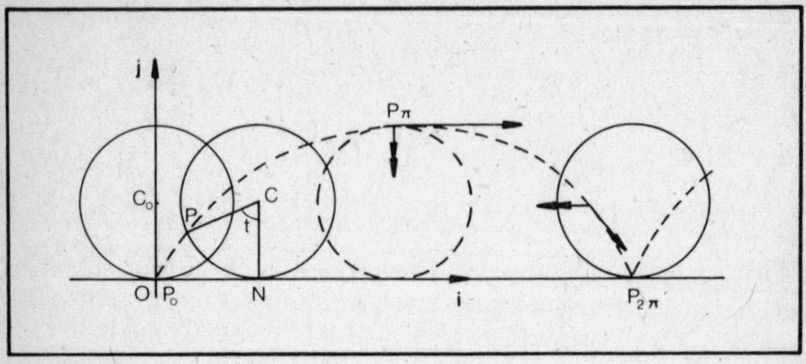

Figure 118

The path traced out by a point on the circumference of a rolling wheel is called a **cycloid**. For convenience let the wheel have radius 1 unit and let it roll along the x-axis with angular velocity 1 radian per second (figure 118). This means that CP turns through 1 radian every second, so that at time t, C_0P_0 has rotated to CP. P_0N is the same length as arc PN, i.e. t units, since CN = 1 and $\angle$PCN = t radians.
The position vector of P, $\mathbf{OP} = \mathbf{r}$ is given by

$$\mathbf{r} = \begin{pmatrix} x \\ y \end{pmatrix} = \mathbf{OC} + \mathbf{CP} = \begin{pmatrix} t \\ 1 \end{pmatrix} + \begin{pmatrix} -\sin t \\ -\cos t \end{pmatrix} = \begin{pmatrix} t - \sin t \\ 1 - \cos t \end{pmatrix}.$$

The following table gives the values of displacement, velocity and acceleration at times $t = 0, \frac{1}{2}\pi, \pi, 1\frac{1}{2}\pi$ and 2π.

t	0	$\frac{1}{2}\pi$	π	$1\frac{1}{2}\pi$	2π
$\mathbf{r} = \begin{pmatrix} x \\ y \end{pmatrix} = \begin{pmatrix} t - \sin t \\ 1 - \cos t \end{pmatrix}$	$\begin{pmatrix} 0 \\ 0 \end{pmatrix}$	$\begin{pmatrix} \frac{1}{2}\pi - 1 \\ 1 \end{pmatrix}$	$\begin{pmatrix} \pi \\ 2 \end{pmatrix}$	$\begin{pmatrix} 1\frac{1}{2}\pi + 1 \\ 1 \end{pmatrix}$	$\begin{pmatrix} 2\pi \\ 0 \end{pmatrix}$
$\mathbf{v} = \begin{pmatrix} \dot{x} \\ \dot{y} \end{pmatrix} = \begin{pmatrix} 1 - \cos t \\ \sin t \end{pmatrix}$	$\begin{pmatrix} 0 \\ 0 \end{pmatrix}$	$\begin{pmatrix} 1 \\ 1 \end{pmatrix}$	$\begin{pmatrix} 2 \\ 0 \end{pmatrix}$	$\begin{pmatrix} 1 \\ -1 \end{pmatrix}$	$\begin{pmatrix} 0 \\ 0 \end{pmatrix}$
$\mathbf{a} = \begin{pmatrix} \ddot{x} \\ \ddot{y} \end{pmatrix} = \begin{pmatrix} \sin t \\ \cos t \end{pmatrix}$	$\begin{pmatrix} 0 \\ 1 \end{pmatrix}$	$\begin{pmatrix} 1 \\ 0 \end{pmatrix}$	$\begin{pmatrix} 0 \\ -1 \end{pmatrix}$	$\begin{pmatrix} -1 \\ 0 \end{pmatrix}$	$\begin{pmatrix} 0 \\ 1 \end{pmatrix}$

The position vector **r** gives the locus of the path, and the velocity vector (single arrow) is along the tangent. The acceleration (double arrow) is on the 'inside' of the curve changing the direction of the path. The velocity is zero when $t = 0, 2\pi, \ldots$ and the curve meets the x-axis in a pointed 'cusp'.

Two-dimensional motion with constant acceleration

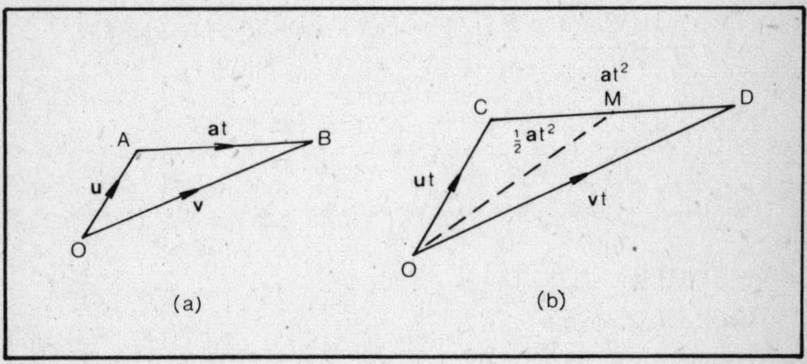

Figure 119

If a particle is moving with velocity **u** and experiences a constant acceleration **a** for a time t seconds, its velocity is increased by an amount $\mathbf{a}t$. The new velocity

$$\mathbf{v} = \mathbf{u} + \mathbf{a}t \qquad (1)$$

and the quantities are added vectorially (figure 119a).

Since $\mathbf{v} = \dfrac{d\mathbf{r}}{dt}$, where **r** is the displacement,

$$\mathbf{v} = \frac{d\mathbf{r}}{dt} = \mathbf{u} + \mathbf{a}t \Rightarrow \mathbf{r} = \mathbf{u}t + \tfrac{1}{2}\mathbf{a}t^2 + \mathbf{b}. \qquad (2)$$

If the particle starts at the origin, $t = 0$, $\mathbf{r} = \begin{pmatrix} 0 \\ 0 \end{pmatrix} \Rightarrow \mathbf{b} = \begin{pmatrix} 0 \\ 0 \end{pmatrix}$.

Figure 119(b) shows the triangle of figure 119(a) enlarged by a scale factor t and $\mathbf{OM} = \mathbf{u}t + \tfrac{1}{2}\mathbf{a}t^2$, which from (2) is equal to **r**, so that **OM** represents the displacement.

$$\mathbf{OM} = \tfrac{1}{2}\mathbf{OC} + \tfrac{1}{2}\mathbf{OD} \Rightarrow \mathbf{r} = \tfrac{1}{2}\mathbf{u}t + \tfrac{1}{2}\mathbf{v}t = \tfrac{1}{2}(\mathbf{u} + \mathbf{v})t. \qquad (3)$$

These formulae are similar to those for one-dimensional motion and we can find an equation similar to $v^2 = u^2 + 2ax$.

$$v^2 = \mathbf{v} \cdot \mathbf{v} = (\mathbf{u} + \mathbf{a}t) \cdot (\mathbf{u} + \mathbf{a}t) = \mathbf{u} \cdot \mathbf{u} + 2\mathbf{u} \cdot \mathbf{a}t + \mathbf{a} \cdot \mathbf{a}t^2$$
$$= u^2 + 2\mathbf{a} \cdot (\mathbf{u}t + \tfrac{1}{2}\mathbf{a}t^2)$$
$$v^2 = u^2 + 2\mathbf{a} \cdot \mathbf{r}. \qquad (4)$$

The formulae are the same as in one dimension, but with multiplication replaced by the scalar product of vectors, and can simply be extended to use in three dimensions. An application of them can be seen in chapter 20.

In three dimensions with reference to axes Ox, Oy and Oz,

displacement is represented by $\mathbf{r} = x\mathbf{i} + y\mathbf{j} + z\mathbf{k}$;

velocity by $\mathbf{v} = \dot{\mathbf{r}} = \dfrac{d\mathbf{r}}{dt} = \dfrac{dx}{dt}\mathbf{i} + \dfrac{dy}{dt}\mathbf{j} + \dfrac{dz}{dt}\mathbf{k} = \dot{x}\mathbf{i} + \dot{y}\mathbf{j} + \dot{z}\mathbf{k}$;

acceleration by $\mathbf{a} = \dot{\mathbf{v}} = \ddot{\mathbf{r}} = \ddot{x}\mathbf{i} + \ddot{y}\mathbf{j} + \ddot{z}\mathbf{k}$.

Key terms

In one dimension

$v = u + at$;
$x = ut + \frac{1}{2}at^2 = \frac{1}{2}(u + v)t$;
$v^2 = u^2 + 2ax$.

In two or more dimensions

$\mathbf{v} = \mathbf{u} + \mathbf{a}t$;
$\mathbf{r} = \mathbf{u}t + \frac{1}{2}\mathbf{a}t^2 = \frac{1}{2}(\mathbf{u} + \mathbf{v})t$;
$v^2 = u^2 + 2\mathbf{a} \cdot \mathbf{r}$.

Chapter 17
Laws of Motion

So far the motion of a particle has been discussed, but this chapter will consider the forces applied to a particle and the subsequent motion. The work is based on the laws of motion formulated by Sir Isaac Newton and have been the basis of mechanics since that time.

The laws of motion are:

(i) Every body remains at rest or in uniform motion unless acted upon by some external forces.

(ii) The rate of change of momentum of a moving body is proportional to the applied force and takes place in the direction of that force.

(iii) To every action there is an equal and opposite reaction.

These laws cannot be proved directly but have been found to agree very closely with observation and experiment. The first law is, in effect, a definition of force as that cause which changes or tends to change the state of rest or uniform motion of a body. The third law is particularly important when considering internal forces within a body or contact between rigid bodies.

The second law gives a very important formula used in dynamics.

Let $\mathbf{F}$ be a force acting on a particle of mass m, moving with a velocity $\mathbf{v}$. Its **momentum** is $m\mathbf{v}$ and hence the rate of change of momentum is

$$\frac{d}{dt}(m\mathbf{v}) = m\frac{d\mathbf{v}}{dt} \quad \text{(if } m \text{ is constant).}$$

Hence by Newton's second law

$$F \propto m\frac{d\mathbf{v}}{dt}.$$

This can be written as $\mathbf{F} = km\mathbf{a}$, where $\mathbf{a}$ is the acceleration and k is the constant of proportionality.

By choosing an appropriate system of units we can make $k = 1$. This necessitates choosing a unit of force such as the newton.

The newton is defined as that force which gives a mass of 1 kg an acceleration of 1 m s^{-1}. Thus in SI units we have

$$F = ma.$$

Example 1 Find the force required to increase the velocity of a mass of 6 kg from 15 m s^{-1} to 25 m s^{-1} in $\frac{1}{2}$ min.

Let the acceleration be a. This can be found using $v = u + at$,

$$25 = 15 + 30a.$$

Hence the acceleration $a = \frac{1}{3}$ m s^{-2}.

By Newton's second law, $F = ma$, and thus the force is

$$6 \times \tfrac{1}{3}\text{ N} = 2\text{ N}.$$

The forces applied to a particle or body may be of many different types. They may be associated with the body as

(i) contact forces, i.e. reactions due to the contact of one body with another,
(ii) frictional forces,
(iii) tensions in strings, rods or springs.

Non-contact forces are also possible and the most familiar is that of the weight of a body, that is, the gravitational attraction of the earth. Other examples include magnetic forces, electric fields and pressure forces from fluids.

Forces are vector quantities in that they possess magnitude and direction and thus can be combined together using the law of vector addition.

Forces acting at a point can be formed into a single resultant. For example, if a body is moving under the action of a force of 750 N against a resistance of 300 N, then the effective force is 450 N in the direction of motion. These two forces are parallel and hence their resultant is simply their difference since they act in opposite directions.

If the forces are not parallel they can still be combined by using a parallelogram of forces (vector addition). Figure 120(a) shows two forces of 2 N and 3 N acting at 60° to one another. Their resultant is given by R, where

$$R^2 = 2^2 + 3^2 - 2 \times 2 \times 3 \cos 120° \quad \text{(cosine rule).}$$
$$R^2 = 4 + 9 + 12 \cos 60°$$
$$= 13 + 6 = 19$$
$$R = \sqrt{19} = 4{\cdot}36\text{ N}.$$

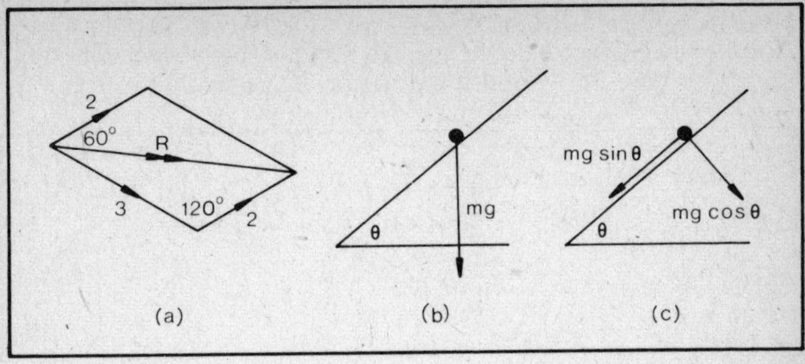

Figure 120

Forces can also be resolved and this is often necessary when considering motion in a particular direction. For example, when considering a particle of mass m on an inclined plane, it may be more useful to consider the resolved parts of the weight along and perpendicular to the plane rather than taking just the single vertical force mg downwards. Figure 120(b) and (c) show this force and its resolution on a plane inclined at an angle θ to the horizontal.

When a surface is said to be **smooth** there is assumed to be no force between the surface and any body placed on it opposing any motion or implied motion. In this case the only contact force between the surface and the body is a force normal to the surface, that is, a **normal reaction**.

If the surface is not smooth then there exists a force tending to prevent motion or opposing any actual motion. This is called a **frictional force** and when the body is in motion or on the point of moving the amount of friction is in a constant ratio to the normal reaction and is dependent on the nature of the two surfaces in contact.

If F is the frictional force between two surfaces and R is the normal reaction, then

$$\frac{F}{R} = \mu \quad \text{or} \quad F = \mu R,$$

where μ is the coefficient of dynamic friction.

Example 2 Find the time taken for a particle of mass 5 kg to slide 19·6 m from rest down a rough inclined plane of inclination $\tan^{-1}(\frac{3}{4})$ to the horizontal if the coefficient of friction is $\frac{1}{2}$.

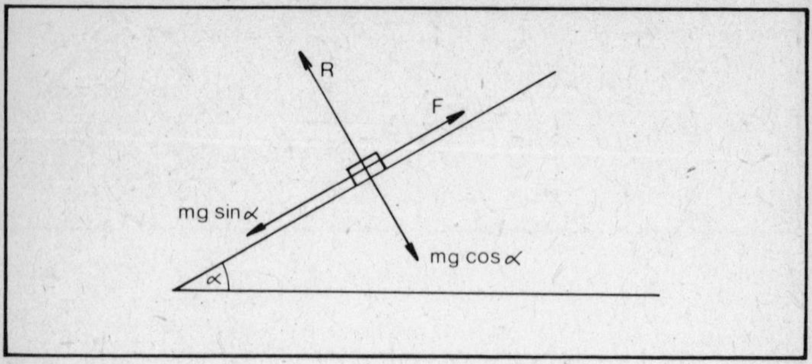

Figure 121

Let the normal reaction be R and the frictional force between the particle and the plane be F as shown in figure 121. Let α be the inclination of the plane. Thus $\tan \alpha = \frac{3}{4}$. Since there is no motion perpendicular to the plane

$$R - 5g \cos \alpha = 0.$$

Now $F = \mu R = 5\mu g \cos \alpha = 5 \times \frac{1}{2}g(\frac{4}{5}) = 2g.$

Consider the motion down the plane. Let the acceleration be a. Using Newton's second law

$$5g \sin \alpha - F = 5a,$$
$$5g(\tfrac{3}{5}) - 2g = 5a,$$
$$a = \frac{g}{5} \text{ m s}^{-2}.$$

Let t seconds be the time taken to travel 19·6 m down the plane. Using $s = ut + \frac{1}{2}at^2$:

$$19{\cdot}6 = 0 + \frac{1}{2}\left(\frac{g}{5}\right)t^2 \Rightarrow t^2 = 20.$$

Taking the positive square root, the time taken to descend is 4·47 s.

It is important to remember that the force used must always be the resultant force in the direction of motion and that the acceleration must be in the direction of this force.

Example 3 A man of mass 60 kg stands in a lift. Find the reaction between the man and the floor when the lift is (i) ascending (ii) descending with an acceleration of 2 m s^{-2}.

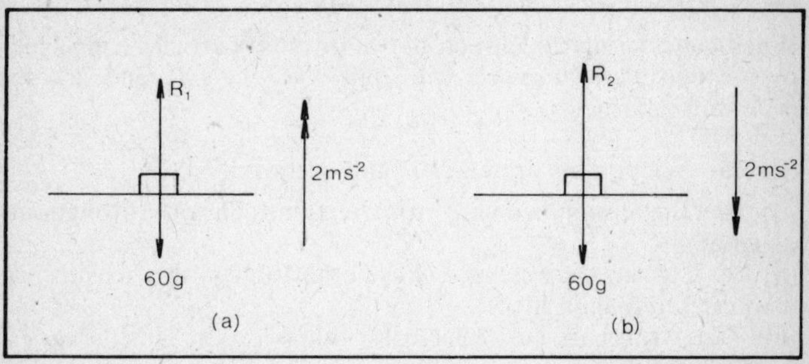

Figure 122

(i) Let the reaction be R_1 when the lift is ascending. Using Newton's second law

$$R_1 - 60g = 60 \times 2 \qquad (1)$$
$$R_1 = 120 + 588 = 708 \text{ N.}$$

(ii) Let the reaction be R_2 when the lift is descending. Using Newton's second law

$$60g - R_2 = 60 \times 2 \qquad (2)$$
$$R_2 = 588 - 120 = 468 \text{ N.}$$

The reactions are 708 N when the lift is ascending and 468 N when it is descending. Notice that in equations (1) and (2) the effective force is given in the direction of motion.

Remember that, by Newton's third law, the force exerted by the floor of the lift on the man will be equal and opposite to that exerted by the man on the floor of the lift. This means that if the lift has a mass of 600 kg and we consider the forces applied to the lift to produce an acceleration of 2 m s^{-2} upwards we have an equation of motion

$$T - 600g - R_1 = 600 \times 2,$$

where T is the tension in the cable.

$$T - 600g - 708 = 1200.$$
$$T = 1200 + 708 + 5880,$$
$$T = 7788 \text{ N}.$$

The tension in the cable is 7788 N.

Motion of connected particles

When considering the motion of two or more particles connected by inextensible strings, we can apply Newton's second law to each of the particles.

Several assumptions are made in this section:

(i) that the tension in a light string remains constant throughout its length,
(ii) if a string passes over a smooth pulley the tension is unaltered in magnitude,
(iii) the strings are inextensible, for otherwise the tension varies with the extension.

Most questions refer to light, inextensible strings passing over smooth pulleys.

Example 4 Masses of 5 kg and 3 kg are connected by a light inextensible string passing over a fixed smooth pulley. Find the acceleration of the masses and the tension in the string.

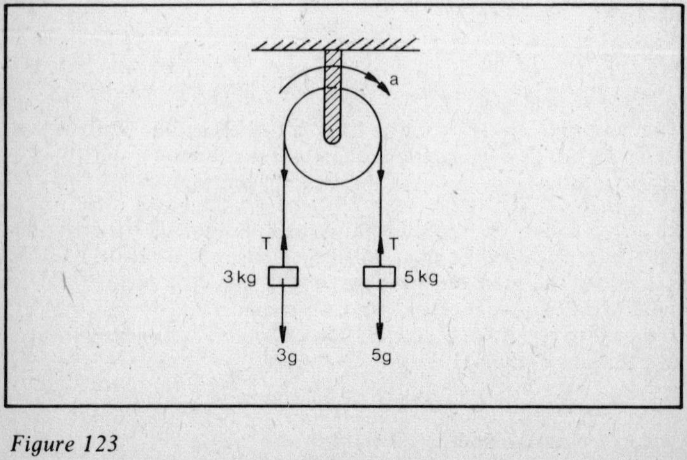

Figure 123

Let the tension throughout the string be T (in newtons) and the acceleration of the system be a (in m s^{-2}) as shown in figure 123. Clearly the acceleration of each of the masses is the same since the string is inextensible. Applying $F = ma$ to the two masses, remembering that the 3 kg will move upwards while the 5 kg mass will move downwards.

The equations of motion are:

$$\text{3 kg mass:} \quad T - 3g = 3a. \qquad (1)$$
$$\text{5 kg mass:} \quad 5g - T = 5a. \qquad (2)$$

Eliminating T from equations (1) and (2) we have

$$2g = 8a \Rightarrow a = (g/4) \text{ m s}^{-2}.$$

Substituting in equation (1) gives $T = (15g/4)$N.

Hence the acceleration of the masses is 2·45 m s^{-2} and the tension in the string is 36·75 N.

Example 5 A particle of mass 5 kg rests on the surface of a rough plane inclined at an angle $\alpha = \tan^{-1}\frac{3}{4}$ to the horizontal and is connected by a light inextensible string passing over a smooth pulley at the top of the plane to a mass of 5 kg hanging vertically. Find the acceleration of the masses if the coefficient of friction is $\frac{1}{4}$.

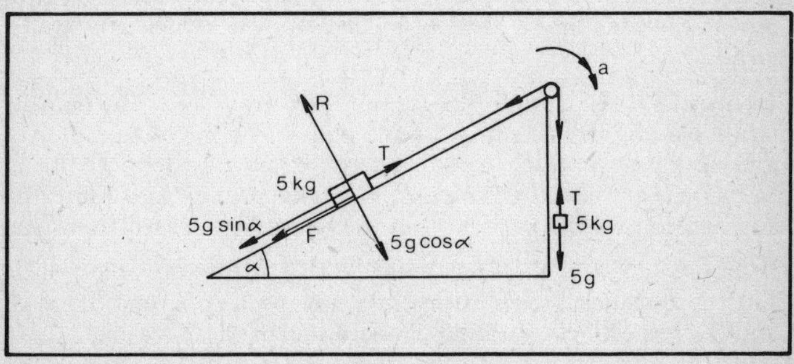

Figure 124

Let the tension in the string be T and the acceleration of the system be a with the 5 kg mass hanging vertically moving downwards. Let the frictional force due to the motion of the mass on the plane be F (figure 124).

The equations of motion are:

$$\text{5 kg mass hanging} \quad 5g - T = 5a. \tag{1}$$
$$\text{5 kg mass on the plane} \quad T - 5g \sin\alpha - F = 5a. \tag{2}$$
$$R - 5g \cos\alpha = 0 \tag{3}$$

Equations (2) and (3) were formed by considering directions along and perpendicular to the plane respectively.

Also $$F = \mu R = 5\mu g \cos\alpha = \tfrac{5}{4}g \cos\alpha. \tag{4}$$

From equations (2) and (4)

$$T - 5g \sin\alpha - \tfrac{5}{4}g \cos\alpha = 5a.$$

Since $\tan\alpha = \frac{3}{4}$, $\cos\alpha = \frac{4}{5}$ and $\sin\alpha = \frac{3}{5}$

$$T - 3g - g = 5a \quad \text{or} \quad T - 4g = 5a. \tag{5}$$

Hence equations (1) and (5) are

$$5g - T = 5a,$$
$$T - 4g = 5a.$$

Adding gives $g = 10a$ or $a = \dfrac{g}{10}$ m s^{-2}.

Hence the acceleration of the system is 0·98 m s^{-2} and the mass will move up the plane. This technique can be extended to cover systems consisting of different parts, each possessing different accelerations, for example, a particle placed on the face of a movable wedge or a system of pulleys some of which are free to move.

Example 6 A mass of 3 kg is connected by a light inextensible string passing over a smooth fixed pulley to a movable pulley of mass 2 kg over which passes a second light inextensible string to which are attached masses of 2 kg and 4 kg. Find the accelerations of the masses when the system is released from rest.

In figure 125, A is the fixed pulley and B is the movable pulley. Let the tensions in the string over the pulleys A and B be T and T_1 respectively. Let the acceleration of the 3 kg mass be a upwards and the acceleration of the 2 kg and 4 kg masses be f **relative to pulley B**.

Thus the actual acceleration of the 2 kg mass is $(a - f)$ downwards (or $f - a$ upwards) and that of the 4 kg mass is $(a + f)$ downwards. Hence applying $F = ma$ to each mass and pulley B, we obtain the equations of motion:

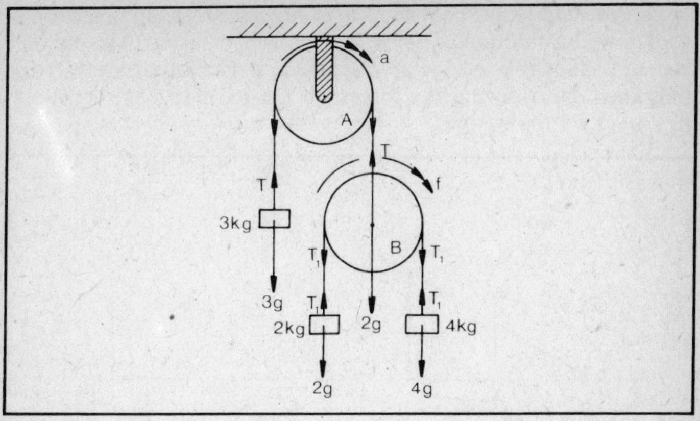

Figure 125

$$\begin{aligned} \text{3 kg mass} \quad & T - 3g = 3a. && (1)\\ \text{On pulley B} \quad & 2T_1 + 2g - T = 2a. && (2)\\ \text{2 kg mass} \quad & 2g - T_1 = 2(a - f). && (3)\\ \text{4 kg mass} \quad & 4g - T_1 = 4(a + f). && (4) \end{aligned}$$

Multiplying equation (3) by 2 and adding equation (4):

$$8g - 3T_1 = 8a \Rightarrow 3T_1 = 8g - 8a.$$

In equation (2):

$$\tfrac{2}{3}(8g - 8a) + 2g - T = 2a,$$

$$\text{or} \quad 22g - 3T = 22a. \qquad (5)$$

Hence from equations (1) and (5):

$$3T - 9g = 9a,$$
$$22g - 3T = 22a.$$

Adding gives $13g = 31a$ or $a = \frac{13}{31}g$ m s^{-2}.

From equations (3) and (4):

$$2g = 2a + 6f,$$
$$6f = 2g - 2 \times \tfrac{13}{31}g = \tfrac{36}{31}g,$$
$$f = \tfrac{6}{31}g \text{ m s}^{-2}.$$

Hence the acceleration of the 3 kg mass is $\frac{13}{31}g$ m s^{-2} upwards, the acceleration of the 4 kg mass is $\frac{19}{31}g$ m s^{-2} downwards and the acceleration of the 2 kg mass is $\frac{7}{31}g$ m s^{-2} downwards.

In considering some problems it is necessary to use Newton's third law. For example, when the motion of a particle on the face of a movable wedge is discussed, it is best to draw two diagrams, one showing the forces on the particle and the other the forces on the wedge.

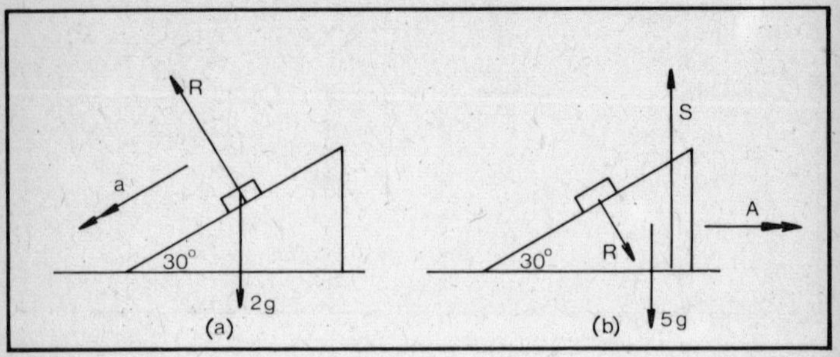

Figure 126

Figure 126 shows a particle of mass 2 kg on the face of a smooth wedge of mass 5 kg and slope 30°.

Let R be the normal reaction on the 2 kg mass normal to the surface and S the normal reaction of the supporting surface of the wedge.

Let the acceleration of the wedge horizontally be A and the acceleration of the 2 kg mass **relative to the wedge** be a. Hence the actual acceleration of the particle is $a \sin 30°$ vertically downwards and $(a \cos 30° - A)$ horizontally (to the left). The equations of motion can be written:

For particle vertically $2g - R \cos 30° = 2a \sin 30°$, (1)
horizontally $R \sin 30° = 2(a \cos 30° - A)$. (2)

For wedge vertically $S - R \cos 30° - 5g = 0$, (3)
horizontally $R \sin 30° = 5A$. (4)

These four equations contain four unknowns and thus can be solved to find any of the quantities a, A, S or R. For example, to find the values of a and A (the accelerations).

Equation (4) gives $R = 10A$. (5)
Equation (1) becomes $2g - 5\sqrt{3}A = a$, (6)
and (2) gives $7A = a\sqrt{3}$. (7)

Hence, eliminating A:

$$2g - 5\sqrt{3}\left(\frac{a\sqrt{3}}{7}\right) = a.$$

$$2g = a + \tfrac{15}{7}a \Rightarrow a = \tfrac{7}{11}g.$$

Thus from equation (7):

$$A = \frac{a\sqrt{3}}{7} = \frac{7\sqrt{3}}{77}g = \frac{\sqrt{3}}{11}g.$$

The acceleration of the wedge is $\frac{\sqrt{3}g}{11}$ m s^{-2} and that of the particle relative to the wedge is $\frac{7g}{11}$ m s^{-2}.

Variable forces

If the force is variable then the acceleration will also be variable and calculus is required to solve the problems. The equation of motion of a particle of mass m moving subject to a force F is (by Newton's second law)

$$m\frac{dv}{dt} = F. \tag{1}$$

Integrating with respect to t:

$$mv = \int F\,dt.$$

If F is a function of t, this integral can be evaluated. Alternatively, if F is a function of distance s, the acceleration can be written as $m\frac{dv}{ds}$ and the equation of motion is

$$mv\frac{dv}{ds} = F, \tag{2}$$

$$\Rightarrow \tfrac{1}{2}mv^2 = \int F\,ds.$$

If F is a function of v, the differential equation (2) will represent the equation of motion and can be solved by separating the variables (see chapter 10).

Example 7 A particle of mass m falls from rest under gravity in a medium whose resistance varies as the velocity. Find an expression for the velocity after time t.

Let the resistance be kv per unit mass where k is a constant and v is the velocity.
The equation of motion is:

$$\frac{dv}{dt} = g - kv. \tag{1}$$

Rearranging gives $$\frac{1}{g - kv}\frac{dv}{dt} = 1.$$

Integrating with respect to t gives

$$\int \frac{dv}{g - kv} = \int dt.$$

Thus $$-\frac{1}{k}\ln(g - kv) = t + C.$$

Now when $t = 0$, $v = 0 \Rightarrow C = -\dfrac{1}{k}\ln g$.

Thus $$\frac{1}{k}\ln\left(\frac{g}{g - kv}\right) = t,$$

$$\frac{g}{g - kv} = e^{kt},$$

$$g - kv = ge^{-kt},$$

and $$v = \frac{g}{k}\left(1 - e^{-kt}\right).$$

Notice that as $t \to \infty$, $v \to \dfrac{g}{k}$. This is the value obtained for v when the acceleration is zero in equation (1). It is called the **limiting or terminal velocity**.

Example 8 The acceleration of a particle is $4 - v^2$ where v is its speed. If it reaches a speed u from rest after travelling a distance of 30 m, find u.

The equation of motion is

$$v\frac{dv}{ds} = 4 - v^2,$$

$$\frac{v}{4 - v^2}\frac{dv}{ds} = 1.$$

Integrating with respect to s:

$$\int \frac{v}{4 - v^2}\, dv = \int ds,$$

and
$$-\tfrac{1}{2}\ln(4 - v^2) = s + C.$$

When $s = 0$, $v = 0 \Rightarrow C = -\tfrac{1}{2}\ln 4$.

$$\tfrac{1}{2}\ln\left(\frac{4}{4 - v^2}\right) = s,$$

$$\frac{4}{4 - v^2} = e^{2s},$$

$$v^2 e^{2s} = 4(e^{2s} - 1),$$

$$v^2 = 4(1 - e^{-2s}).$$

When $s = 30$, $v = u$ and

$$u^2 = 4(1 - e^{-60}),$$

$$u = 2(1 - e^{-60})^{1/2}.$$

Key terms

Newton's first law Every body will continue in its state of rest or uniform motion in a straight line unless acted upon by some external force.

Newton's second law $\mathbf{F} = m\mathbf{a}$.

Newton's third law To every action there is an equal and opposite reaction.

If F is the frictional force between two surfaces and R is the normal reaction, then

$$F = \mu R,$$

where μ is the coefficient of dynamical friction.

When variable forces are used, Newton's second law becomes

$$m\frac{d\mathbf{v}}{dt} = \mathbf{F} \quad \text{or} \quad m\mathbf{v}\frac{d\mathbf{v}}{ds} = \mathbf{F}.$$

Chapter 18
Work, Power and Energy

When a constant force **F** moves its point of application through a displacement **r**, the work done by the force is **F** . **r**.

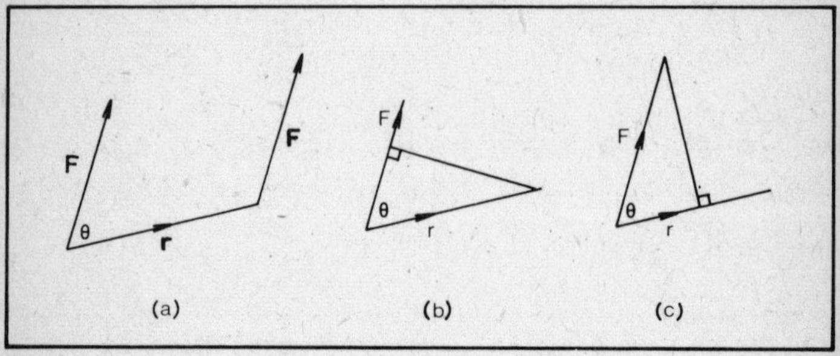

Figure 127

In scalar terms this gives the work done as $Fr \cos \theta$, where θ is the angle between F and r (see figure 127a).

Thus the work done may be regarded as the product of the force and the projection of the displacement on it (figure 127b) or as the product of the component of the force in the direction of the displacement and the distance moved (figure 127c).

Since the work done involves the product of force and distance, the SI unit of work is the newton metre (Nm) = joule (J).

Definition A joule is the work done when a force of 1 newton moves its point of application 1 metre in the direction of the force.

Two special cases should be remembered:

(i) if **F** is perpendicular to **r**, **F** . **r** = 0 and thus no work is done;
(ii) if **F** is parallel to **r**, then the work done is Fr (if they are in the same sense) or $-Fr$ (if they are in the opposite sense).

For a number of forces $\mathbf{F}_1, \mathbf{F}_2, \mathbf{F}_3, \ldots, \mathbf{F}_n$ having displacements $\mathbf{r}_1, \mathbf{r}_2, \mathbf{r}_3, \ldots, \mathbf{r}_n$, the total work done is the vector sum of the work done by each force.

$$\text{Work done} = \mathbf{F}_1 \cdot \mathbf{r}_1 + \mathbf{F}_2 \cdot \mathbf{r}_2 + \cdots + \mathbf{F}_n \cdot \mathbf{r}_n = \sum \mathbf{F} \cdot \mathbf{r}.$$

If a constant force moves through different displacements this becomes

$$\text{Work done} = \mathbf{F} \cdot \mathbf{r}_1 + \mathbf{F} \cdot \mathbf{r}_2 + \cdots + \mathbf{F} \cdot \mathbf{r}_n = \mathbf{F} \cdot \sum \mathbf{r}.$$

If a number of forces all move through the same displacement then

$$\text{Work done} = \mathbf{F}_1 \cdot \mathbf{r} + \mathbf{F}_2 \cdot \mathbf{r} + \cdots + \mathbf{F}_n \cdot \mathbf{r} = (\sum \mathbf{F}) \cdot \mathbf{r},$$

i.e. the forces could be replaced by their resultant.

Example 1 A particle of mass m is pulled up a rough inclined plane by a force P applied through a string parallel to the plane. Find expressions for the work done by P, the frictional force, the normal reaction and the weight.

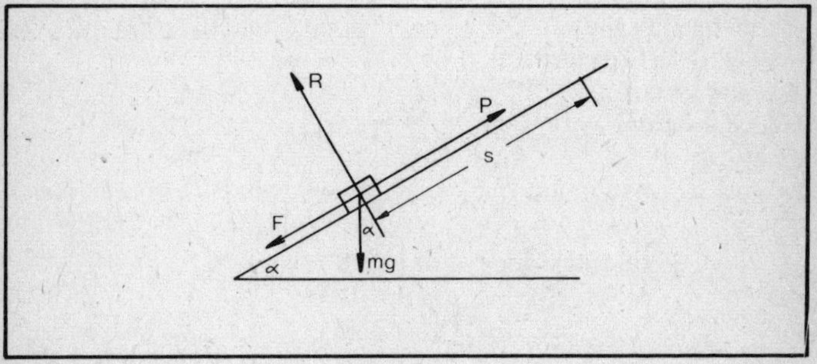

Figure 128

Let α be the inclination of the plane to the horizontal. The forces are shown in figure 128.

Since the work done is $\mathbf{F} \cdot \mathbf{r}$ or $Fr \cos \theta$, where θ is the angle between F and r we have

$$\begin{aligned}
\text{Work done by } P &= Ps \\
\text{Work done by } R &= Rs \cos 90° = 0 \\
\text{Work done by } F &= Fs \cos 180° = -Fs \\
\text{Work done by } mg &= mgs \cos(90° + \alpha) = -mgs \sin \alpha.
\end{aligned}$$

Notice that two of the results are negative. We say that Fs is the work done **against** the friction. Also, since R and s are perpendicular the normal reaction R does no work. This is verified by the answer of zero.

Example 2 A constant force $\mathbf{F} = \mathbf{i} + 2\mathbf{j} + 3\mathbf{k}$ moves its point of application from $\mathbf{r}_1 = \mathbf{i} + \mathbf{j} + \mathbf{k}$ to $\mathbf{r}_2 = \mathbf{i} - 2\mathbf{j} + 4\mathbf{k}$. Find the work done by the force.

Now the displacement vector is $\mathbf{r}_2 - \mathbf{r}_1 = -3\mathbf{j} + 3\mathbf{k}$. Hence the work done by F is $(\mathbf{i} + 2\mathbf{j} + 3\mathbf{k}) \cdot (-3\mathbf{j} + 3\mathbf{k}) = 3$ units.

Work done by a variable force

Consider a variable force F moving along a general path. If we regard the path as the aggregate of a large number of small sections, say n, then for successive displacements $\delta\mathbf{r}_1, \delta\mathbf{r}_2, \delta\mathbf{r}_3, \ldots, \delta\mathbf{r}_n$ where the force has values of $\mathbf{F}_1, \mathbf{F}_2, \ldots, \mathbf{F}_n$ over these sections we have

$$\text{Total work done} \simeq \mathbf{F}_1 \cdot \delta\mathbf{r}_1 + \mathbf{F}_2 \cdot \delta\mathbf{r}_2 + \cdots + \mathbf{F}_n \cdot \delta\mathbf{r}_n,$$

assuming that the force $\mathbf{F}_i$ remains approximately constant over the small displacement $\delta\mathbf{r}_i$.

Hence the total work done $\simeq \sum_{i=1}^{n} \mathbf{F}_i \cdot \delta\mathbf{r}_i$.

As $n \to \infty$, $\delta\mathbf{r}_i \to 0$ and

$$\text{Total work done} = \lim_{n\to\infty} \sum_{i=1}^{n} \mathbf{F}_i \cdot \delta\mathbf{r}_i = \int \mathbf{F} \cdot d\mathbf{r},$$

the integral being over the whole path.

A typical example is the motion of a particle suspended by an elastic string. The tension provides the variable force.

An **elastic string** supporting a mass m will obey **Hooke's law**. This states that the tension in an elastic string is directly proportional to the extension beyond the natural length provided that the elastic limit is not exceeded. This can be expressed in the form

$$T = \frac{\lambda}{l} x,$$

where l is the natural length, x is the extension and λ is a constant.

λ is a characteristic of the string itself and is called the **modulus of elasticity**. In fact, λ is the tension required to stretch the string to double its natural length, and hence is measured in the same units as the tension.

The work done in stretching an elastic string from an extension x_1 to x_2 is found by considering the work done in increasing the extension from x to $x + \delta x$ and summing in the limit. We assume that T remains constant over this small interval.

$$\text{Work done} = \int_{x_1}^{x_2} \frac{\lambda}{l} x \, dx = \frac{\lambda}{l} \left[\frac{x^2}{2}\right]_{x_1}^{x_2}$$

$$= \frac{\lambda}{2l}(x_2^2 - x_1^2).$$

Example 3 Find the work done in stretching an elastic string from a length of 0·5 m to 0·7 m given that its natural length is 0·4 m and its modulus is 56 N.

The extension increases from 0·1 m to 0·3 m and hence

$$\text{Work done} = \frac{\lambda}{2l}(x_2^2 - x_1^2) = \frac{56}{0{\cdot}8}(0{\cdot}3^2 - 0{\cdot}1^2)$$

$$= 5{\cdot}6 \text{ J}.$$

Example 4 Find the work done when a particle is moved 8 m by the action of a force F of magnitude $(10 - x)$, where x is the distance of the particle from the initial position in the direction of motion.

$$\text{Work done} = \int F \, dx$$

$$= \int_0^8 (10 - x) \, dx = \left[10x - \frac{x^2}{2}\right]_0^8$$

$$= 80 - 32 = 48 \text{ J}.$$

Power

Power is defined as the rate at which work is done. The SI unit of power is the **watt**, and is equal to 1 joule per second.

It follows that, if a force P N moves its point of application along its line of action at a constant speed v m s^{-1}, the work done per second is Pv, that is, the power is Pv watts. This is a special case of the general result

$$\text{Power} = \frac{d}{dt}(\text{work}) = \lim_{\delta t \to 0} \frac{\delta(\text{work})}{\delta t}$$

$$= \lim_{\delta t \to 0} \frac{\mathbf{F} \cdot \delta \mathbf{r}}{\delta t} = \mathbf{F} \cdot \mathbf{v}.$$

Thus the power of an engine at an instant is the scalar product of the force it exerts and the velocity with which it moves.

Example 5 Find the tractive force produced by an engine when it is travelling at a constant speed of 30 m s^{-1} and the engine is working at the rate of 450 kW.

Since the tractive force, say P N, moves its point of application in the direction of the velocity we have:

$$\text{Power} = Pv,$$
$$450\,000 = 30P \Rightarrow P = 15\,000 \text{ N}.$$

The tractive force produced by the engine is 15 000 N.

Example 6 An engine of mass 30 Mg pulls six coaches of total mass 120 Mg up a slope of 1 in 150 against a resistance equal to the weight of 200 kg. If the engine can work at 300 kW find the maximum speed of the train up the slope.

The component of weight down the slope is $\frac{150 \times 1000}{150} g$ N.

The resistances are $200g$ N.
Since at maximum speed the pull of the engine will equal the total resisting force we have:

$$\text{Pull of the engine} = 1200g \text{ N}.$$

At a speed of v m s^{-1} the work done $= 1200gv$ J s^{-1}.
Now the power is 300 kW so

$$1200gv = 300 \times 1000.$$

Thus the maximum speed is $\frac{300 \times 1000}{1200 \times 9{\cdot}8}$ m s^{-1} = 25·5 m s^{-1}.

Energy

The energy of a body is its capacity for doing work. The SI unit for energy is the joule.

A body may possess energy in many different forms such as heat or mechanical energy and it is possible to convert from one form to another (the principle on which steam engines work). In dynamics we consider only the mechanical forms, namely kinetic and potential energy.

The **kinetic energy** of a body is that which it possesses due to its motion and is measured by the work done in bringing it to rest.

Consider a particle of mass m moving with velocity v. If it is brought to rest under the action of a force F, the retardation a is given by $F = ma$.

If it comes to rest in a distance s, we have, using $v^2 = u^2 + 2as$,

$$0 = v^2 + 2as \Rightarrow -as = \tfrac{1}{2}v^2.$$

So the work done is $-Fs = -mas = \frac{1}{2}mv^2$. The kinetic energy of the body is $\frac{1}{2}mv^2$.

The **potential energy** of a body is that which it possesses by virtue of its position and is measured by the work done in moving from its actual position to some standard position.

The surface of the earth is often taken as the reference level and hence the potential energy possessed by a mass m at a height h above the earth's surface is mgh.

The potential energy of an elastic string when it is subjected to an extension x is thus

$$\int_0^x \frac{\lambda}{l} x \, dx = \frac{\lambda}{l}\left[\frac{x^2}{2}\right]_0^x = \frac{\lambda x^2}{2l}.$$

Example 7 A pump is required to deliver 3 m^3 of water per minute through a circular pipe 6 cm in diameter having raised it from a depth of 20 m. Find the power of the pump if friction is neglected.

The water is given potential and kinetic energy, the sum of which gives the work done per minute.

The potential energy gained per minute

= work done per minute in raising the water 20 m

$= 3 \times 10^3 \times 9{\cdot}8 \times 20$ J

$= 588 \times 10^3$ J,

(assuming that 1 m^3 of water has a mass of 10^3 kg).

Let the velocity of the water leaving the pipe be v m s^{-1}. The volume of water leaving the pipe per minute is given by

$$\begin{aligned} \text{volume} &= v \times (\text{cross-sectional area}) \times 60 \text{ m}^3 \\ &= v \times \pi(0{\cdot}03)^2 \times 60 \\ &= 0{\cdot}17v. \end{aligned}$$

But the volume issuing per minute = 3 m^3.

$$\therefore \qquad 0{\cdot}17v = 3 \Rightarrow v = 17{\cdot}68 \text{ m s}^{-1}.$$

The kinetic energy given to the water per minute

$$\begin{aligned} &= \tfrac{1}{2} \times 3 \times 10^3 \times (17{\cdot}68)^2 \text{ J} \\ &= 468{\cdot}9 \times 10^3 \text{ J}. \end{aligned}$$

The total gain in energy per minute = $1056{\cdot}9 \times 10^3$ J.

$$\begin{aligned} \text{Power required} &= 1056{\cdot}9 \times \frac{10^3}{60} \text{ W} \\ &= 17{\cdot}6 \text{ kW}. \end{aligned}$$

Conservation of energy

If a particle of mass m is allowed to fall from rest at a height h above the ground, its speed v after falling a distance s is $\sqrt{2gs}$.

The kinetic energy of the particle on reaching this point is

$$\tfrac{1}{2}mv^2 = mgs.$$

But mgs is the potential energy lost during this part of the motion and hence the kinetic energy gained is equal to the potential energy lost.

Or, we can note that:

$$\begin{aligned} \text{Kinetic energy} &= mgs, \\ \text{Potential energy} &= mg(h - s). \end{aligned}$$

Thus the total energy = mgh which is constant.

In this case the sum of the kinetic and potential energies remains constant. This is an example of the **principle of conservation of energy** which states that the total amount of energy in the universe is constant although it may be converted into different forms.

If we exclude forces in our system which cause a change in the form of energy (impacts, friction, etc.) then we can use a special form of the general principle called the **principle of energy**. This states that, if the forces acting on a system of particles are conservative, the sum of the kinetic and potential energies is constant.

Conservative forces are such that the work done by them in moving a particle depends only on the initial and final positions and not on the path taken. This excludes frictional forces and impacts, but includes gravitational forces.

Example 8 A particle of mass m is attached to one end of a light inextensible string of length a, the other end of which is tied to a fixed point O. When the particle is hanging freely it is given a horizontal velocity of $\sqrt{5ga}$ so that it performs circular motion in a vertical plane with centre O. Find the speed of the particle at P when the string makes an angle of 60° with the upward vertical.

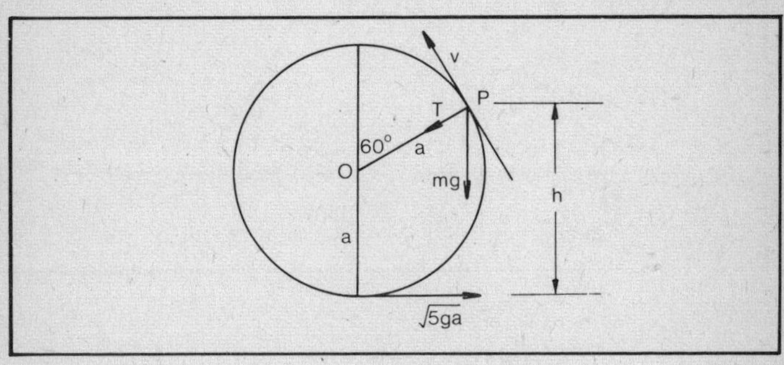

Figure 129

Since the tension, T, in the string is always perpendicular to the direction of motion, it does no work and hence the sum of the potential and kinetic energies is constant.

Let the speed of the particle at P be v. Regarding the point of projection as the zero potential energy level we have

$$\tfrac{1}{2}mv^2 + mgh = \tfrac{1}{2}m(5ga) + 0,$$
$$v^2 + 2ga(1 + \cos 60°) = 5ga,$$
$$v^2 = 2ga.$$

Thus the speed of the particle when it reaches the point P is given by $\sqrt{2ga}$.

Remember that the principle of energy can only be used in a conservative system. If the system is such that non-conservative forces are included (such as friction) then we cannot use this principle. In this case we use the **principle of work** which states that the change in kinetic energy of a particle during a displacement is equal to the work done, i.e.

Final K.E. – initial K.E. = work done.

Example 9 A particle of mass m is projected from a point A up a plane of inclination θ with an initial velocity of u. Find how far up the plane the particle will travel before coming to rest if (i) the plane is smooth, (ii) the plane is rough with a coefficient of friction μ.

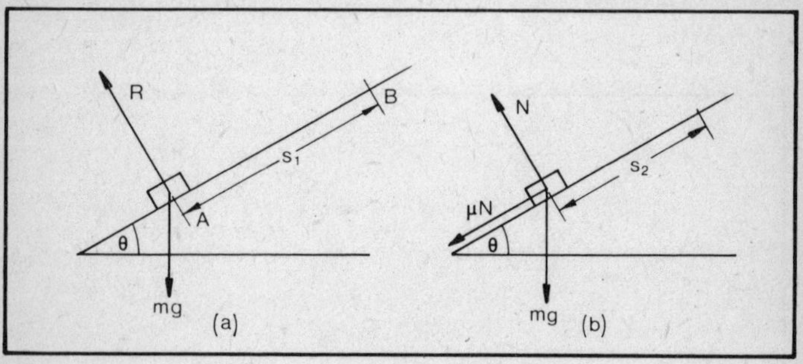

Figure 130

(i) If the plane is smooth then the only forces acting are the weight and a reaction normal to the plane (figure 130a).

Let the particle come to rest at a point B, a distance s_1 up the plane, giving a vertical rise of $s_1 \sin\theta$. Since this is a conservative system we can use the principle of conservation of energy, i.e.

$$\text{K.E.} + \text{P.E. at A} = \text{K.E.} + \text{P.E. at B.}$$

$$\tfrac{1}{2}mu^2 + 0 = 0 + mgs_1 \sin\theta$$

$$s_1 = \frac{u^2}{2g} \operatorname{cosec}\theta.$$

(ii) If the plane is rough, energy is lost and so we use the principle of work. In this case there is a non-conservative force μN acting in addition to the normal reaction N and the weight. Work is done **against** both gravity and friction.

$$\text{Work done} = \text{change in K.E.},$$

thus
$$-mgs_2 \sin\theta - \mu N s_2 = 0 - \tfrac{1}{2}mu^2$$
$$s_2(mg\sin\theta + \mu N) = \tfrac{1}{2}mu^2$$
$$s_2 = \frac{mu^2}{2(mg\sin\theta + \mu N)}.$$

Example 10 A particle of mass 2 kg is attached to one end of an elastic string of natural length 0·5 m whose other end is fixed to a point O. The modulus of elasticity of the string is $24g$ N. If the particle is released from O and falls vertically find the extension in the string when it first comes to rest.

The motion is in two parts: the motion of a free particle under gravity until the natural length is reached and the motion when the string is being stretched.

We can consider the motion as a whole since there are no impulses etc. Since the particle starts and finishes at rest, there is no change in kinetic energy and thus the work done is zero. Thus the work done by gravity is equal to the work done in stretching the string.

Let the particle come to rest with an extension x beyond the natural length.

$$2g(x + \tfrac{1}{2}) = \frac{24gx^2}{2 \times 0{\cdot}5},$$
$$24x^2 - 2x - 1 = 0,$$
$$(6x + 1)(4x - 1) = 0.$$

Neglecting the negative solution leaves $x = \frac{1}{4}$. Hence the particle comes to rest after an extension of $\frac{1}{4}$ m.

Key terms

Work done by a constant force is $\mathbf{F} \cdot \mathbf{r}$.

Work done by a variable force is $\int \mathbf{F} \cdot d\mathbf{r}$.

In an elastic string the **tension** T is given by $T = \dfrac{\lambda}{l}x$, where x is the extension, λ is the modulus of elasticity and l is the natural length of the string.

Work done in stretching an elastic string from an extension x_1 to x_2 is

$$\frac{\lambda}{2l}(x_2^2 - x_1^2).$$

Potential energy stored in an elastic string is $\dfrac{\lambda x^2}{2l}$, where x is the extension.

Power is the rate of doing work, measured in watts.

Kinetic energy of a body is $\frac{1}{2}mv^2$.

Potential energy of a body is mgh.

Principle of work is

$$\text{Final K.E.} - \text{initial K.E.} = \text{work done.}$$

Principle of energy gives

$$\text{K.E.} + \text{P.E. at point A} = \text{K.E.} + \text{P.E. at point B}$$

in a conservative field of force.

Chapter 19
Impulse and Impact

When a constant force **F** acts for a time t, the force is said to exert an **impulse**. The impulse of the force is defined as the product $\mathbf{F}t$. The SI unit of impulse is the newton second (Ns).

If the force is variable, the total impulse will be approximately given by the sum of the impulses over successive small intervals of time δt.

$$\mathbf{I} \simeq \mathbf{F}_1\,\delta t_1 + \mathbf{F}_2\,\delta t_2 + \cdots + F_n\,\delta t_n = \sum \mathbf{F}\,\delta t.$$

In the limit as $\delta t \to 0$

$$\mathbf{I} = \int_0^t \mathbf{F}\,dt.$$

In most problems about impacts the force is usually large and acts for a short interval of time. For example, the blow of a hammer, a cricket bat striking a ball or a sudden jerk in a string. If the force is variable it is not always possible to measure the different magnitudes and we measure the impulse by the effect it produces.

Thus if **I** is the impulse of a force **F** which acts for a time t,

$$I = \int_0^t \mathbf{F}\,dt = \int_0^t m\,\frac{d\mathbf{v}}{dt}\,dt = \Big[m\,\mathbf{v}\Big]_0^t,$$

$$\mathbf{I} = m(\mathbf{v} - \mathbf{u}),$$

where **u** and **v** are the initial and final velocities respectively. Hence, since this is true if **F** is constant or variable.

Impulse = change in momentum produced.

Remember that any change in position is neglected since this will be small if the time interval is small. Also impulse is a vector quantity.

Note that it can be resolved into two perpendicular directions. Thus an impulse of magnitude I acting in a direction θ to the line Ox is equivalent to an impulse $I\cos\theta$ along Ox together with an impulse $I\sin\theta$ perpendicular to Ox.

Example 1 A footballer receives a pass of a ball of mass 1 kg which, just before it reaches him is travelling horizontally with a speed of 6 m s^{-1}. He immediately kicks the ball to another player so that it again moves horizontally with a speed of 6 m s^{-1}, but at 60° to its original direction. Find the magnitude of the impulse given to the ball.

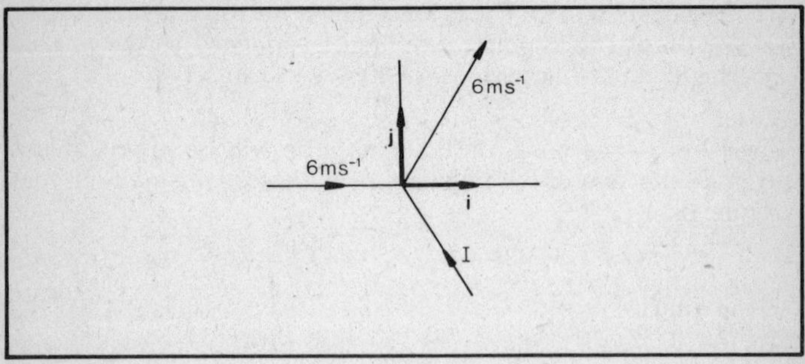

Figure 131

With the footballer as origin, set up unit vectors **i**, **j** along and perpendicular to the original direction of motion as shown in figure 131.

The initial velocity is $6\mathbf{i}$ and the final velocity is $3\mathbf{i} + 3\sqrt{3}\,\mathbf{j}$.

Now
$$\begin{aligned} \mathbf{I} &= m\mathbf{v} - m\mathbf{u} \\ &= 1(3\mathbf{i} + 3\sqrt{3}\,\mathbf{j}) - 1(6\mathbf{i}) \\ &= -3\mathbf{i} + 3\sqrt{3}\,\mathbf{j}. \end{aligned}$$

The impulse on the ball $= |\mathbf{I}| = \sqrt{3^2 + (3\sqrt{3})^2} = 6$ N s.

Impact of water

When rain falls on the ground, or water from a hose pipe is directed normally at a fixed surface, the momentum of a certain quantity of water is destroyed each second (assuming the water does not rebound). This momentum destroyed per second measures the impulse on the surface in that second and gives a measure of the average force on the surface. This average force is the force acting per second that would destroy that particular amount of momentum and is called the **thrust**.

Example 2 Water issues from a hose pipe of cross-sectional area 8 cm^2 with a velocity of 4 m s^{-1}. If it strikes a wall at right angles find the thrust on the wall.

The volume of water issuing per second $= (8 \times 10^{-4}) \times 4$ m^3.
The mass of water is $(8 \times 10^{-4}) \times 4 \times 10^3 = 3{\cdot}2$ kg.
The momentum destroyed per second is $3{\cdot}2 \times 4 = 12{\cdot}8$ N s.
The thrust on the surface is 12·8 N.

Impact of inelastic bodies

If two bodies A and B collide then, by Newton's third law, the action of A on B is equal and opposite to the action of B on A. Thus the impulses are equal and opposite and it follows that the changes in momentum of A and B are equal and opposite.

This can be expressed as the **principle of conservation of momentum** which states that if the vector sum of the external forces acting on a system is zero then the total momentum is constant. Remember that momentum is a vector and hence when using this principle the direction must be given its correct sign.

Example 3 A particle of mass $2m$ moving with a speed of 10 m s^{-1} strikes and sticks to another particle of mass m moving in the same direction with a speed of 4 m s^{-1}. Find their common speed after impact.

Let V be their common speed after impact. We can use the principle of conservation of momentum.

Total momentum before impact = total momentum after impact.

$$2m \times 10 + m \times 4 = 3m \times V,$$
$$\Rightarrow 24m = 3mV \quad \text{or} \quad V = 8 \text{ m s}^{-1}.$$

Remember that energy is **not** conserved during impacts because some is changed into different forms. In this example:
Total kinetic energy before impact $= \frac{1}{2}(2m)10^2 + \frac{1}{2}(m)4^2 = 108m$ J.
Total kinetic energy after impact $= \frac{1}{2}(3m)8^2 = 96m$ J.
Loss in energy $= 108m - 96m = 12m$ J.

Example 4 A gun of mass M is free to recoil horizontally and fires a shell of mass m with its barrel inclined at θ to the horizontal. Find the initial direction of motion of the shell to the horizontal.

When the explosion takes place, the gases formed exert a force on the shell and drive it from the barrel and, at the same time, exert

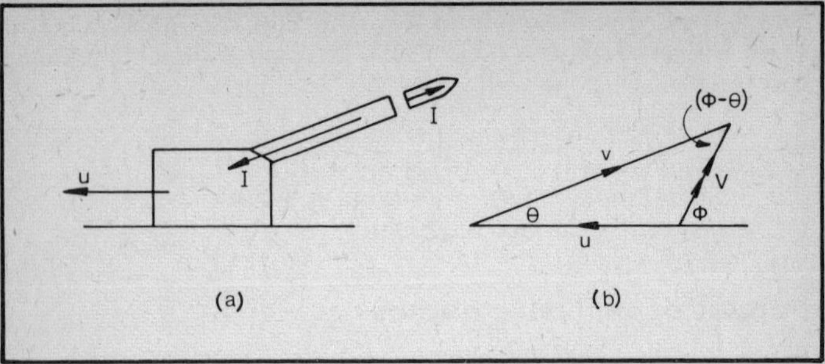

Figure 132

an equal and opposite force on the gun itself. This will cause the gun to recoil horizontally with speed u. As the shell leaves the barrel it is subject to a velocity v relative to the barrel and a velocity u horizontally. Its resultant velocity V is shown in figure 132(b).

The horizontal momentum of the shell is equal to the horizontal momentum of the gun.

$$Mu = mV \cos \phi. \qquad (1)$$

From the vector triangle (figure 132b) using the sine rule

$$\frac{\sin(\phi - \theta)}{u} = \frac{\sin \theta}{V}. \qquad (2)$$

From equations (1) and (2)

$$M \sin(\phi - \theta) = m \sin \theta \cos \phi.$$

Thus $\quad M \sin \phi \cos \theta - M \cos \phi \sin \theta = m \sin \theta \cos \phi.$

Dividing by $\cos \theta \cos \phi$

$$M \tan \phi - M \tan \theta = m \tan \theta,$$

$$\text{or} \quad M \tan \phi = (M + m)\tan \theta.$$

Hence the shell initially moves at an angle of ϕ to the horizontal where

$$\tan \phi = \left(\frac{M + m}{m}\right)\tan \theta.$$

Impulsive tension in strings

Let two particles A and B which are connected by a light inextensible string be placed on a smooth horizontal table with the string taut. Let an impulse **I** be applied to one of them, say B. This will cause an impulsive tension in the string $\mathbf{I}_1$, which has equal and opposite effects on A and B.

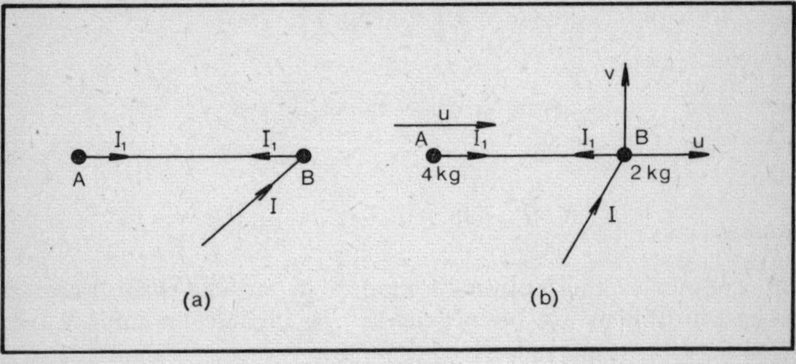

Figure 133

Since A has only one impulse applied to it, it will begin to move along AB. However, B has two impulses **I** and $\mathbf{I}_1$ applied to it and its direction of motion will be at an angle to AB.

Example 5 Two particles A and B of masses 4 kg and 2 kg respectively are joined by a light, taut, inextensible string and lie on a smooth horizontal table. If B is given an impulse **I**, of magnitude 6 N s in a direction making an angle of 60° with AB, find the velocities of A and B immediately after the blow.

Let B move initially with speeds u and v in directions along and perpendicular to AB (see figure 133b).

Since impulse = change in momentum,

For A $\qquad I_1 = 4u, \qquad (1)$

where I_1 is the magnitude of $\mathbf{I}_1$. (Note that the velocity of A is equal to the component of the velocity of B along AB since the string is taut.)

For B, along AB $\qquad 6 \cos 60° - I_1 = 2u. \qquad (2)$

Pendicular to AB $\qquad 6 \sin 60° = 2v. \qquad (3)$

Equations (1) and (2) give $\quad 6 = 12u \Rightarrow u = \frac{1}{2}\text{ m s}^{-1}$.

From equation (3) $\quad 6 \times \dfrac{\sqrt{3}}{2} = 2v \Rightarrow v = \dfrac{3\sqrt{3}}{2}\text{ m s}^{-1}$.

The velocity of B is $\sqrt{(\frac{1}{2})^2 + \left(\frac{3\sqrt{3}}{2}\right)^2} = \sqrt{7}$ m s^{-1} at an angle of $\tan^{-1}(3\sqrt{3})$ with AB produced.
The velocity of A is $\frac{1}{2}$ m s^{-1} along AB.

Note that since the internal impulsive tension $\mathbf{I}_1$ has no effect on the whole system, this problem could have been solved by applying the principle to the complete system, i.e.

In the direction of AB:

$$6 \cos 60° = 2u + 4u \Rightarrow u = \tfrac{1}{2} \text{ m s}^{-1}.$$

Perpendicular to AB:

$$6 \sin 60° = 2v + 0 \Rightarrow v = \frac{3\sqrt{3}}{2} \text{ m s}^{-1}.$$

In chapter 17 the motion of connected particles was discussed. These problems are often extended to include an impact. The following example is fairly typical.

Example 6 Two particles A and B of masses 3 kg and 5 kg are connected by a light inextensible string which passes over a smooth fixed pulley. After 8 seconds the 5 kg mass strikes the floor without rebounding. Find the speed with which the system begins to move just after the string becomes taut again.

Let a be the common acceleration of the system and T the tension in the string.
The equations of motion are:

$$5g - T = 5a,$$
$$T - 3g = 3a.$$

Adding gives $\qquad 2g = 8a \Rightarrow a = \dfrac{g}{4}.$

After 8 s the common velocity, v, is given by

$$v = 0 + \left(\frac{g}{4}\right)8 = 2g.$$

At this moment the heavier mass hits the floor and the string then becomes slack. The 3 kg mass now moves as a free particle under gravity until it momentarily comes to rest and then falls to the point where the string becomes taut again. At this point it will again have a speed of $2g$. When the string becomes taut it causes an impulse in the string which sets the 5 kg mass in motion again. If I is the impulse and V is the common velocity after we have:

$$I = 5V - 0,$$
$$-I = 3V - 3(2g).$$

Adding gives $\qquad 0 = 8V - 6g \Rightarrow V = \frac{3}{4}g.$

Hence just after the string becomes taut the speed is $\frac{3}{4}g$ m s^{-1}.

Impact of elastic bodies

When two spheres collide and separate again the spheres are slightly compressed and then resume their original shape. This property of a body to regain its shape after compression is called **elasticity**.

In this section we consider the spheres to be smooth (so that the reaction between them lies along the common normal at the point of contact) and that the spheres collide directly (i.e. along the common normal).

Clearly we can use the **principle of conservation of momentum**, since there are no external impulses. But another equation is also required if the problem is to be solved. This is provided by **Newton's experimental law** which states that if two bodies collide directly, the relative velocity after impact is in a constant ratio to their relative velocity before impact.

This constant is denoted by e and is called the **coefficient of restitution**. The value of e can only be determined by experiment and one should remember that the law itself is only approximate.

If $e = 1$, the bodies are said to be perfectly elastic, while if $e = 0$ the bodies are inelastic.

Newton's law can be expressed in the form

$$\mathbf{v}_2 - \mathbf{v}_1 = -e(\mathbf{u}_2 - \mathbf{u}_1),$$

where $\mathbf{u}_1$, $\mathbf{u}_2$, $\mathbf{v}_1$ and $\mathbf{v}_2$ are the velocities before and after impact respectively.

Consider figure 134 which shows two spheres of masses m_1 and m_2 moving with velocities $\mathbf{u}_1$ and $\mathbf{u}_2$ ($\mathbf{u}_1 > \mathbf{u}_2$). Let $\mathbf{v}_1$ and $\mathbf{v}_2$ be their velocities after they have collided directly and let e be the coefficient of restitution.

The two results needed for this problem are (i) conservation of momentum and (ii) Newton's experimental law.

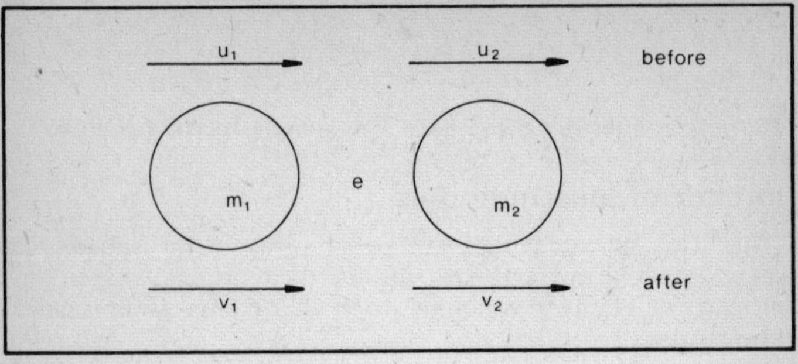

Figure 134

Conservation of momentum gives:

$$m_1\,\mathbf{u}_1 + m_2\,\mathbf{u}_2 = m_1\,\mathbf{v}_1 + m_2\,\mathbf{v}_2\,.$$

Newton's law gives:

$$\mathbf{v}_2 - \mathbf{v}_1 = -e(\mathbf{u}_2 - \mathbf{u}_1).$$

Remember the velocities are vectors and careful attention must be paid to the signs in each case.

Example 7 A sphere of mass 4 kg moving with a speed of 5 m s^{-1} overtakes and collides directly with a similar sphere of mass 2 kg moving in the same direction with a speed of 2 m s^{-1}. If the coefficient of restitution is $\frac{1}{2}$, find the speeds of the two spheres after impact.

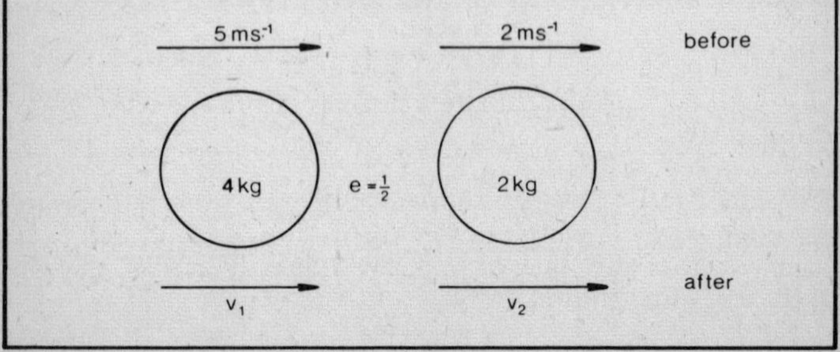

Figure 135

Let v_1 and v_2 be the speeds of the 4 kg and 2 kg spheres respectively after impact as shown in figure 135. We shall assume that the original direction of motion be taken as the positive direction and the motion takes place in a straight line.

By conservation of momentum,

$$4v_1 + 2v_2 = 5 \times 4 + 2 \times 2 = 24,$$

thus $$2v_1 + v_2 = 12. \tag{1}$$

By Newton's experimental law,

$$v_2 - v_1 = -\tfrac{1}{2}(2 - 5),$$

thus $$2v_2 - 2v_1 = 3. \tag{2}$$

Adding equations (1) and (2) gives

$$3v_2 = 15 \Rightarrow v_2 = 5 \text{ m s}^{-1}.$$

Hence $$v_1 = 3\tfrac{1}{2} \text{ m s}^{-1}.$$

Thus the speed of the 4 kg mass is $3\frac{1}{2}$ m s^{-1} and that of the 2 kg mass is 5 m s^{-1}.

Example 8 A sphere of mass 2 kg travelling at 5 m s^{-1} meets a similar sphere of equal mass moving in the opposite direction at 3 m s^{-1}. If the coefficient of restitution is $\frac{1}{4}$, show that one sphere is brought to rest by the impact and find the speed of the other sphere. Find also the loss in kinetic energy during the impact.

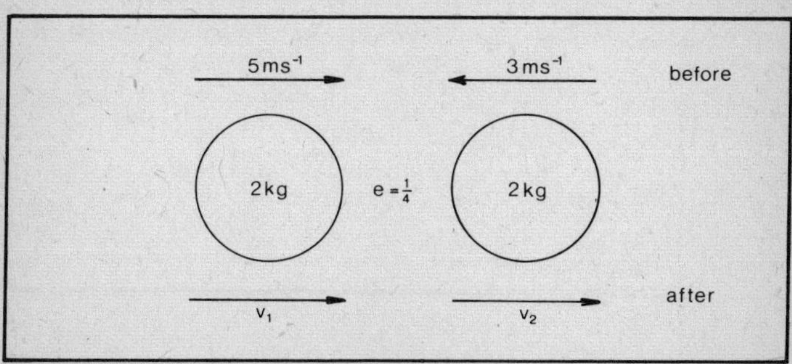

Figure 136

Let v_1 and v_2 be the speeds of the first and second spheres respectively after impact is shown in figure 136. Again the positive direction is taken from left to right.
By Newton's law,

$$v_1 - v_2 = -\tfrac{1}{4}\{5 - (-3)\},$$

thus $$v_1 - v_2 = -2. \quad (1)$$

By conservation of momentum:

$$2v_1 + 2v_2 = 5 \times 2 - 3 \times 2,$$

or $$v_1 + v_2 = 2. \quad (2)$$

Solving equations (1) and (2) by adding and subtracting we obtain $v_1 = 0$ and $v_2 = 2 \text{ m s}^{-1}$.

Hence the first sphere is brought to rest by the impact and the second sphere is given a speed of 2 m s^{-1} having its direction reversed.

The loss in K.E. of the system is

$$\tfrac{1}{2}(2)5^2 + \tfrac{1}{2}(2)3^2 - \tfrac{1}{2}(2)2^2 - 0 = \tfrac{1}{2}(50 + 18 - 8) = 30 \text{ J}.$$

Impact of a sphere on a fixed plane

Consider a sphere falling on to a fixed smooth horizontal plane (figure 137a). If u is the velocity before impact, v the velocity after impact and e is the coefficient of restitution then by Newton's experimental law

$$v = -eu.$$

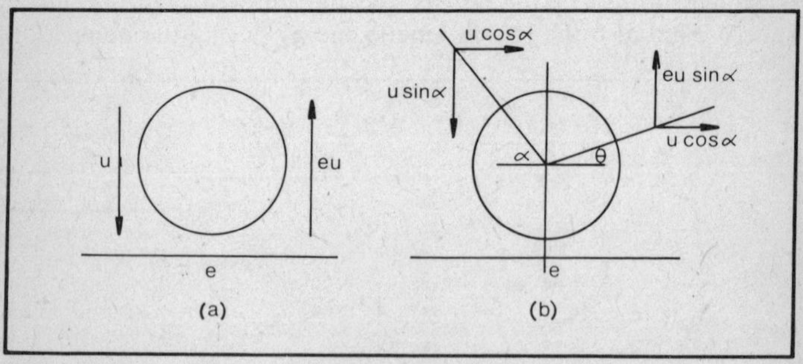

Figure 137

For a sphere which approaches a plane at an angle or falls vertically on to an inclined plane we need to remember that the speed of the sphere parallel to the plane will be unaltered, since there is no force between them in this direction. Figure 137(b) shows a sphere striking a smooth plane at an angle α to the horizontal. If u is the velocity of approach and e is the coefficient of restitution then by applying Newton's law **normal** to the plane we can obtain the components of the velocity after impact.

Note that the speed parallel to the plane is constant (i.e. $u \cos \alpha$) and that the speed normal to the plane changes from $u \sin \alpha$ to $eu \sin \alpha$ reversing its direction.

If θ is the angle between the final direction of motion and the horizontal then,

$$\tan \theta = \frac{eu \sin \alpha}{u \cos \alpha} = e \tan \alpha.$$

Key terms

The **impulse of a constant force**, $\mathbf{F}$, is $\mathbf{F}t$, where t is the time for which it acts.

The **impulse of a variable force F** is $\int \mathbf{F}\, dt$.

In both cases

impulse = change in momentum produced.

The **principle of conservation of momentum** applied to the impact of two masses m_1 and m_2 with initial and final velocities $\mathbf{u}_1$, $\mathbf{u}_2$, $\mathbf{v}_1$, $\mathbf{v}_2$ respectively is

$$m_1 \mathbf{u}_1 + m_2 \mathbf{u}_2 = m_1 \mathbf{v}_1 + m_2 \mathbf{v}_2 .$$

Newton's experimental law is

$$\mathbf{v}_2 - \mathbf{v}_1 = -e(\mathbf{u}_2 - \mathbf{u}_1),$$

where $\mathbf{u}_1$, $\mathbf{u}_2$, $\mathbf{v}_1$ and $\mathbf{v}_2$ are velocities along the common normal at the point of impact.

Chapter 20
Motion in Two Dimensions

Relative velocity

How long will a car travelling at 50 km hr^{-1} take to overtake one travelling at 40 km hr^{-1}? (By overtaking, we mean the time at which the cars are overlapping.) Why is it that if the cars are travelling in opposite directions this overlapping time is so much shorter? The answers to these questions are best explained by referring to the **relative velocities** of the cars.

If the cars travel in the same direction, their relative velocity is 10 km hr^{-1}. To overtake, the faster car must travel the total length of both cars (8 metres say) at an 'apparent' speed of 10 km hr^{-1} ($2{\cdot}\dot{7}$ m s^{-1}) and this will take 2·88 seconds.

If the cars travel in opposite directions, their relative velocity is 90 km hr^{-1}. They will overlap for a time given by 8 metres divided by 90 km hr^{-1} (25 m s^{-1}) which is 0·32 seconds. The idea of relative velocity can be further illustrated by saying that two cars crashing head on both travelling at 50 km hr^{-1} have the same effect as a car travelling at 100 km hr^{-1} crashing into a stationary car, because the relative velocities are the same, 100 km hr^{-1}. The idea of relative velocity gives a powerful method of solving certain problems in two or three dimensions.

Example 1 A ship S is travelling due east at 20 knots and a ship T is travelling due north at 25 knots. At noon they are converging and both 50 nautical miles (n.m.) from the intersection point of their lines of travel. What is their closest distance apart and when does this occur?

Figure 138(a) shows the velocity of S, $\mathbf{v}_S$, represented by **OA** and the velocity of T, $\mathbf{v}_T$, represented by **OB**. Figure 138(b) shows the positions of S and T at noon. We shall consider the problem from S's point of view and measure velocities and displacements relative to S.

The velocity of T relative to S, ${}_T\mathbf{v}_S$, is given by $\mathbf{v}_T - \mathbf{v}_S$.
We reduce S to rest by imposing a velocity of $-\mathbf{v}_S$.
Using **i**, **j** components $\mathbf{v}_T = 25\mathbf{j}$ and $\mathbf{v}_S = 20\mathbf{i}$, so

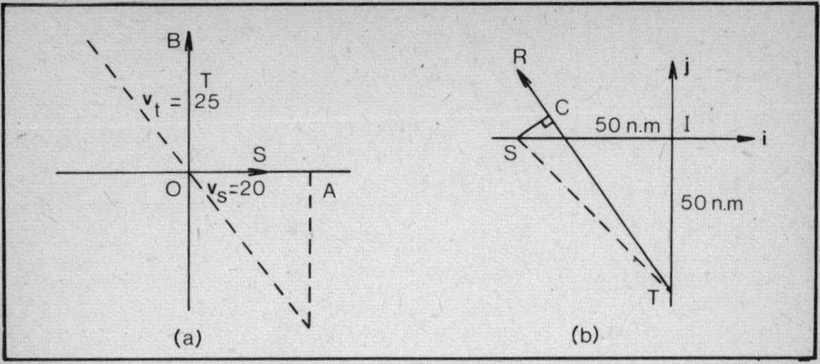

Figure 138

$$_T\mathbf{v}_S = \mathbf{v}_T - \mathbf{v}_S = 25\mathbf{j} - 20\mathbf{i} = \begin{pmatrix} -20 \\ 25 \end{pmatrix}.$$

T appears to S to be moving towards her on a roughly NW course to cut across her bows. If we regard S as stationary, T follows the line TR in the direction of its velocity relative to S. Its closest point to S can be found by drawing SC at right angles to TR (figure 138b).

$$\angle ITS = 45° \quad \text{and} \quad \angle ITC = \tan^{-1}(\tfrac{20}{25}) = 38{\cdot}7° \Rightarrow \angle CTS = 6{\cdot}3°.$$
$$SC = ST \times \tan 6{\cdot}3° = 50\sqrt{2} \times 0{\cdot}11 = 7{\cdot}81 \text{ n.m.}$$

This occurs after T has 'apparently' travelled a distance CT with an 'apparent' (relative) velocity of $\begin{pmatrix} -20 \\ 25 \end{pmatrix}$ (speed of 32·02 knots). The time taken to reach their closest approach is

$$TC \div 32{\cdot}02 = ST \cos 6{\cdot}3° \div 32{\cdot}02 = 70{\cdot}28 \div 32{\cdot}02 = 2{\cdot}195 \text{ hr.}$$
$$= 2 \text{ hr } 11{\cdot}7 \text{ min.}$$

Closest approach when t = noon + 2 hr 11·7 min ≃ 14·12 hr.

Example 2 An aircraft J on a course of 027° travelling at 600 km hr^{-1} is 20 km due west of an aircraft K travelling at the same height on a course of 300° at 1000 km hr. If a 'near miss' is classified as closer than 1 km, does this constitute a 'near miss'?

Figure 139(a) shows the aircraft starting at J_0 and K_0 (20 km apart) and their directions of travel.

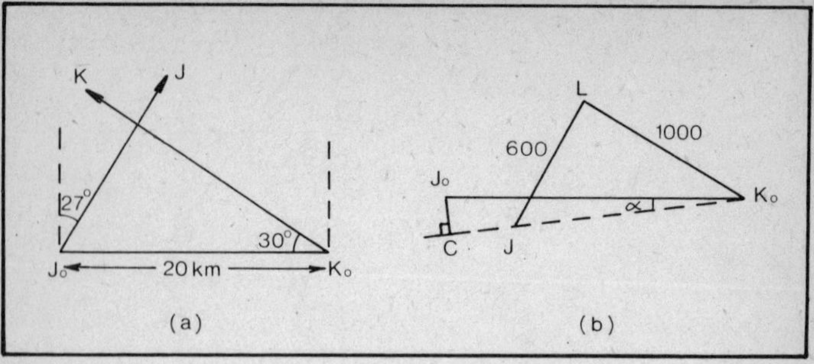

Figure 139

$$\mathbf{v}_K = \begin{pmatrix} -1000 \cos 30° \\ 1000 \sin 30° \end{pmatrix} = \begin{pmatrix} -866 \\ 500 \end{pmatrix};$$

$$\mathbf{v}_J = \begin{pmatrix} 600 \sin 27° \\ 600 \cos 27° \end{pmatrix} = \begin{pmatrix} 272{\cdot}4 \\ 534{\cdot}6 \end{pmatrix}.$$

Velocity of K relative to J, ${}_K\mathbf{v}_J = \mathbf{v}_K - \mathbf{v}_J = \begin{pmatrix} -1138 \\ -34{\cdot}6 \end{pmatrix}$.

In figure 139(b), K_0C is the direction of ${}_K\mathbf{v}_J$ and α is given by

$$\tan \alpha = \frac{34{\cdot}6}{1138} = 0{\cdot}0304 \Rightarrow \alpha = 1{\cdot}741°.$$

Closest approach $\quad CJ_0 = J_0K_0 \times \sin \alpha = 20 \times 0{\cdot}0304$
$\qquad = 0{\cdot}608$ km.

This constitutes a 'near miss' so how long has the navigator to change course?

The time taken to reach closest approach is $CK_0 \div |{}_K\mathbf{v}_J|$

$$= \frac{20 \cos 1{\cdot}741°}{\sqrt{(1138^2 + 34{\cdot}6^2)}} = \frac{19{\cdot}991}{1138{\cdot}5} = 0{\cdot}01756 \text{ hr} = 1{\cdot}054 \text{ min}$$

$$\simeq 63 \text{ seconds!!}$$

Example 3 A particle A travels with velocity $2\mathbf{i} + 2\mathbf{j}$ starting at (0, 4) and a particle B travels with velocity $\mathbf{i} + 3\mathbf{j}$ from (4, 4) at the same time, find when they are closest together and their distance apart at this time. (Distances in metres and velocities in m s^{-1}).

Figure 140 shows the starting positions A_0 and B_0 (4, 4). Velocity of A relative to B is

$$\mathbf{v}_A - \mathbf{v}_B = \begin{pmatrix}2\\2\end{pmatrix} - \begin{pmatrix}1\\3\end{pmatrix} = \begin{pmatrix}1\\-1\end{pmatrix}.$$

Relative to B, A follows the path A_0A. The closest approach is clearly at C which takes place at time $t = 2$ ($\mathbf{A_0C} = 2\mathbf{i} - 2\mathbf{j}$) and the closest distance is

$$CB_0 = \sqrt{8}\text{ m.}$$

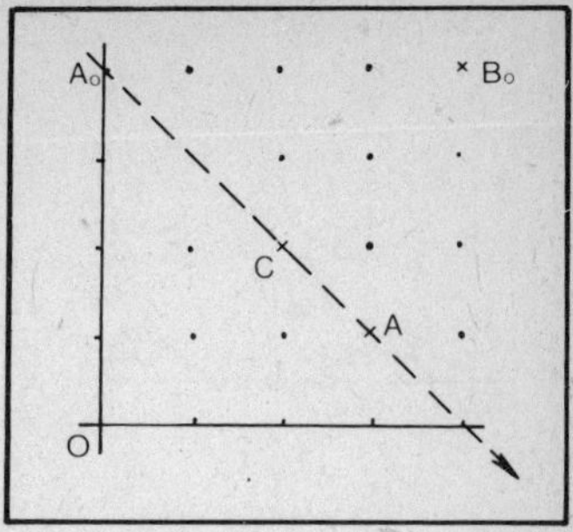

Figure 140

Projectiles

The motion of a particle under the action of its gravitational force gives rise to a branch of mathematics known as **projectiles**. Particles are projected with varying speeds and the motion takes place close to the surface of the earth, so that the force of gravity can be taken to be constant. The motion is two-dimensional, since the only force acting on the particle is in the vertical plane (towards the centre of the earth) containing the vector representing the velocity of projection. In reality, there is a frictional force due to air resistance which usually depends on the square of the speed of the particle. In the study of projectiles this is neglected, although in the flight of a golf ball the air resistance is so large (owing to the dimpled surface of the ball) that it cannot be neglected.

Example 4 A ball is projected with a velocity of 20 m s^{-1} at an angle of 40° above the horizontal. Find how far it will travel and how long it is in the air.

Taking x- and y-axes whose origin is at the point of projection, (figure 141a) the displacement of the ball is represented by the vector $\begin{pmatrix}x\\y\end{pmatrix}$ and the starting velocity by $\mathbf{v} = \begin{pmatrix}20\cos 40°\\20\sin 40°\end{pmatrix}$. The only force acting on the particle is its weight which produces an acceleration vector of $-9{\cdot}8\mathbf{j}$ m s^{-2}.

This acceleration is constant, so the formulae for constant acceleration can be used. $\mathbf{v} = \mathbf{u} + \mathbf{a}t$ gives

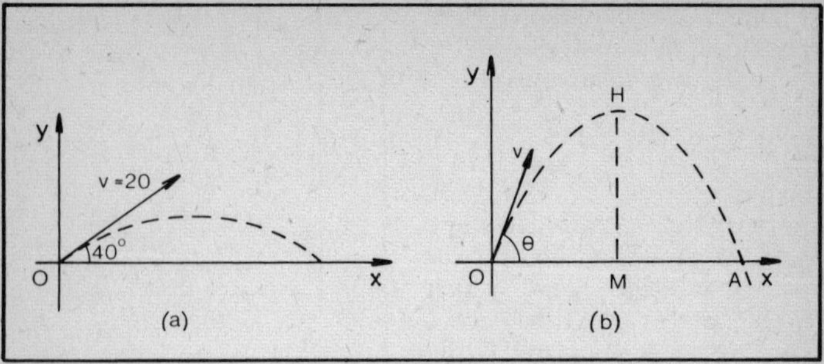

Figure 141

$$\begin{pmatrix}\dot{x}\\ \dot{y}\end{pmatrix} = \begin{pmatrix}20\cos 40^\circ\\ 20\sin 40^\circ\end{pmatrix} + \begin{pmatrix}0\\ -9{\cdot}8\end{pmatrix}t,$$

where $\begin{pmatrix}\dot{x}\\ \dot{y}\end{pmatrix}$ is the velocity at time t.

$\dot{x} = 20\cos 40^\circ = 15{\cdot}3\ \text{m s}^{-1}$ and is constant for all t.
$\dot{y} = 20\sin 40^\circ - 9{\cdot}8t = 12{\cdot}9 - 9{\cdot}8t\ \text{m s}^{-1}$.

$$\mathbf{r} = \mathbf{u}t + \tfrac{1}{2}\mathbf{a}t^2 \Rightarrow \begin{pmatrix}x\\ y\end{pmatrix} = \begin{pmatrix}15{\cdot}3\\ 12{\cdot}9\end{pmatrix}t + \tfrac{1}{2}\begin{pmatrix}0\\ -9{\cdot}8\end{pmatrix}t^2$$

$$\Rightarrow x = 15{\cdot}3t \quad \text{and} \quad y = 12{\cdot}9t - 4{\cdot}9t^2.$$

The ball hits the ground again (assuming it was projected from ground level) when $y = 0 \Rightarrow t = \dfrac{12{\cdot}9}{4{\cdot}9} = 2{\cdot}63$ seconds.

The x-displacement is called the horizontal **range** of the ball and when $t = 2{\cdot}63$ s, $x = 15{\cdot}3 \times 2{\cdot}63 = 40{\cdot}2$ m. In general, if the particle is projected at an angle of θ° above the horizontal (figure 141b) with a speed of $v\ \text{m s}^{-1}$, the equations are

$$\begin{pmatrix}\dot{x}\\ \dot{y}\end{pmatrix} = \begin{pmatrix}v\cos\theta\\ v\sin\theta\end{pmatrix} + \begin{pmatrix}0\\ -g\end{pmatrix}t,$$

where $g = 9{\cdot}8\ \text{m s}^{-1}$.

At the highest point of the flight $\dot{y} = 0 \Rightarrow t = \dfrac{v\sin\theta}{g}$.

$$\begin{pmatrix} x \\ y \end{pmatrix} = \begin{pmatrix} v\cos\theta \\ v\sin\theta \end{pmatrix} t + \tfrac{1}{2}\begin{pmatrix} 0 \\ -g \end{pmatrix} t^2 \Rightarrow \begin{array}{ll} x = (v\cos\theta)t, & (1) \\ y = (v\sin\theta)t - \tfrac{1}{2}gt^2. & (2) \end{array}$$

$y = 0 \Rightarrow t = 0$ (start) or $t = \dfrac{2v\sin\theta}{g}$ (when the particle returns to its starting level). The time to reach the highest point is half that to reach A. The horizontal range OA is given by

$$\begin{aligned} OA = x = (v\cos\theta)t = v\cos\theta \times \frac{2v\sin\theta}{g} &= \frac{2v^2\sin\theta\cos\theta}{g} \\ &= \frac{v^2\sin 2\theta}{g}. \end{aligned}$$

For a given velocity, the range $R = \dfrac{v^2\sin 2\theta}{g}$ is a maximum when $\sin 2\theta = 1 \Rightarrow \theta = 45°$.

The path of a projectile

Eliminating t from (1) and (2) gives

$$y = v\sin\theta \times \frac{x}{v\cos\theta} - \tfrac{1}{2}g \times \frac{x^2}{v^2\cos^2\theta} = x\tan\theta - \frac{gx^2\sec^2\theta}{2v^2}. \quad (3)$$

This represents a parabola with line of symmetry HM (figure 141b). This confirms that the time to reach H is half that to reach A and that OM = MA.

The velocity at A, $\mathbf{v}_A = \begin{pmatrix} v\cos\theta \\ -v\sin\theta \end{pmatrix}$ if $\mathbf{u} = \begin{pmatrix} v\cos\theta \\ v\sin\theta \end{pmatrix}$.

Example 5 How fast must a batsman run to take 'one for the throw' against a fielder who is 50 metres from the wicket and can throw at 25 m s^{-1} (90 km hr^{-1} = 56 miles hr^{-1}). The batsman must cover 22 yards before the cricket ball can travel 50 m.

Assuming the ball is thrown from the same height as it could possibly hit the wicket, the range, $R = \dfrac{v^2\sin 2\theta}{g}$ gives

$$50 = \frac{25^2\sin 2\theta}{9{\cdot}8} \Rightarrow \sin 2\theta = \frac{490}{625} = 0{\cdot}784$$

$$\Rightarrow 2\theta = 51{\cdot}6° \quad \text{or} \quad 124{\cdot}4° \Rightarrow \theta = 25.8° \quad \text{or} \quad 62{\cdot}2°.$$

There are two possible angles of projection to cover 50 m with a starting speed of 25 m s. Clearly the smaller angle gives the shorter time.

$$\text{Time of flight } T = \frac{2v \sin \theta}{g} = \frac{50 \sin 25{\cdot}8^\circ}{9{\cdot}8} = 2{\cdot}22 \text{ seconds.}$$

The batsman must cover 22 yards in 2·22 seconds, that is, run as fast as a top-class sprinter. In reality, the batsman covers less than 22 yd (he extends his bat to gain his ground) and a direct hit from such a distance is rare.

If the wicket-keeper has to remove the bails, the batsman has an extra half-second perhaps, so a more realistic speed might be 20 yd in 2·72 s, i.e. 100 yd in 13·6 s.

If the fielder hits the stumps direct at a point 1 metre below the level from which the ball was thrown, can the time of flight be found?

The projectile passes through a point where $x = 50$ and $y = -1$.

$$\text{Equation (3) gives } -1 = 50 \tan \theta - \frac{9{\cdot}8 \times 50^2 \sec^2 \theta}{2 \times 625}.$$

Writing $\sec^2 \theta = 1 + \tan^2 \theta$ gives a quadratic in $\tan \theta$

$$-1 = 50 \tan \theta - 19{\cdot}6(1 + \tan^2 \theta).$$

$$19{\cdot}6 \tan^2 \theta - 50 \tan \theta + 18{\cdot}6 = 0.$$

$$\tan \theta = \frac{50 \pm \sqrt{2500 - 4 \times 19{\cdot}6 \times 18{\cdot}6}}{39{\cdot}2} = 2{\cdot}099 \quad \text{or} \quad 0{\cdot}452$$

$\Rightarrow \theta = 64{\cdot}5$ or $24{\cdot}3$ and the smaller angle $\Rightarrow$ shorter time.

$$\text{Equation (1) gives} \quad t = \frac{x}{v \cos \theta} = \frac{50}{25 \cos 24{\cdot}3^\circ} = 2{\cdot}194 \text{ s.}$$

The larger angle gives a higher trajectory and a longer time. In general, for a projectile to pass through a given point there are two possible angles of projection.

Range on an inclined plane

In figure 142 to find the distance OA, the range up a plane inclined at an angle above the horizontal, let the particle be projected with speed v at an angle θ above the horizontal. It is most convenient to find the time taken to reach A by using a direction perpendicular to OA. In this direction the starting velocity is $v\sin(\theta-\alpha)$ and the acceleration $-g\cos\alpha$. Since the displacement in this direction (perpendicular to OA) at A is zero, $x = ut + \frac{1}{2}at^2$

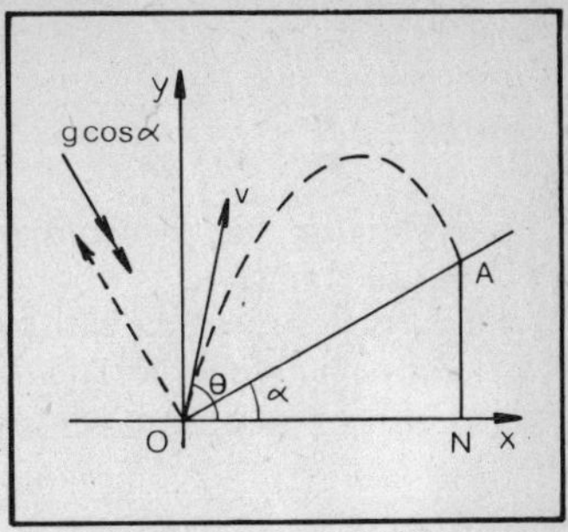

Figure 142

$$\Rightarrow 0 = v\sin(\theta-\alpha)t - \tfrac{1}{2}g(\cos\alpha)t^2 \Rightarrow t = \frac{2v\sin(\theta-\alpha)}{g\cos\alpha}.$$

$$\mathrm{ON} = x = v(\cos\theta)t \quad \text{and} \quad \mathrm{OA} = \mathrm{ON}\sec\alpha$$

$$\Rightarrow \mathrm{OA} = v\cos\theta \times \frac{2v\sin(\theta-\alpha)}{g\cos\alpha}\sec\alpha = \frac{2v^2\sin(\theta-\alpha)\cos\theta}{g\cos^2\alpha}.$$

Motion in a circle

Consider a particle moving in a circle in the x-y plane with radius r and let the particle start from A$(r, 0)$ at time $t = 0$.

At time t, the particle is at P where $\mathrm{OP} = \mathbf{r} = \begin{pmatrix} x \\ y \end{pmatrix} = \begin{pmatrix} r\cos\theta \\ r\sin\theta \end{pmatrix}$.

The velocity $\mathbf{v} = \begin{pmatrix} \dot{x} \\ \dot{y} \end{pmatrix} = \begin{pmatrix} -r\sin\theta\,\dot{\theta} \\ r\cos\theta\,\dot{\theta} \end{pmatrix} = r\dot{\theta}\begin{pmatrix} -\sin\theta \\ \cos\theta \end{pmatrix}$ which is at right angles to **OP**, since $\mathbf{r}\,.\,\mathbf{v} = r^2\dot{\theta}(-\cos\theta\sin\theta + \sin\theta\cos\theta) = 0$, which shows that the velocity is directed along the tangent to the circle with magnitude $|\mathbf{v}| = v = r\dot{\theta}$.

$\dot{\theta}\left(=\dfrac{d\theta}{dt}\right)$ is the rate at which $\angle$AOP is changing and is the angular velocity of the particle. This is usually denoted by ω and $v = r\omega$ for a particle moving on a circle of radius r, whether ω is constant or not.

The acceleration $\mathbf{a} = \dot{\mathbf{v}} = \ddot{\mathbf{r}} = \begin{pmatrix} \ddot{x} \\ \ddot{y} \end{pmatrix} = \begin{pmatrix} -r\cos\theta\,\dot{\theta}^2 - r\sin\theta\,\ddot{\theta} \\ -r\sin\theta\,\dot{\theta}^2 + r\cos\theta\,\ddot{\theta} \end{pmatrix}$

$$= -r\dot{\theta}^2\begin{pmatrix} \cos\theta \\ \sin\theta \end{pmatrix} + r\ddot{\theta}\begin{pmatrix} -\sin\theta \\ \cos\theta \end{pmatrix}.$$

The acceleration has two components, $-r\dot{\theta}^2$ along the radius and $r\ddot{\theta}$ along the tangent.
If the particle is moving with constant speed, then $\ddot{\theta} = 0$ and the acceleration has one component along the radius, towards the centre of the circle, of magnitude $r\omega^2$.

Since $v = r\omega$, $a = r \times \dfrac{v^2}{r^2} = \dfrac{v^2}{r}$.

Motion in a vertical circle under gravity

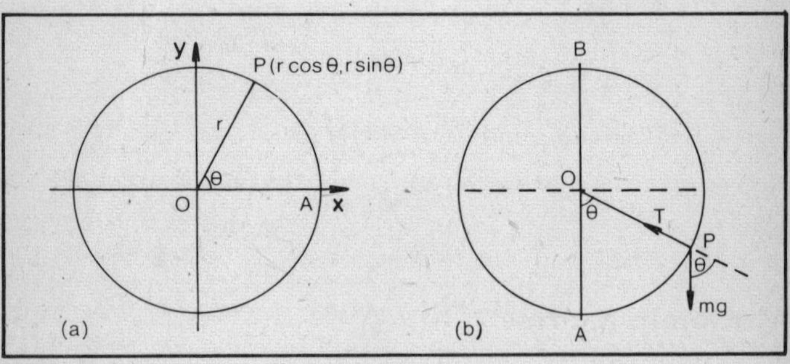

Figure 143

A particle of mas m is fixed to the end of an inextensible light string of length a whose other end is secured to a fixed point O (figure 143b). The particle is projected horizontally from A, and we shall find the necessary speed of projection for the particle to reach B.

By the principle of conservation of energy, the kinetic energy at A is converted into potential energy at B.

$$\tfrac{1}{2}mv^2 = 2mga \Rightarrow v^2 = 4ga \Rightarrow v = 2\sqrt{ga}.$$

If the particle P is projected from A with speed $2\sqrt{ga}$, at what point will the string become slack?

The acceleration of P is $a\dot{\theta}^2$ towards the centre and $a\ddot{\theta}$ along the tangent. Using $F = ma$,

Radially $\quad T - mg\cos\theta = ma\dot{\theta}^2$.

Tangentially $\quad -mg\sin\theta = ma\ddot{\theta} \Rightarrow 2mg\cos\theta = ma\dot{\theta}^2 + c. \qquad (1)$

$$\theta = 0,\ v^2 = 4ga \Rightarrow a^2\omega^2 = 4ga \Rightarrow 2mg = 4mg + c \Rightarrow c = -2mg.$$

Equation (1) becomes $ma\dot{\theta}^2 = 2mg\cos\theta + 2mg$,

$$\Rightarrow T = 3mg\cos\theta + 2mg.$$

$T = 0$ when $\cos\theta = -\frac{2}{3} \Rightarrow \theta = 131{\cdot}8°$.

At this point with the string slack the particle behaves like a projectile with initial speed $v = a\dot{\theta} = \sqrt{\frac{2}{3}ga}$ at an angle of $\cos^{-1}\frac{2}{3}$ above the horizontal.

Its greatest height is $\dfrac{v^2\sin^2\theta}{2g} = \dfrac{2}{3}ga \times \dfrac{5}{9} \times \dfrac{1}{2g} = \dfrac{5a}{27}$.

The particle's greatest height above O is $\dfrac{2}{3}a + \dfrac{5a}{27} = \dfrac{23a}{27}$.

The energy equation result assumes that the string remains taut until the particle reaches B. If T becomes zero just as the particle reaches B, then at B, $mv^2/a = mg \Rightarrow v^2 = ga$. The particle needs to be projected with a speed of $\sqrt{5ga}$ for it to reach B.

Conical pendulum

A particle of mass 1 kg on the end of a string of length 1 m describes a horizontal circle so that the string traces out the

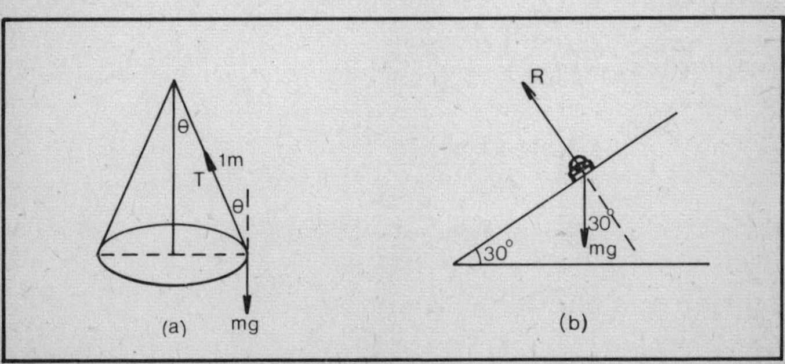

Figure 144

surface of a cone. If it performs 1 revolution per second find its velocity and the tension in the string (figure 144a). The acceleration is towards the centre of the circle and is $a\omega^2$, where $a = \sin\theta$ and $\omega = 2\pi$ radians per second.

$$v = a\omega = 2\pi \sin\theta.$$

Horizontally $F = ma \Rightarrow T\sin\theta = ma\omega^2 = 4\pi^2 \sin\theta$

$$T = 4\pi^2 = 39{\cdot}5 \text{ N}.$$

Vertically there is no acceleration so $T\cos\theta = mg = 9{\cdot}8$ N.

$$T = 39{\cdot}5 \Rightarrow \cos\theta = 9{\cdot}8 \div 39{\cdot}5 = 0{\cdot}2481 \Rightarrow \theta = 75{\cdot}6^\circ.$$

$$v = 2\pi \sin\theta = 6{\cdot}09 \text{ m s}^{-1}.$$

Example 6 Find the speed that a racing car has to be driven around a circular corner of radius 100 m banked at 30° if there is no tendency for the car to side-slip (figure 144b).

With no sideways friction the only forces acting are the car's weight and the reaction normal to the plane.

$$\uparrow \quad R\cos 30^\circ = mg \qquad \rightarrow R\sin 30^\circ = \frac{mv^2}{r}.$$

Dividing $\dfrac{v^2}{gr} = \tan 30^\circ \Rightarrow v^2 = gr\tan 30^\circ = 565{\cdot}8$

$$\Rightarrow v = 23{\cdot}79 \text{ m s}^{-1} = 85{\cdot}63 \text{ km hr}^{-1}.$$

If the car travels faster it will tend to slide up the banking and if it travels slower it will tend to slide down the banking.

Key terms

Relative velocity of A relative to B, $_A\mathbf{v}_B = \mathbf{v}_A - \mathbf{v}_B$.

Projectiles Range $= \dfrac{v^2 \sin 2\theta}{g}$.

Greatest height $= \dfrac{v^2 \sin^2\theta}{2g}$.

Time of flight $= \dfrac{2v\sin\theta}{g}$.

Motion in a circle $v = r\omega = r\dot{\theta}$

Acceleration $a = r\omega^2 = r\dot{\theta}^2 = v^2/r$ towards centre; $r\dot{\omega} = r\ddot{\theta}$ along tangent.

Index